Biologically-Inspired Radar and Sonar

Related titles on radar:

Advances in Bistatic Radar Willis and Griffiths
Airborne Early Warning System Concepts, 3rd Edition Long
Bistatic Radar, 2nd Edition Willis
Design of Multi-Frequency CW Radars Jankiraman
Digital Techniques for Wideband Receivers, 2nd Edition Tsui
Electronic Warfare Pocket Guide Adamy
Foliage Penetration Radar: Detection and characterisation of objects under trees Davis
Fundamentals of Ground Radar for ATC Engineers and Technicians Bouwman
Fundamentals of Systems Engineering and Defense Systems Applications Jeffrey
Introduction to Electronic Warfare Modeling and Simulation Adamy
Introduction to Electronic Defense Systems Neri
Introduction to Sensors for Ranging and Imaging Brooker
Microwave Passive Direction Finding Lipsky
Microwave Receivers with Electronic Warfare Applications Tsui
Phased-Array Radar Design: Application of radar fundamentals Jeffrey
Pocket Radar Guide: Key facts, equations, and data Curry
Principles of Modern Radar, Volume 1: Basic principles Richards, Scheer and Holm
Principles of Modern Radar, Volume 2: Advanced techniques Melvin and Scheer
Principles of Modern Radar, Volume 3: Applications Scheer and Melvin
Principles of Waveform Diversity and Design Wicks *et al.*
Pulse Doppler Radar Alabaster
Radar Cross Section Measurements Knott
Radar Cross Section, 2nd Edition Knott *et al.*
Radar Design Principles: Signal processing and the environment, 2nd Edition Nathanson *et al.*
Radar Detection DiFranco and Ruby
Radar Essentials: A concise handbook for radar design and performance Curry
Radar Foundations for Imaging and Advanced Concepts Sullivan
Radar Principles for the Non-Specialist, 3rd Edition Toomay and Hannan
Test and Evaluation of Aircraft Avionics and Weapons Systems McShea
Understanding Radar Systems Kingsley and Quegan
Understanding Synthetic Aperture Radar Images Oliver and Quegan
Radar and Electronic Warfare Principles for the Non-specialist, 4th Edition Hannen
Inverse Synthetic Aperture Radar Imaging: Principles, algorithms and applications Chen and Marotella
Stimson's Introduction to Airborne Radar, 3rd Edition Griffiths, Baker and Adamy
Test and Evaluation of Avionics and Weapon Systems, 2nd Edition McShea
Angle-of-Arrival Estimation Using Radar Interferometry: Methods and applications Holder

Biologically-Inspired Radar and Sonar

Lessons from nature

Alessio Balleri
Cranfield University

Hugh Griffiths
University College London

Chris Baker
Aveillant Ltd

SciTech Publishing
an imprint of the IET

theiet.org

SciTech Publishing
an imprint of the IET

Published by SciTech Publishing, an imprint of the IET
www.scitechpub.com
www.theiet.org

First published 2017

ISBN 978-1-61353-235-5 (hardback)
ISBN 978-1-61353-236-2 (PDF)

Typeset in India by MPS Limited
Printed in the UK by CPI Antony Rowe Ltd

Contents

About the editors **xi**
Foreword **xiii**

1 Introduction **1**
Alessio Balleri, Hugh Griffiths and Chris Baker

1.1 Motivations 1
1.2 Scope of the book 2

2 Biosonar-inspired signal processing and acoustic imaging from echolocating bats **5**
James A. Simmons, Jason E. Gaudette and Michaela Warnecke

2.1 Introduction 5
2.1.1 Engineered vs biological solutions to design 6
2.1.2 Varieties of biosonar 8
2.1.3 Technical challenges 9
2.2 Computational model of biosonar: spectrogram correlation and transformation (SCAT) receiver 15
2.2.1 Time–frequency representation in FM biosonar 15
2.2.2 Determination of echo delay – target range images 19
2.2.3 Incorporation of the echo spectrum – focused target-shape images on the range axis 22
2.2.4 Defocusing of images for suppressing clutter 27
2.3 Principles of biosonar imaging by SCAT 30
Acknowledgements 31
References 32

3 Enhanced range resolution: comparison with the matched filter **37**
Krasin Georgiev, Alessio Balleri, Andy Stove and Marc W. Holderied

3.1 Introduction 37
3.2 Description of the spectrogram correlation and transformation model 39
3.2.1 Cochlear block 40
3.2.2 Temporal block 41
3.2.3 Spectral block 42
3.2.4 Model output 42

3.3 The baseband spectrogram transformation receiver 42
3.4 Response of the BSCT to two closely spaced ideal reflectors 45
3.4.1 Central lobe suppression 48
3.5 Experimental setup and data collection 50
3.5.1 General settings and equipment 50
3.5.2 Simulations 51
3.5.3 Phantom targets 51
3.5.4 Physical targets 51
3.6 Results 52
3.7 Conclusion 58
References 58

4 Air-coupled sonar systems inspired by bat echolocation 61
James F.C. Windmill and Francesco Guarato

4.1 Introduction 61
4.2 What is a sonar system? 62
4.3 Bioinspired (biomimetic) design of sonar systems: emitters and receivers 65
4.3.1 Biomimetic emitters 67
4.3.2 Biomimetic receivers 68
4.4 Bat-inspired sonar systems and localisation methods 73
4.4.1 Bat-inspired signal processing (cochlea models, discrimination of closely spaced objects and estimation of time delays using chirps, bat-inspired waveforms, target identification) 83
4.5 Conclusions 84
References 85

5 Analysis of acoustic echoes from bat-pollinated plants 89
Alessio Balleri, Hugh Griffiths, Chris Baker and Marc Holderied

5.1 Introduction 89
5.2 Analysis of the signature of *C. scandens* corollas 92
5.3 Signature of *C. scandens* flowers 94
5.3.1 Power reflection as a potential cue 97
5.4 Analysis of a *R. auriculatum* inflorescence 98
5.5 Conclusion 104
References 105

6 The biosonar arms race between bats and insects 109
Thomas R. Neil and Marc W. Holderied

6.1 Introduction 109
6.2 Bat biosonar 109
6.2.1 Implementing the sonar equation 110
6.2.2 Adaptive bat biosonar 111

6.3 Prey defences 112
6.3.1 Passive defences 112
6.3.2 Acoustic decoys 114
6.3.3 Active defences and ear evolution in insects 114
6.3.4 Active responses to ultrasound in moths 118
6.3.5 Defences in other insects 123
6.4 Bat countermeasure to insect defences 125
6.4.1 Allotonic frequency hypothesis 125
6.4.2 Stealth aerial hawking 126
6.5 Conclusion 128
References 128

7 Biologically-inspired coordination of guidance and adaptive radiated waveform for interception and rendezvous problems 137
Alessio Balleri, Alfonso Farina and Alessio Benavoli

7.1 Introduction 137
7.2 Theoretical framework 138
7.2.1 Gaussian linear chirp 140
7.3 Two-dimensional case study 141
7.4 Simulation results 143
7.5 Conclusion 151
References 153

8 Cognitive sensor/processor system framework for target tracking 155
Kristine L. Bell, Chris Baker, Graeme E. Smith, Joel T. Johnson and Muralidhar Rangaswamy

8.1 Introduction 155
8.2 Framework 157
8.2.1 Cognitive sensor/processor system framework 157
8.2.2 Cognitive single target tracking 159
8.2.3 Cognitive MAP-PF single target tracking 163
8.2.4 Summary 166
8.3 Distributed sensor example 168
8.3.1 Model 168
8.3.2 Implementation 171
8.3.3 Simulation results 171
8.4 Software defined radar example 174
8.4.1 Model 176
8.4.2 Implementation 179
8.4.3 Collected data results 179
8.5 Conclusions 181
8.6 Appendix A: Bayesian Cramér–Rao lower Bound 182
References 183

9 The biosonar of the Mediterranean Bottlenose dolphins: analysis and modelling of echolocation signals **189**
Maria Greco and Fulvio Gini

9.1 Introduction 189
9.2 Data acquisition 191
9.2.1 The hydrophone 191
9.2.2 The amplifier 192
9.2.3 Digital card 192
9.3 Biosonar model 193
9.4 Signal estimation 196
9.4.1 Exponential pulse 196
9.4.2 Gaussian pulse 197
9.5 Estimation results 199
9.5.1 Exponential pulse 199
9.5.2 Gaussian pulse 201
9.5.3 Audio band 204
9.6 Conclusions 206
References 206

10 Human echolocation – spatial resolution and signal properties **209**
Liam Norman and Lore Thaler

10.1 Introduction and background 209
10.2 Acoustic properties of human sonar emissions 210
10.3 Environmental factors 212
10.4 Localising objects in space 212
10.4.1 Localising objects in depth (ranging): signal properties 212
10.4.2 Localising objects in depth (ranging): spatial resolution of human echolocators 214
10.4.3 Localising objects in the horizontal plane (azimuth): signal properties 217
10.4.4 Localising objects in the horizontal plane: spatial resolution of human echolocators 217
10.4.5 Localising objects in the vertical plane (elevation) 218
10.5 Discriminating object size, shape and material 219
10.6 Concluding comments and future directions 222
References 224

11 Polarization tensors and object recognition in weakly electric fish **229**
William R.B. Lionheart and Taufiq K. Ahmad Khairuddin

11.1 Electrosensing in fish and electrical imaging by humans 229
11.2 Algorithms for electrical imaging 231
11.3 Polarization tensors 234
11.4 Generalized polarization tensor 235

11.5 The second rank polarization tensor 236
11.6 PT in electrosensing fish 237
11.7 PT in metal detection by human technology 241
11.8 Conclusions 245
References 247

Postscript **251**

Index **253**

About the editors

Dr Alessio Balleri is a Senior Lecturer with Cranfield University at the Defence Academy of the United Kingdom, Shrivenham. He obtained a PhD degree in Electronic and Electrical Engineering from University College London in 2010 and a Laurea Degree (5 legal years, summa cum laude) from University of Pisa in 2004. His research interests include radar and sonar system design, biologically-inspired radar and sonar systems, radar and sonar target classification, target feature extraction and modelling of radar clutter.

Prof. Hugh Griffiths holds the THALES/Royal Academy of Engineering Chair of RF Sensors at University College London. His research interests include radar sensor systems and signal processing as well as antennas and antenna measurement techniques. He has published over 500 papers and technical articles on these subjects. He received the IET A F Harvey Research Prize in 2012 and the IEEE Picard Medal in 2017. He is Editor-in-Chief of the journal *IET Radar, Sonar and Navigation*.

Prof. Chris Baker is the Chief Technology Officer for Aveillant. He has been actively engaged in radar systems research since 1984 and is the author of over 250 publications in this field. He is a visiting Professor at the University of Cape Town, Wright State University, Strathclyde University, Birmingham University, Ohio State University and Cranfield University and is a fellow of the IET and IEEE.

Foreword

This book is without doubt an important landmark, as we move from sonar and radar technologies dominated by Cold War scenarios, to respond to more recent challenges of shallow coastal waters, asymmetric warfare and digitally-empowered terrorism. Responses to this include the interface and exchange of knowledge between the sonar and radar communities, and the utilization of improved signal generation, processing and control facilities to produce waveforms that mimic those generated by various fauna. This in turn has led to consideration of how and why an organism has produced such signals and processes their echoes.

A given organism can appear to have remarkable capabilities that other organisms or man-made systems lack. These capabilities might enable it to meet challenges that confound others, or exploit a niche that others find uninviting. The topic of this book concerns capabilities in sonar and radar, of which a classic example from nature is that of dolphins. They have remarkable sonar capabilities, which cannot be understood if one analyses only components of the biological system, such as the power and bandwidth of the sonar pulse: man-made sonar can have far more impressive power and bandwidth, but perform less well. This conundrum is reconciled by considering the whole organism, including the way the sonar is mounted on an agile platform (its body) that allows the target to be interrogated from a rapidly adapting series of ranges and orientations, an adaptation that is driven by a brain that has evolved through tens of millions of years of facing just such challenges.

However achieving such a level of understanding is a formidable research challenge, and we only know how far from real understanding we are when our predictions and technologies fail to match up to the animal's capabilities. This is a major weakness of biomimetic systems: we might copy the pulses emitted by animals, but without understanding the role that these pulses play in the organism as whole, from platform to detection to processing to perception to adaptation, we will not match the echolocation capability of the animal.

Bioinspiration instead looks at the challenges faced by an organism, and matches these to our technological tool-set. We become problem-solvers, not mimics.

There is of course a mismatch in terms of the timescales. Here again, we must not simply restrict our vision to imagine we are competing with evolution through natural selection over timescales of tens of millions of years. This manner of evolution towards a solution is somewhat wasteful and time-consuming. Whilst it is possible, depending on the trait, the strength of the selection pressure and the breadth of genetic variation, for evolutionary change to occur much more rapidly, the output of natural selection is not necessarily the solution required by technologists. By addressing the problem through bioinspiration, rather than the solution through biomimetics, it is possible to find these supra-optimal solutions.

With science and technology, we can test variants against a carefully constructed set of challenges, designed to answer predetermined questions that allow the next generation variant to be designed. This gives greater speed, though runs the risk of not accounting in tests for all the factors our device will face 'in the field', in the way that systems derived through natural selection do automatically. However scientific testing allows the exploration of effective solutions which might not have survived down the generations to reach us, such as a hypothetical game-changer in a predator that wipes out its food source: undesirable for biological systems but desirable for some radar and sonar.

As the bioinspired approach becomes more established, where next? A significant factor to address is that we are not only comparing our solutions and methodologies with those that arose through evolution by natural selection over millions of years. Complex organisms can pass solutions to the next generation through example, education and training; we can trial and adapt behaviour, testing a wide range of variants in much shorter timescales, and these can be copied (by friend and foe alike). Adaptation and learning can develop solutions on a much faster timescale than through natural selection, provided our prior evolution has equipped us to do that. Responses on such timescales can be critical, because extinctions can arise through catastrophe as well as the slower process of being out-competed.

This brings me to my final point, in that when we take inspiration from nature to produce better solutions (technological, behavioural, etc.), we must not simply restrict our viewpoint to the idea of highly evolved complex organisms which exhibit extraordinary capabilities in the face of challenges that would defeat others, those capabilities having arisen through the slow process of evolution by natural selection, and the complex processes of learning and mimicry, and the role of key drivers such as co-operation, synergism and ecological balance. Bacteria, for example, can rapidly adapt to the environment through genetic processes, creating challenges we currently have no idea how to solve, and which present us with an existential threat. Our sewage systems mix animal and human faeces, and if an *Escherichia coli* that is resistant to antibiotics enters that sewage system from the gut of a cow, and resides in close proximity to human faeces containing *E. coli* that was susceptible to antibiotics, then after four hours the *E. coli* that can infect humans can gain resistance. Adaptation in this manner, via the transfer of genetic material, can occur in a matter of hours, not the millions of years we associate with complex creatures; and consideration of the individual alone might fail to identify advantages imparted to the group (e.g. the pod, or the biofilm). The generation of solutions through bio-inspiration has barely begun to understand such pathways, by which the solutions we see in nature, have developed. As we move from biomimetics to bioinspiration, with its greater emphasis on understanding the challenges and the whole biological system that responds to them, we will see that such understanding allows more complete problem-solving through bioinspiration.

In this context I am delighted to see the book '*Biologically-Inspired Radar and Sonar: Lessons from Nature*'. The growth of sonar and radar in facing the challenges of the last century compressed in time the requirement for a sensory advantage (against competitors and prey who can themselves adapt) for hunting, navigation, etc. The editors have selected from the best of research across bats, humans, dolphins and fish, taking in prey species such as insects. The authors have judiciously walked that difficult line in bioinspiration research: assessing similarities between anthropogenic challenges and those found in nature; hypothesizing without anthropomorphising; and then rigorously testing hypotheses whilst liberating themselves of the analogy to nature in order to make use of the full power of digital signal processing. The result is a pioneering compendium across species and radiations, which should underpin the new era in sonar and radar that must come as we face the challenges of the new century, which differ significantly from those of the World Wars and Cold War on which our legacy sonar and radar were established.

Timothy G. Leighton, FREng FRS
Professor of Ultrasonics and Underwater Acoustics
Institute of Sound and Vibration Research
Faculty of Engineering and the Environment
University of Southampton, UK

Chapter 1
Introduction

Alessio Balleri[1], Hugh Griffiths[2] and Chris Baker[3]

1.1 Motivations

Radar and sonar systems play a key role in many modern defence and civilian sensing applications. They are used to accomplish a large variety of tasks which include the detection and classification of targets, the acquisition of intelligence information, imaging, autonomous navigation and collision avoidance. Although radar and sonar systems have been available, deployed and continuously improved for nearly 100 years, improvements in technology and the pressing appetite for increased performance have resulted in many new technical challenges and, as a result, radar and sonar still attract a lot of interest from both academia and industry.

Nature presents examples of active sensing which are unique, sophisticated and incredibly fascinating. There are animals that sense the environment actively, for example through echolocation, which have evolved their capabilities over millions of years and that, as a result of evolution, have developed unique in-built sensing mechanisms that are often the envy of synthetic systems. Examples of these are nectar feeding bats that can detect and land at the corolla of tiny flowers with an outstanding precision in a highly cluttered environment and dolphins that can detect and recognise object shapes on the sea-floor with a high level of accuracy. Active sensing is used in nature to carry out many different tasks such as navigation, collision avoidance and selection, identification and attack of prey. The interaction between predator and prey has often resulted in counter-measure solutions and tactical target prosecution strategies which present unique and fascinating similarities with their counterpart solutions in radar and sonar Electronic Warfare (EW).

Because natural sensing systems have evolved and been optimised over such an extended time period, there may be lessons that radar and sonar practitioners can learn from looking at how nature operates and at the solutions evolution has provided. Darwin's principle of Natural Selection, the 'survival of the fittest', tells us that

[1]Centre for Electronic Warfare, Information and Cyber, Cranfield University, Defence Academy of the UK, Shrivenham SN6 8LA, UK
[2]Department of Electronic and Electrical Engineering, University College London, London WC1E 6BT, UK
[3]Aveillant Ltd, 300 Science Park, Milton Road, Cambridge CB4 0XL, UK

the most successful species will survive and propagate further, whilst the less successful will die out.[1] It is the result of this evolution that we seek to study, since an understanding may allow us to devise better algorithms in our own sensors and in the associated processing. This is of particular importance in a research era where significant resources are spent to drive future sensors towards an increasing level of cognition.

The aim of this book is to present some of the recent work that has been carried out to investigate if nature can provide useful lessons, and how sophisticated sensing techniques used in nature can be applied to radar and sonar systems to improve their performance. Biologically-inspired radar and sonar is a unique and wide topic that embraces many different disciplines. It presents many positive aspects and at the same time requires a lot of care in its treatment. The uniqueness and multidisciplinarity of the topic is such that considerable interest can be easily attracted across readers from many different academic backgrounds. Indeed, it is a topic that can so easily and so nicely facilitate multi-disciplinary collaborations. In our previous experience of giving talks and seminars worldwide, there has probably not been a single time when a new idea, something completely different, has not come up in the resulting discussions. At the same time, this is a topic that requires particular care and precision in the execution of the research and in the presentation of the results. Data is often scarce and experiments are often very expensive and time consuming. One needs to ensure, very meticulously, that the execution is perfectly planned as easy repetitions are often not possible. Also, great care must be taken in the presentation of the results, in the definition of the hypothesis and, just as importantly, in clarifying the research goals. This is because it can be easy to fall in the trap of trying to replicate nature or draw conclusions on how we believe nature actually operates but, in fact, the main goal is the improvement of radar and sonar sensors. If a hypothesis is made that eventually leads to a solution that works better than existing solutions in practice then that goal has been achieved even if, after all, it is found that the same hypothesis does not always obey the rules of nature.

Because of the characteristics and uniqueness of biologically-inspired sensor research, it would be unrealistic to try and attempt to include all topics in a book alone. However, we hope this book will be an excellent starting point and a key guide to those who wish to further expand their knowledge in this research direction as it includes the relevant research streams that relate to Radar and Sonar.

1.2 Scope of the book

The book is organised in 10 principal chapters that cover some of the main recent aspects of the topic. These are grouped in five main sections.

Chapters 2–5 describe the case of echolocating bats. Chapter 2 introduces the bio-neural mechanisms that are believed to be used by the bat to process target

[1] C. Darwin, *On the Origin of Species*, John Murray, Albemarle Street, London, 1859, https://en.wikisource.org/wiki/On_the_Origin_of_Species_(1859), accessed on 22 February 2016.

echoes together with a comparison with the case of radar and sonar. Chapter 3 describes a signal processing algorithm, inspired by a model of the bat auditory processing, that can be applied to radar waveforms and investigates potential advantages with respect to range resolution performance. Chapter 4 presents a review of the biomimetic sonar systems developed in the last few years with a particular emphasis to biomimetic transmitter and receiver design and target localisation solutions. Chapter 5 describes the fascinating case of nectar feeding bats that, by visiting flowers of bat-pollinated plants, facilitate pollen transfers between individual plants. The chapter investigates the signatures of the corolla of these types of flowers to identify features that might ease target classification by the bat with the goal to understand strategies for target classification.

Chapter 6 presents key aspects of the biosonar arms race between insects and bats to highlight differences and similarities with radar and sonar EW. The chapter describes both passive solutions, such as acoustic absorption by wing scales or acoustic decoys employed to divert a bat, and sonar jamming active techniques.

Chapters 7 and 8 describe very recent signal processing solutions that, deriving inspiration from the diversity and strategies available in nature, have been developed to address key challenges for adaptive and cognitive sensors. Chapter 7, in particular, derives inspiration from the neurobiological processes of human cognition and develops a theoretical framework to examine artificial cognition and its application to target tracking. Chapter 8 derives inspiration from the bat and develops an algorithm that guides an airborne radar interceptor towards a target by jointly developing an optimal guidance and automatically adapting and optimising the transmitted waveform on a pulse-to-pulse basis.

Chapter 9 is concerned with the case of the Dolphin and presents an analysis of typical echolocation calls produced by these mammals both in the audio domain and at ultrasound. The calls are modelled with Gaussian and exponential pulses and their parameters are estimated to inform their application to man-made sonar.

Chapter 10 presents the extraordinary case of echolocating humans and contains an excellent review that will allow readers to develop a general understanding of the challenges tackled in human echolocation research. The chapter describes the properties of typical human sound emissions and discusses the localisation performance, in distance, azimuth and elevation, and object recognition performance achieved by echolocating humans.

Finally, Chapter 11 presents the case of weakly electric fish that can achieve active electro sensing using an electric discharge organ and shows how similar techniques can be used for target detection in heavy clutter. The chapter introduces the concept of polarization tensor as a tool that provides a low-dimensional information of a target and shows how this can be employed to improve object discrimination performance in metal detection.

Chapter 2

Biosonar-inspired signal processing and acoustic imaging from echolocating bats

James A. Simmons[1], Jason E. Gaudette[2] and Michaela Warnecke[3]

2.1 Introduction

No biological system more immediately evokes thinking about technology than the sonar sense of bats [1,2]. This chapter describes a novel approach to sonar signal-processing that has been identified from experiments carried out with echolocating bats. Bats use their biosonar to perceive target distance (range), target azimuth (crossrange), and target shape using wideband signals, and target velocity and fluttering motion using narrowband signals [2–4]. Being able to perceive these components in real-time while flying through surroundings that vary from simple to complex makes them a very valuable source of engineering inspiration. From a technological perspective, the parallels between biosonar and pulse-echo sensing by man-made radar and sonar systems are very compelling [5]. Moreover, progress towards understanding biosonar has benefitted from applying radar and sonar concepts to biology [5–8]. Furthermore, the historical development of radar and sonar technology overlapped with the discovery and advancement of research on biosonar [1,2,9,10]. Broadcast transmission, echo reception, signal-processing, display of images, and guidance of motion are recognizable functions both in biosonar [2,5] and man-made systems [9,10]. These functions correspond fairly well to specific stages in a block diagram of an engineered system. In biosonar, however, these functions are spread across several stages of biological and perceptual processing because they are interrelated and are carried out in a distributed fashion by overlapping biological structures [2]. The content of this chapter follows a signal-processing theme across stages in the bat's auditory system that comprises its sonar imaging system. At several stages in this story, operations seem similar to practices used in some man-made systems, only to have their outputs treated quite differently at higher stages.

[1]Department of Neuroscience, Brown University, Providence, RI 02912, USA
[2]Sensors and Sonar Systems Department, Naval Undersea Warfare Center, Newport, RI 02841, USA
[3]Department of Psychological and Brain Sciences, Johns Hopkins University, Baltimore, MD 21218, USA

Most bats emit brief, wideband, frequency-modulated (FM) signals of which they change the bandwidth, frequency content, duration, and interpulse interval in response to changes in their surroundings [2]. These bats exploit the available bandwidth of echoes to form images that sharply depict the features of both large and small objects [4,8,11]. Other species of bats instead rely primarily on narrowband signals, mainly constant-frequency (CF) signals with relatively long durations, and they concentrate on the kinds of information conveyed by narrowband echoes, such as velocity from Doppler shifts and modulations caused by target fluttering motions [2] (see Figure 2.1). Using the assumption of matched-filter reception, i.e., crosscorrelation [3,4] across many species of bats, biosonar sounds cover a large gamut of signal types that have been explored over more than a half-century for their technological potential by radar and sonar engineers [6]. In this chapter, we examine the biosonar system of one species of bats for lessons about design principles.

2.1.1 Engineered vs biological solutions to design

Guided by biosonar, bats are remarkably successful at achieving high levels of real-time, flexible performance in complex dynamic environments [1,2,7]. Man-made systems generally do not achieve such a wide range of capabilities, especially when required to operate in real time. Clearly, the behaviours and capabilities of these animals offer lessons for improving technology. Results of recent research have revealed new knowledge about the performance of echolocating bats in specific tasks, such as target ranging and target shape classification, in conjunction with the underlying neural mechanisms that enable this performance [5,8]. These mechanisms do not resemble processing methods commonly found in radar or sonar systems [9,10]. Instead, they are more closely related to the mechanisms of perception and action in nervous systems [2,7,11].

For example, consider the problem of clutter interference: the presence of clutter surrounding a target of interest is a significant challenge for localization and classification by man-made sonar and radar systems [9,10]. A common engineering approach for rejecting clutter is to narrow the beam of the sonar or radar so that echoes arriving from objects off to the sides are excluded while echoes from objects located straight ahead are accepted. Although, using this approach, the acoustic clutter is removed and potential targets of interest can be isolated, the resulting loss of information about the surrounding objects compromises guidance through the scene. Moreover, obtaining a comprehensive view of the scene requires sequential scanning of that scene with the narrow beam, which is time-consuming and limits the capacity for real-time operation while the beam is moved from one point to the next across all directions. Instead, the solution used by bats is to emit sounds with a broad beam, which ensonifies the entire scene with each broadcast [11,12]. The bat aims its beam at the target and thereby tracks it from one broadcast to the next. Being on the axis of the beam, the target returns echoes containing the full bandwidth of the incident sounds. As demonstrated experimentally, full-band echoes from the target are received and transduced to yield images that are sharply focused [12], thus

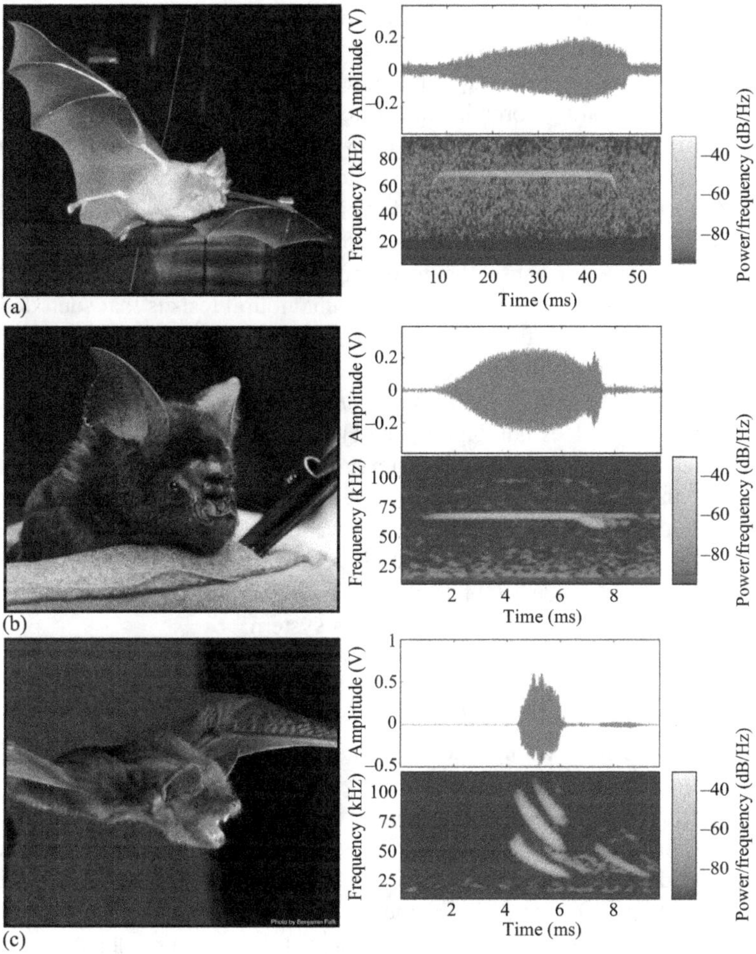

Figure 2.1. Sample of species and call diversity in echolocating bats. (a) Photograph of the greater horseshoe bat, Rhinolophus ferrumequinum, *in flight (left). This bat has a two-part distinctive noseleaf: the upper part is pointed, while the lower section is horseshoe-shaped.* Rh. ferrumequinum *sends out long (20–60 ms) FM–CF–FM sounds between 70 and 80 kHz (right). (b) Photograph of the Formosan leaf-nosed bat,* Hipposideros armiger terasensis. *This bat's noseleaf has a complex, multifold structure, and it sends out comparatively short (6–8 ms) CF–FM sounds at around 60 kHz (right). (c) Photograph of the big brown bat,* Eptesicus fuscus, *in flight (left). This bat is common to northern America and echolocates through its open mouth.* E. fuscus *sends out 0.5 to 20 ms FM sounds that cover frequencies from 100 kHz down to 20 kHz. Note horizontal time scale differences. We thank Benjamin Falk for permission to use his photograph*

facilitating target localization and classification [3,4]. Because the target is locked onto the beam to stabilize the spectrum of incident sounds, any changes in echo spectra can be attributed to the target itself, not to the target's direction. At the same time, because the beam is broad, objects located off to the sides – clutter – are strongly ensonified, too, but due to the transmitting organ's aperture, the ensonification of off-axis objects is by lowpass versions of the broadcasts. Echoes from the clutter thus arrive with predominantly lowpass spectra. These off-side echoes are received and transduced along with echoes from the on-axis target, but the off-axis object's lowpass spectra cause the resulting images to be blurred, or defocused [12,13]. Here, the most important finding of behavioural tests is that such defocused images do not intrude into the bat's perception of targets of interest, which instead are registered as focused images. The dimension of *focused* vs *defocused* images intrinsically distinguishes *targets* from *clutter* in the bat's perceptions. The point is that the bat's solution does not depend on (1) sequential narrow-beam scanning and (2) narrow-beam exclusion of clutter echoes. Instead, the bat uses a broad beam and still receives information from clutter echoes, sending that information into other parts of its processing pathway to guide flight and adapt the transmitted waveforms to prevailing conditions [5,14–16]. Multiple experiments have been carried out on the ability of bats to guide flight in dense, extended clutter [17–22]; their results dramatically illustrate the effectiveness of this system.

2.1.2 Varieties of biosonar

Biosonar systems offer a whole suite of technical challenges for the design of bioinspired sonar or radar systems. There are roughly a thousand species of echolocating bats, divided into over a dozen prominently represented families that differ in their geographical distribution [2]. The bats in these families use several different combinations of broadband FM and narrowband CF biosonar signals. They also process FM or CF echoes with different auditory mechanisms to exploit the different kinds of information they contain [2,6]. Man-made systems, too, use broadband and narrowband signals in the same way – for example, to determine target range from echo delay (broadband) or target velocity from echo Doppler shift (narrowband) [9,10]. Consider, for example, the detection of flying insect prey. Detection can be achieved achieved by transmitting narrowband CF sounds and using sharply tuned receiving filters that accept only echoes of the same frequency (give or take any Doppler shifts). At the same time, flutter-induced Doppler shifts are conveyed in the echoes. Many species of bats transmit narrowband, shallow-sweeping or CF sounds when they are searching for prey, especially in open-space conditions where no other echoes compete with the echoes from potential prey [2]. However, after detecting a target or when flying in proximity to clutter, these same bats adapt their call characteristics and instead use wideband FM sweeps. In this situation detecting a target entails separating it from clutter, which involves more information than just the presence or absence of narrowband echoes. Once target detection has occurred, the bat's task is to determine if the target is indeed an insect and worth trying to capture. Recognition of active prey can either be done by using multidimensional high-resolution acoustic images derived from wideband broadcasts [8], which is the strategy used by big brown bats, or by exploiting some specific salient acoustic feature of the echoes, most easily by emitting narrowband sounds, a strategy used by horseshoe bats [2]. The acoustic signature of a flying insect lies in the fluttering movements of its wings, which typically occur at wingbeat rates of 10–30 Hz [23,24]. Wingbeat flutter is easily detectable by bats that emit long-duration CF sounds, such as horseshoe bats [2,24–26]. Because the duty-cycle of the sounds is high (60%–80%), there is almost continuous coverage of multiple periods of the insect's wingbeats. Moreover, each sound impinging on the insect lasts long enough to encompass several wingbeats just within one echo, and each of which has amplitude and Doppler-shift fluctuations that convey the fluttering of an insect to the bat. CF-emitting bats will immediately attack targets that return flutter-modulated echoes but ignore targets that return stationary echoes [2,25,26]. These bats even have a unique behaviour – Doppler shift compensation – that acts to facilitate detection of flutter [2]. They adjust the precise frequency of their broadcasts to compensate for the Doppler shift caused by their own flight velocity, bringing the frequency of echoes returning to the bat from the flying insect into a fixed, narrow frequency region where their inner ears have particularly high sensitivity and also frequency resolution. This region is called the "acoustic fovea", an analogy with the fovea of the eye's retina. Like a magnifying lens for sound, the acoustic fovea focuses the bat's hearing on the frequencies that carry the target's wingbeat modulations. This narrowband scheme concentrates on picking out one acoustic feature – the modulation in CF echoes – as the cue to identify the target as prey.

On the other hand, echoes of typical broadband FM biosonar sounds are not as efficient in picking up the modulations created by the insect's wingbeats within individual echoes because the duty cycle of the sounds is low (~5%–10%). Each wideband broadcast is typically no more than 10 ms long, too short to capture more than a fragment of one wingbeat cycle. Moreover, during the bat's approach to evaluate the target for final capture, the FM sounds shorten in proportion to the decreasing range, which makes flutter detection even more difficult unless successive images are integrated to follow the changing shape. No single acoustic feature of echoes is reliable for this task. Instead, FM bats process echoes to create percepts (images) that depict the target's location and shape in spatial terms [3,4,8,27]. They perceive the flutter as rhythmic changes in shape from one "look" to the next. This is a more comprehensive process that involves all of the echo's acoustic features, not just modulations of one selected feature.

2.1.3 Technical challenges

Here, we concentrate on extracting technological lessons we find from wideband FM biosonar [5,8]. We use as our model the big brown bat (*Eptesicus fuscus*), a common North American insect-eating species [28] that captures beetles and moths [29] either in aerial interception manoeuvres guided by sonar or by using sonar to find prey on surfaces such as vegetation or the ground [30]. Figure 2.2(a) shows a big brown bat flying in a laboratory test of guidance and orientation in dense, extended clutter [17–22]. These bats frequently fly through vegetation to reach

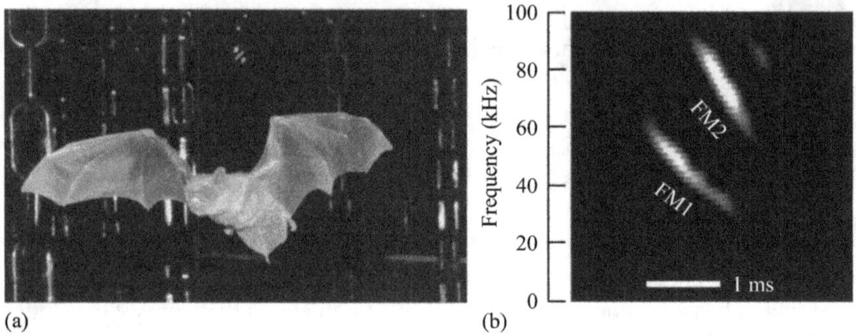

Figure 2.2. (a) Big brown bat flying through a corridor between rows of vertically hanging plastic chains in tests of their guidance and reactions to dense surrounding clutter. (b) FM biosonar sound emitted during flight and picked up by a microphone next to the array of chains. This signal is short in duration and covers 25–100 kHz in two prominent downward-sweeping harmonics (FM1, FM2). Bats exploit the full 75-kHz bandwidth of these sounds to determine distance, or target range, from echo delay and to reconstruct a target's shape from the echo spectrum

feeding areas, and they also chase each other in and out of the branches of trees and bushes [30] and use their biosonar to locate objects in three dimensions using information about echo delay, echo amplitude, and echo spectrum in relation to the broadcast [4].

2.1.3.1 Wide signal bandwidth

The first challenge posed by biosonar comes from the wide bandwidth of biosonar sounds in relation to their location in the baseband. The signals transmitted by our example species, the big brown bat, are brief ultrasonic FM sounds in the 20–100 kHz band [1,26,27,31,32]. Figure 2.2(b) shows the spectrogram of this bat's FM biosonar signal recorded during a flight in clutter (Figure 2.2(a)). This sound contains two prominent harmonics; the first harmonic ("FM1") sweeps from 50 to 25 kHz and the second harmonic ("FM2") sweeps from about 100 kHz down to 50 kHz. The acoustic waveform is generated in the bat's larynx and projected out through the open mouth, which filters the sound to the frequency band shown in the spectrogram [33]. The presence of two or more harmonics widens the transmitted bandwidth, which the bat exploits for higher-resolution imaging [8]. With transmitted frequencies of 25–100 kHz, the sounds actually have a bandwidth (~75 kHz) greater than their centre frequency (~60 kHz). By this formulation, the percentage bandwidth exceeds 100%. Broadcasts used in man-made sonar systems are narrowband or only moderately wideband by comparison; due to transducer limitations, their bandwidths are rarely more than a third of an octave around the centre frequency. Radar signals can have very wide absolute bandwidths but their centre frequencies are so high that the percentage bandwidth is also much lower than for the big brown bat. In both cases, the ambiguity (i.e., crosscorrelation) functions for echoes have envelopes-widths reciprocally related to bandwidth but fine cycle-by-cycle structure related to the reciprocal of centre frequency [9]. Because the structure underlying the envelope consists of many cycles, the usual procedure is to take the envelope to be the ambiguity plot. This concentrates echo energy to estimate delay from the envelope's peak, but it also discards information carried by the fine structure, which often relates to the target's shape. Because the bat's bandwidth and centre frequency are similar, the fine cycle structure underneath the envelope of the ambiguity function has about the same time scale as the envelope itself. If reflections are altered in phase at the instant of echo formation, changes in phase could be registered either by coherent processing or by rectification followed by envelope detection. The mammalian auditory system uses half-wave instead of full-wave rectification, so the potential for coherent processing is retained and used to form images [3,8]. Furthermore, the role of the harmonics goes beyond merely brute-force widening the broadcast bandwidth [8,11,12]. They fill the specific role of providing the substrate in the sounds for controlling the content of biosonar images as will be described below.

2.1.3.2 Compact size

The next challenge is the small size of the big brown bat [28]. The entire biosonar system – including the transmitting antenna (open mouth), the receiving antennas (external ears), the transducer apparatus (the inner ear), and the processing and display hardware (auditory pathways in the brain) – fits on or in the bat's head, which has dimensions of only a few centimetres. The processing apparatus – the brain – occupies a volume of about 1 cm^3, only part of which actually includes the auditory system. Man-made systems often contain an array of multiple receiving elements that stretches over an effective receiving antenna aperture of many wavelengths. The sharp angular resolution of the system is based on the width of this compound aperture, which has a narrow receiving beam that can be steered in different directions by adjusting the delays of the signals coming off the various array elements. The method of excluding clutter (mentioned above) by keeping the beam narrow to avoid ensonifying the surrounding clutter is a classic example of the utility of extended, multi-element arrays [9,10]. In contrast, the bat's mouth and ears have dimensions comparable to only one or two average wavelengths [28]. Figure 2.2(a) illustrates the configuration of the transmitting and receiving structures on the bat's head. The big brown bat uses wavelengths from 17 mm (at 20 kHz) to 3.4 mm (at 100 kHz), while the bat's open mouth is about 10 mm wide and up to 5 mm high [34]. Each external ear is about 14 mm high and 5 mm wide, and the length axis of each ear is tilted to the side. Figure 2.3 illustrates the emitted beam [35] and the receiving beams for the left and right ears [36] at 25 and 40 kHz (in FM1) as well as 60 and 80 kHz (in FM2). These broad beams are consistent with the small size of the effective emitting and receiving antenna apertures available to the bat. Most critically, the spacing of the left and right ear-canal openings is about 14 mm, which defines the aperture of the ears if they are considered as the two receiving elements

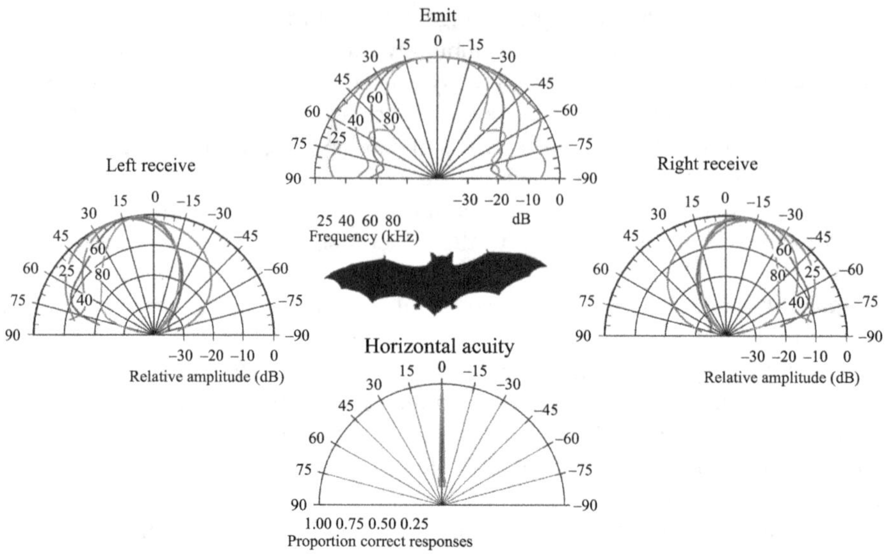

Figure 2.3. Widths of emitting and receiving beams for the big brown bat at 25, 40, 60, and 80 kHz for comparison with the horizontal (crossrange) acuity of the bat in tests of two-point resolution or resolution of the width of five-point gratings

in an array. For comparison, Figure 2.3 also shows the horizontal (i.e., crossrange) angular acuity of the big brown bat [37]. While the emitting and receiving beams are broad (±30°–90°), the angular acuity of the bat's perception is only about 1°–2°, which implies the use of an aperture width greater than the size of the bat's head. By registering echo delay and target shape separately at each ear and then comparing (i.e., crosscorrelating by a method described below) the binaural information, such angular acuity is achievable. The bat may also use a biological version of the monopulse technique, although its time–frequency signal representation implies different computations than the man-made systems of this type. An obvious technological goal is to design man-made systems with antennas and processing elements that have sizes comparable to the wavelengths of the signals, instead of the current lengths of multi-element arrays that are much longer than the wavelengths, while achieving the angular acuity associated with longer array size.

2.1.3.3 Adaptability

The third challenge is that the bat's echolocation is very versatile and can be used for a multitude of tasks. It serves the bat for all aspects of orientation, guidance, and object perception in the immediate surroundings. The operating range of biosonar depends on the strength of broadcasts, the distance to reflecting objects, their target strengths, and the capacity of the receiver to use the energy in echoes to achieve detection [9,10]. For big brown bats, small targets such as insects are detectable at roughly 5–10 m, while large surfaces such as the ground or screens of vegetation would be detectable as far away as 20–40 m [38]. Consequently, the bat's effective surroundings extend as far away as 10–40 m depending on the nature of the objects comprising the scene. Considering these aspects, it is important to keep in mind the dynamic nature of the scene, too: at a flight velocity of 5 m/s, the bat will cross its operating range for insect-hunting in only a second or two, and the entire scene, including distant clutter, in 4–8 s. Another aspect of versatility is evident when bats routinely fly from simple surroundings, such as an open area while hunting for flying insects, to an area with several reflecting sources, such as when trying to capture an insect from the ground, near the surface of a pond, or on vegetation [30] as well as within a seriously cluttered space, such as when flying through vegetation on the way to a feeding area. This challenge also includes bats flying in close proximity to other bats or even during bat-to-bat chases, in which bats need to deal with conspecific echolocation behaviour.

In short, engineering design often favours use of a different system for different conditions – open-area search, target tracking, interception, guidance in clutter, target classification – whereas the bat instead uses the same system for all of its needs.

The adaptability of biosonar is most visible in the signals themselves [2]. Big brown bats change the duration, and frequency spectrum of their FM sweeps, as well as the interval between broadcasts according to the distance to objects of interest and the composition of the scene [14–22]. The bat's solution to environmental challenges is at least in part to alter the sounds, although the underlying processing system also has to accommodate these altered signals. The difficulty here is that the auditory mechanisms of biosonar are deeply concealed. Furthermore, they have been studied extensively primarily in static tests of neural responses, and only somewhat in truly dynamic conditions of natural tasks [2,39]. The second part of this chapter outlines several principles of processing gleaned from neural data in conjunction with results of behavioural experiments on biosonar performance [5,8,11]. The adaptability of the bat's sounds typically takes place on a sound-to-sound basis [14,15] and the receiver must, in effect, accommodate these changes "on the fly". A sonar or radar system can be designed to perform optimally for any one task [9], such as echo detection or echo-delay resolution [40]. The necessary changes in signal processing can be identified from the processing algorithms and the content of the resulting images. However, achieving very good performance across multiple tasks is a different problem that is not so obviously solved with one specific receiver design. While big brown bats appear to have near-optimal echo delay accuracy [41] and very likely better delay resolution [42,43], how do these benchmark measurements fit into the wider design problem of achieving successful performance in complex, dynamic environments?

2.1.3.4 Real-time operation

The fourth challenge is the bat's ability to operate in real time. Video recordings and acoustic tracking studies document big brown bats making surprisingly complex manoeuvres and using clearly-defined rules for steering when capturing prey or operating in clutter [13,17–22,26,30,44–48]. The time course of whole interceptions is only a second or two, and steering manoeuvres involve action in fractions of a second [2]. A good estimate of the fastest reaction speed in bats is derived from head-aim tracking of a moving target; the big brown bat points its sonar beam at the target's location for the most recent echo with an accuracy of about 1°–2° [49]. Because big brown bats rely on biosonar for finding and intercepting prey, as well as for guidance when getting from place to place, there is little room for compromising on real-time performance. In effect, an off-line, slow sonar means a dead bat. The design of a receiver that has both versatility and real-time capability in a wide variety of situations concentrates our attention on how bat sonar "works".

2.1.3.5 Rejection of clutter interference

The fifth challenge is the ability of bats to fly and even to find prey in cluttered environments such as vegetation. Here, clutter is defined as multiple objects located off to the sides in relation to the bat's direction of acoustic gaze, which is fixated onto the target by head-aim tracking, or objects located farther away, in the background. Because echoes from clutter often arrive in the same window of time as echoes from the target, they can interfere with perception of the target's location and distinguishing features. Echoes from the ground, a common problem for airborne radar, or multipath reverberation, a common problem for underwater sonar, fit this description of clutter. To assess the ability of bats to guide flight in sonar scenes with dense, extended clutter, big brown bats were flown through arrays consisting of multiple rows of closely spaced plastic chains hanging from the ceiling of a test room (Figure 2.2) [17,18,20,22,50]. In such conditions, each broadcast is followed by a series of echoes from all of the chains that extend from the bat's momentary position farther into the room. Figure 2.4 shows the spectrogram of a typical FM biosonar broadcast emitted during flight in a chain array followed by a series of echoes from successive rows of chains that arrive at different delays according to the distances to each chain (~6 ms delay per meter of range) [50]. The array of chains is a good surrogate for the complexity of echoes from natural vegetation [51,52]. The length of the flight room was 8 m, and the

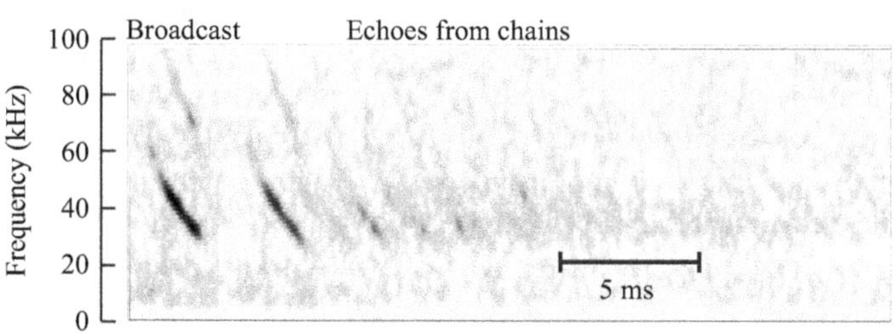

Figure 2.4. Spectrogram showing an FM biosonar broadcast (left) followed by a long series of echoes returning to the bat following this broadcast. The bat is flying through an array of vertically hanging plastic chains in tests of clutter rejection and flight guidance (Figure 2.2(a)) [50]. The sounds were recorded by placing a miniature radiotelemetry microphone on the bat's head to pick up these sounds

chains are very strong reflectors [20], so all of the echoes are audible to the bat [38]. In other words, the bat receives strong echoes not just the nearest row of chains but from all the chains comprising the entire scene – row after row of chains all the way to the end wall of the room. In this situation, because the bat's transmitted beam is very wide, encompassing virtually all of the scene that the bat faces, echoes from potential obstacles along the upcoming flight path will necessarily be arriving in the same time window as the echoes from offside or distant chains, and the bat has to peer into the corridor to determine that it is free of collision hazards [22]. Furthermore, the proximity of the nearest chains requires the bat to emit sounds at short interpulse intervals to manoeuvre through the array of obstacles. The short intervals result in pulse-echo, or range, ambiguity [9,10] because echoes of one broadcast still are arriving when the next broadcast is emitted. The bat adapts its pulse emission rate to the dimensions of the scene by emitting successive sounds at alternating long and short intervals [22]. The resulting pattern consists of pairs or triplets of closely spaced sounds interspersed with longer intervals [14–22]. The sheer number of cluttering echoes still is large, however, too large for an interpulse-interval solution to cope with all of the instances of ambiguity. The big brown bat employs a further means to keep the wrong echoes from being associated with each broadcast – it changes the frequencies of its FM sweeps in successive sounds belonging to pairs or triplets by a few percent to distinguish echoes that belong to one particular broadcast from those that belong to the other broadcast [50]. The problems of target localization, target shape classification, plus rejection of clutter and ambiguity really dominate the quest for better sonar and radar designs. The ability of bats to perform these tasks is the guiding inspiration for trying to understand how their echolocation "works". The bulk of this chapter describes the computations carried out within the bat's auditory system – its sonar receiver – based on experiments undertaken to shed light on the receiver's mechanisms. The following sections trace the passage of each broadcast and its echoes through the bat's receiver, from the inner ear which transduces the sounds, along the neural pathway of auditory nervous system for processing the representation created during transduction, to the images the bat perceives each time it receives an echo [3–5,8,11–14,53]. They describe the proposed processing operations that take place in the stages leading up to the composition of acoustic "images" in the bat brain.

2.2 Computational model of biosonar: spectrogram correlation and transformation (SCAT) receiver

2.2.1 Time–frequency representation in FM biosonar

The big brown bat's auditory system receives both the FM broadcasts and the echoes through the ears. An internal time–frequency replica of each broadcast is formed and retained in the auditory system while the broadcast itself is propagating through the air, to impinge on and reflect from objects before returning to the bat as echoes. Then, returning echoes are cast in a time–frequency representation that has the same qualities as that of the broadcast. Each echo's time–frequency characteristics are then compared to the internally stored replica of the broadcast to detect any changes that might have been introduced by the target or by the location of other reflecting objects. The computational model of biosonar imaging derived from experiments on big brown bats operates on these spectrogram-like time–frequency representations. First, to determine overall echo delay, the spectrograms undergo crosscorrelation along the time axis of the representation – Spectrogram Correlation (SC) [54]. This is implemented by neuronal delay-lines in the bat's auditory system [8]. Second, to determine target shape, the spectrograms undergo a sieving process that finds frequencies where echoes have relatively lower amplitudes that broadcasts. The numerical values of these frequencies are used to estimate the time-separation of multiple reflections that could interfere with each other to create spectral nulls at these frequencies [8,54]. The result is Spectrogram Transformation (ST) [54], a form of deconvolution that modifies the echo-delay estimates acquired from SC to incorporate the much shorter delay separations between multiple reflecting points in the target – called *glints*. The combined process is SCAT [54]. As the necessary precursor of this process, the bat listens to each broadcast as it is transmitted through the open mouth and sets up a time–frequency replica of that signal to compare to characteristics of the time–frequency representation of returning echoes. Detection of differences between echoes and broadcasts is based on registering microsecond-sized changes in the timing of neural responses making up the time–frequency representations and by locating individual frequencies where such small time changes occur [3,4,8,11,12,50]. Creating a new time–frequency replica for each broadcast allows the bat to change successive broadcast waveforms and adapting to conditions in the environment, without having to rewire the receiver. Instead, the receiver automatically accommodates each new replica. The neural responses that comprise this internal replica of the broadcast propagate across populations of neurons at successive times while the sound travels through the air [8]. Then, as the echoes arrive, they are received and transduced into the same time–frequency format for comparison with the existing replica of the original broadcast that already exists [8,53]. As is necessary in any pulse-echo radar or sonar system [9], the lifetime of each broadcast's replica has to encompass the span of time during which echoes might be received. For bats, this epoch is about 24–30 ms for echoes from small insects out to about 4–5 m of range, and probably as much as 50 ms for other, more distant, elements of the scene [38].

After the bat's broadcasts or echoes are received by the ears, transduction takes place in the inner ear. The cochlea, which transduces mechanical movement into neural signals, employs hundreds of parallel bandpass filters tuned to frequencies in the echolocation band (20–100 kHz for big brown bats) [2,53]. These bandpass filters are organized in the descending order of frequency along the organ of Corti, a structure within the cochlea, and have frequency responses that overlap substantially from one filter to the next. They segment each FM sweep into hundreds of parallel frequency channels that are sent into the auditory neural pathway for subsequent processing, which takes place on the framework provided by the frequency channels, with only local interactions among neighbouring channels. Figure 2.5(a) depicts the time–frequency (i.e., spectrogram) plane of sound

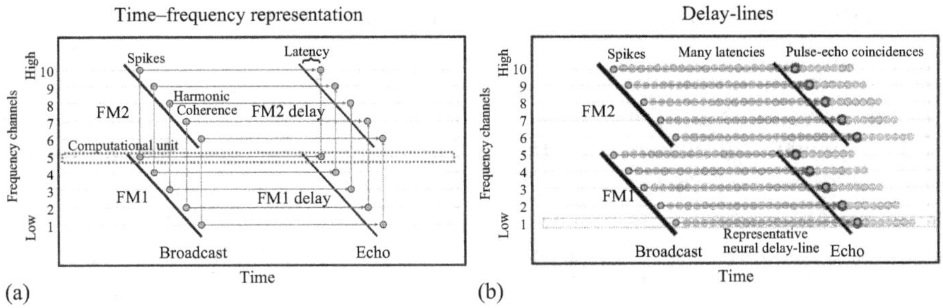

Figure 2.5. *(a) Diagram of the big brown bat's auditory time–frequency representation for the FM sweeps of a broadcast (left) and an echo (right). (b) Same diagram as in (a) but with the neural delay-lines that store and propagate the time–frequency replica of the broadcast for comparison with the incoming echo. Neural responses (circles) register coincidences between responses to the delayed broadcast template and responses to the echo*

transduction, with *time* on the horizontal axis and *frequency* on the vertical axis. The example of a broadcast signal contains two harmonics (FM1, FM2; sloping black lines for the broadcast and for the echo) and sweeps downward in frequency, keeping corresponding harmonic frequencies in a 2:1 ratio. The horizontal scale of the diagram shows the epoch of time that contains a broadcast and its echo. It includes the sounds themselves (shown as sloping lines that trace the FM sweeps) as well as the neural responses (sloping rows of dots) that result from transduction of the FM sweeps. The vertical frequency scale differs from a conventional spectrogram in two ways: First, it is not produced by a series of short-term Fourier transforms (FFTs) operating on overlapping serial segments of the incoming signals. Instead, it is created by passing the FM signals through a bank of parallel bandpass filters tuned to centre frequencies that encompass the entire band of the bat's sounds. An FFT breaks up the FM sweeps into short, overlapping frequency segments, but it does so serially, by executing large numbers of multiplication operations in sequence on matrices of data, whereas the inner ear does the same segmentation simultaneously on all of the frequencies in the signals. The bandpass filters' centre frequencies define the vertical frequency scale. Here, only ten filters are illustrated (filter frequencies are indicated schematically from 1 to 10 in the order of their centre frequencies). The frequency scale defined by the spacing of the filters' centre frequencies in Figure 2.5 is not linear in frequency but exponential [2,53]. To a first approximation, the spacing of the centre frequencies is logarithmic with frequency (from channel 1 to channel 10 in Figure 2.5). Consequently, FM sweeps that appear curvilinear in conventional spectrograms (Figure 2.2(b)) presumably become nearly straight lines on the bat's own internal time–frequency surface. Furthermore, as a departure from conventional signal-processing methods, the cochlear bandpass filters are not simply a feature of the initial auditory "analogue-to-digital" scheme for creating a time–frequency representation of broadcasts and echoes. The existence of parallel frequency channels persist through all subsequent stages of processing. Prior to image formation, these channels are not replaced by summary representations such as the spectrum or correlation function that are more familiar in engineering terms. The hypothesis is that the basic computational elements are the individual frequency channels (an example traced by grey dashed rectangle) plus the sequence of neural circuits housed inside each frequency channel, and replicated from one channel to another.

The echo arrives after the broadcast is sent out, displaced to the right in the diagram of Figure 2.5(a). Like the broadcast, it also appears as two sloping black lines for the two harmonics.

Finally, we hypothesize the following sequence of events to create the neural representation: the two downward FM sweeps of the broadcast on the left simultaneously pass rapidly along the array of bandpass filters (numbered from 10 to 6 for FM2 and from 5 to 1 for FM1) tuned to successive frequencies in those harmonics. Activation of each filter leads to the triggering of neural responses – single spikes (dots in Figure 2.5(a)) – that register the time-of-occurrence of the frequencies in a new, neural spectrogram composed of the sloping rows of spikes. This transformation involves a serious compressive nonlinearity caused by the mechanoelectric properties of the receptor cells that amplify weak inputs and provide progressively less amplification as inputs become stronger [2]. The resulting spike-time representation traces the instantaneous frequency of the sweeps by the single spikes spread across the parallel frequency-tuned computational units (one is outlined by horizontal dashed lines in the diagram). To keep this representation very sharp, the bat's auditory system suppresses the occurrence of any potential second or third spikes that might otherwise occur when each frequency in the broadcast is registered [8]. Subsequent processing thus can be based entirely on the timing of these single spikes without risk of misrepresenting the broadcast itself. When the echo is received, we propose that its harmonics also sweep rapidly along the array of bandpass filters and trigger two new rows of spikes (dots) that trace the frequencies within each harmonic (Figure 2.5(a)). The original acoustic delay of the echo is broken down into a series of elapsed times (horizontal arrows) between the response to the broadcast at each frequency and the response to the echo. Physiological experiments show that the bat's auditory system then extracts these elapsed times, frequency by frequency, and merges them into a single overall delay estimate that represents the target's accurate range [2,3]. As would be expected, delay acuity deteriorates when frequencies are removed from echoes, but the effect is not simply a loss in proportion to the narrowing of bandwidth. For fine delay acuity, this integrative process combines time estimates across frequencies to achieve an accuracy not proportional to the reciprocal of absolute bandwidth or centre frequency, but approximately in proportion to the reciprocal of *relative bandwidth*, or the ratio of centre frequency to bandwidth [55]. In addition, a new process is initiated by weakening or outright removal of some frequencies in echoes that creates new components in the delay images by a process akin to deconvolution [8,54]. Here, the harmonic structure of the sounds plays a decisive role for translating missing frequencies into changes in the images. Any given frequency in FM2 is twice the corresponding frequency in FM1 (2:1 ratio for coherence of harmonics indicated by vertical dashed arrows connecting corresponding harmonic frequencies). The harmonically related frequencies necessarily occur simultaneously in the acoustic signals. However, when the *neural representation* of the harmonics is disrupted, the effects on the images are disproportionate, and they are essential for both target shape perception and clutter rejection. The next sections explain how this takes place.

2.2.2 Determination of echo delay – target range images

Perception of delay is derived from the timing, or latency, of neural responses evoked by each echo in relation to the latency of responses to the preceding broadcast [8,53]. Building on Figure 2.5(a), the diagram in Figure 2.5(a) shows the system of neural delay-lines that store the broadcast template for comparison with the incoming echo. In essence, at each frequency there is a series of neurons that respond to the broadcast with a single spike at different latencies in each cell, collectively covering up to about 50 ms across the whole population of cells. When an echo arrives, the response evoked by the echo is matched up with the ongoing response to the broadcast in each delay-line to register the echo's time-of-arrival in relation to the broadcast. By this means, each frequency contributes an estimate of delay, and these estimates are pooled across frequencies to create an image of echo delay or target range. (Although this chapter describes echo-delay imaging by neural circuits that derive their inputs from only one ear, in fact the bat has two parallel delay-determining pathways that generate delay images for both ears and reconstructs a range/crossrange image from binaural differences in delay [2]. The sharpness of binaural acuity in azimuth (~1°) shows how internal left-right delay comparisons supplant the wide broadcast and receiving beams to create sharp azimuth images.) The diagrams in Figure 2.6 explain the supporting algorithms that convert the bat's time–frequency representation of each broadcast and its echo delay (Figure 2.5(b)) into a target-range image. The parallel frequency channels created by the bat's inner ear (Figure 2.5(a)), depicted by horizontal rows of events and distributed in time, persist throughout the delay-imaging process. The information carried by responses in these frequency channels is integrated across frequencies only at the last stage, at the "moment of image formation".

The two diagrams in Figure 2.6(a) show the same time–frequency plane as in Figure 2.5, with *time* on the horizontal axis and *frequency* on the vertical axis. The associated diagrams in Figure 2.6(b) show the computational process which we propose turns the corresponding echoes from Figure 2.6(a) into images that display target range. Two cases are illustrated in Figure 2.6(a) – reception of the FM broadcast itself followed by reception of a strong echo (left panel) or a weak echo (right panel). In both cases, the echo has a "flat spectrum"; that is, it has the same spectrum as the incident sound, as though it were reflected by a single reflecting point located on the axis of the broadcast beam. In effect, the echo is just a delayed replica of the broadcast, with no coloration of its spectrum. The only difference is the strength of the echo. In the left panel of Figure 2.6(a), after the broadcast is received, passed through the parallel bandpass filters, is transduced, and then

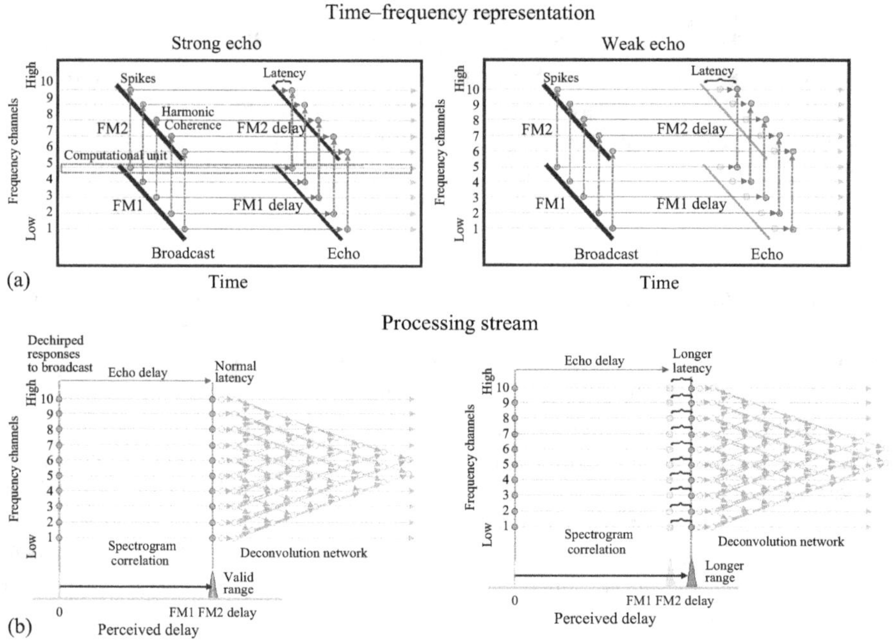

Figure 2.6. Diagram of the SCAT model I: formation of echo delay images by SC for target ranging. (a) In the bat's time–frequency representation, the latency of neural responses (dots) carries information about echo delay, which depends on echo amplitude due to amplitude–latency trading. For the weak echo (right) the longer response latencies cause the echo to appear to arrive at a longer delay. (b) Delay is determined by integrating all of the spectrogram delays at individual frequencies to form an overall delay image (peak) that registers target range. The separateness of delay estimates across frequencies leads to dechirping of the FM sweeps (transition from sloping rows of responses in (a) to vertically aligned rows of responses in (b))

converted into two sloping rows of neural responses that trace the harmonics (FM1, FM2), a strong echo arrives (displaced to the right in the diagram). Like the broadcast, this echo is also transduced and converted into two sloping rows of neural responses. Note that the two harmonics are registered as separate rows of responses in both the broadcast and the echo because the frequency-tuned channels that created the time–frequency representation keep them segregated by their frequencies. No transform is used to produce the neural spectrograms because the bandpass filters do the frequency segmentation all at once mechanically in the cochlea. Neural responses in the auditory pathway occur after a brief latency due to a cascade of cellular events involving mechanical activation of receptor cells in the cochlea followed by the neuronal spike-generation process [5,8]. In the left panel of Figure 2.6(a), for both the broadcast and the strong echo, the spikes (dots) depicted in the diagram are displaced slightly to the right of the echo (black sloping lines) as an indication of this intrinsic delay representing response latency. The neural responses occur independently in each frequency channel, and their latency is dependent not only on the timing but the amplitude of the sound at each frequency. The peculiarity of auditory coding is that weaker levels of sound at each frequency trigger their responses slightly later than stronger levels [56–58]. This dependence of spike latency upon amplitude is called "amplitude–latency trading". In big brown bats, amplitude–latency trading is about 15–20 μs of added latency in the inferior colliculus for every decibel of sound attenuation. From an engineering viewpoint, it seems unreliable to have registration of the timing of different frequencies in sounds also be dependent on their amplitudes. In conventional analogue-to-digital representation, the numerical values that represent the amplitudes just become smaller as sounds are attenuated, not later in time. However, the actual size of amplitude–latency trading – in tens of microseconds for a few decibels of attenuation – does seem minuscule, so the initial reaction is that in practice it is not a serious distortion of the time–frequency representation. Instead, behavioural tests with the big brown bat have shown that even very slight increases in the latency of spikes amounting to only a few microseconds turn out to have dramatic effects on the bat's perception [12]. Amplitude–latency trading highlights why it is so difficult to model biosonar using digital signal-processing transforms. In Figure 2.6(a), the latencies of responses to the strong echo (left panel) are only slightly longer than the latencies to the broadcast itself, so the overall delay estimate based on these neural responses is very close to the actual acoustic delay of the echo. In contrast, in the right panel of Figure 2.6(a), where a weak echo arrives instead of the strong echo (left panel), the amplitude–latency trading effect is larger due to the attenuation of the echo. Like the broadcast, it is also converted into two sloping lines of neural responses (dots), but the latency of these responses is longer because the echo is weaker. Frequency-by-frequency, the elapsed times between neural responses to the broadcast and responses to this weaker echo are longer than the corresponding elapsed times for the stronger echo. Note that the added response latencies are the same for all of the frequencies in the weak echo because the reduction in echo strength occurs equally across the echo spectrum.

The latency difference between responses to the strong echo and the weak echo in Figure 2.6 illustrates how the delay of echoes is determined from the neural broadcast-to-echo elapsed times. In Figure 2.6(b), the left panel shows the processing stream that produces the overall delay estimate for the strong echo and, thus, the bat's image of target range. The first point to appreciate is that by deriving independent estimates of delay at individual frequencies, the bat compresses, or "dechirps" the FM sweeps automatically when the overall delay estimate is compounded across frequencies. This property of biosonar processing is demonstrated in the diagrams of Figure 2.6(b) by the vertical arrangement of the neural responses. The ideal or matched-filter representation of FM signals in chirp radar or sonar is the compressed version of the signals, which corresponds to the crosscorrelation function of the echo with the broadcast [9,10]. We hypothesize that the bat perceives a version of this crosscorrelation function that it produces when it combines individual delay estimates across frequencies (spectrogram correlation, SC). The panel for the strong echo at the left illustrates dechirping by the conversion of the two sloping rows of neural responses in Figure 2.6(a) into a vertical row of neural responses in Figure 2.6(b). The echo, a delayed replica of the broadcast, evokes a similar set of two sloping rows of neural responses that are dechirped into a vertical row of responses in Figure 2.6(b). The resulting tightly compressed vertical row of dots registers the perceived delay of the echo as an image (peak) displayed at the bottom of the panel. In essence, this image would be compounded by vertically integrating the dechirped echo responses across frequencies.

For comparison, in the right panel of Figure 2.6(b), the echo is weaker, so the neural responses to the echo occur at longer latencies. This lengthening of latency becomes a longer perceived echo delay when the dechirped responses are integrated across frequencies to form the perceived delay. Note the displacement of the dark grey image peak slightly to the right, to a longer delay, than the image peak derived from the stronger echo. In accordance with the model's output, we thus propose that neural spectrogram delays are the basis for the bat's perception of target range.

2.2.3 Incorporation of the echo spectrum – focused target-shape images on the range axis

For realistic echoes, the echo spectrum is not flat as in Figure 2.6. Amplitude varies across frequencies owing to atmospheric absorption during sound propagation, the location of the target in the broadcast and receiving beams, and the target's shape. Along with the time–frequency trace of the FM sweeps, the spectrum of echoes is carried into the auditory representation, too, but in a format that is unfamiliar to engineered systems and seems like a defect instead of a useful feature. This part of the story requires some explanation for the reader, as follows: Big brown bats represent the spectrum of echoes – the amplitude of each echo at different frequencies – in terms of the timing of responses that register these frequencies [8]. Through amplitude–latency trading, a new method is introduced in signal processing. The diagrams in Figure 2.7 show how the relative timing of neural discharges at different frequencies affects the bat's perception of objects, both from conventional temporal coding of echo delay to determine target range (see Figure 2.6), and from a novel deconvolution process that use neural response timing to encode spectral information. The format of these diagrams is the same as in Figure 2.6. The top row of panels (Figure 2.7(a)) shows the auditory spectrograms for the FM broadcast and an echo, with three different types of echo – a 1-glint flat spectrum echo (same as in Figure 2.6(a)), an echo from a structured target with two closely spaced glints approximating an insect (e.g., for the head and a wing as different reflecting points), and an echo from a structured target with two glints spaced farther apart in distance, so the reflections are farther apart in delay. The crucial role of amplitude–latency trading as a signal-processing innovation is illustrated by comparing these panels. First, look at the time–frequency representations in Figure 2.7(a). The flat-spectrum echo returned by a 1-glint target as depicted in the left panel in A is essentially a delayed replica of the broadcast (again, same as in Figure 2.6(a)). The corresponding

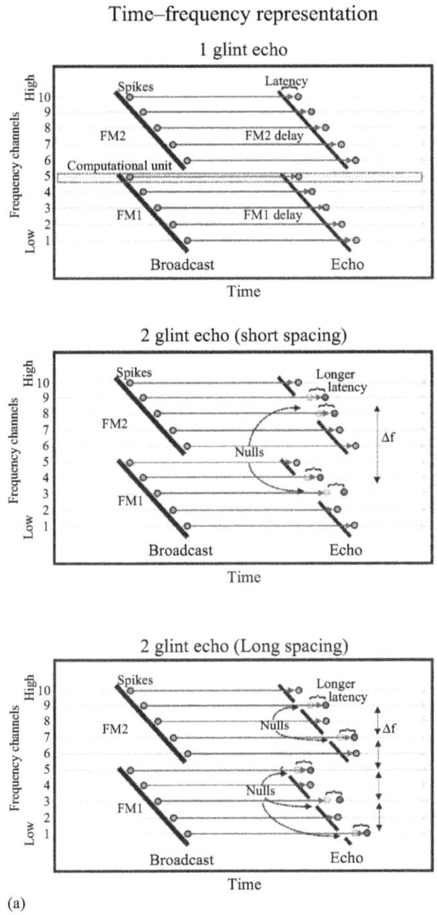

Figure 2.7. Diagram of the SCAT model II: formation of composite biosonar images depicting target shape from ST in tandem with target range from SC.
(a) Time–frequency representations for a 1 glint echo (left; same as Figure 2.6(a), left), a 2-glint echo with short glint spacing (middle), and a 2-glint echo with long glint spacing (bottom). The overlapping glint reflections interfere with each other when passing through the bandpass filters to make the auditory time–frequency representation, creating nulls at specific frequencies according to the time separation of the glint reflections. At the frequencies of the interference nulls, the latencies of responses (dots) are longer than for responses between the nulls (dots), which are used to determine echo delay. Amplitude–latency trading converts the amplitude profile of the echo spectrum into a profile of latencies that portray the nulls as a pattern of responses in time, not amplitude. This amplitude-to-time transformation is unique to the auditory representation of echoes.

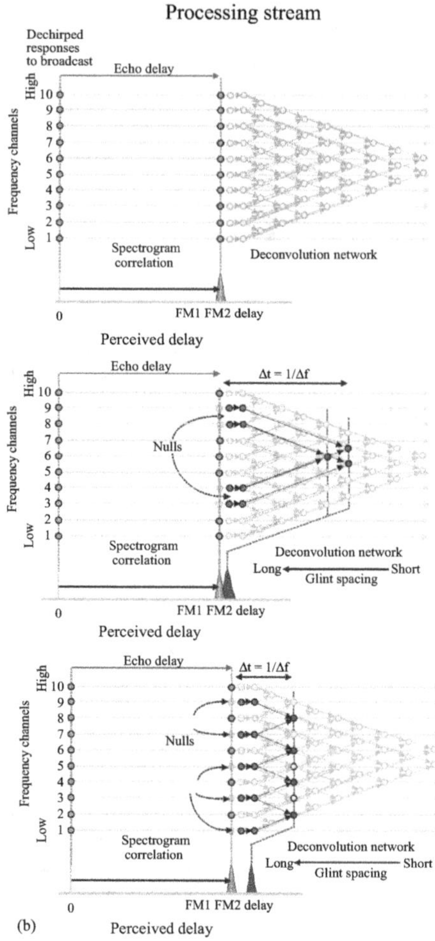

Figure 2.7. (Continued)

(b) First, the longer-latency responses to the nulls (dots) are blocked from contributing to the overall delay estimates (peaks). Second, they are routed into a new neural mechanism – a deconvolution network that uses triangular patterns of connections to determine the frequency spacing of the nulls, which is reciprocally related to the time separation of the glint reflections. By plotting the vertical rows of nodes in the deconvolution network that contain coincidences onto the delay images, just to the right of the overall delay estimates, the separation of the glints becomes part of the delay image, thus expressing target shape in perceptual units of distance, not spectral units of frequency. The grey dashed lines show the relation between the activated nodes of the deconvolution network and the appearance of the additional delay estimate

left panel in Figure 2.7(b) shows how the delay of the 1-glint echo is accurately determined by integrating individual spectrogram delays across frequencies (same as in Figure 2.6(b)). Interference nulls are the distinguishing feature of 2-glint echo spectrograms. They are spaced at frequency intervals along the vertical axis equal to the reciprocal of the time separation of the reflections from the target's two glints. For the short glint spacing (Figure 2.7(a), middle panel), there are only two nulls spaced far apart in frequency because the glint reflections are close together in time. Compare this with the 2-glint echo having a long spacing between glints (Figure 2.7 (a), right). Its spectrogram contains five nulls that are closer together in frequency (narrower interruptions in the FM sweeps) because the glint reflections are farther apart. Based on behavioural tests, the bat's target shape images are built onto the target range images using a scheme that segregates frequencies affected by interference between glint reflections (frequencies at nulls) frequencies that are unaffected by interference (frequencies between nulls). Realistic targets such as insects consist of only a few prominent reflecting glints [23,27], so echo-delay images are a parsimonious basis for portraying target shape, especially if the convergence of left-ear and right-ear delay images into range/crossrange images is taken into account.

In Figure 2.7(b), the images of the 2-glint echoes (middle, right) are formed by adding a spectral null-reading process to the basic mechanism for determining echo delay. The illustrated triangular network of connections describes the neural circuitry for this mechanism, which deconvolves the shape of the spectrum into an image that registers the target's glints.

The image of the 1-glint echo is simple; all of the spectrogram delays are integrated across frequencies to form a sharp image along the axis of perceived delay (same as Figure 2.6(b)). The augmented process applied to the 2-glint echoes illustrates the broader scope for the role of neural response latency as a new tool for extracting information from echoes. The new factor introduced into response timing by the interference nulls in the 2-glint echoes is the effect of amplitude–latency trading. At frequencies between the nulls, where echo amplitude is normal (i.e., unaffected by interference), the individual frequencies in the FM sweeps evoke neural responses (dots) at the same latencies as for the 1-glint echo. These responses trace the sloping sweeps of the harmonics in the 2-glint echoes. In Figure 2.7(b), when the slope of the FM sweeps is dechirped by the vertical realignment of the spectrogram delays across frequencies, the vertical row of neural responses (dots) to frequencies between the nulls is integrated across frequencies to form the perceived delay (peak on perceived delay axis). In contrast, at the frequencies of the nulls themselves, the responses occur at longer latencies (dots displaced to the right) because amplitude–latency trading drags these responses to longer times and misaligns them with respect to the responses between the nulls. This misalignment is detected by inhibitory neural circuits that feed information from each frequency channel to the adjacent channels [8]. If some of the neural responses to nulls in the spectrum are out of temporal alignment with the neural slope of the sweep because the null responses have longer than normal latencies, the model proposes that they are suppressed for purposes of determining echo delay and shunted to another computational pathway for deconvolution. The anomalously longer latencies of neural responses to the frequencies of nulls thus do not distort the overall delay estimates. These estimates depend only on the responses to frequencies between the nulls, which have reliable latencies with respect to overall echo delay. The new computational factor is that at the null frequencies, the longer latencies are fed forward, not into the delay integration process, but into another computational mechanism that operates on the frequencies of the nulls *only*. The triangular network of connections to the right of the diagrams in Figure 2.7(b) depicts this processing stream. In this scheme, the neural responses to nulls (dots) mark the frequencies of the nulls only, not any of the other frequencies. Responses representing nulls enter the triangular deconvolution network which converts the shape of the spectrum into an image of the shape of the target. This network searches for the separation in frequency between the responses affected via amplitude–latency trading (dots) by finding all of the pairwise combinations of different frequencies that yield a match between nulls at different frequencies due to the convergence of the connections. Each of the nodes in the network registers a particular frequency pair, and the pair's frequency separation, by responding to the occurrence of a coincidence between the two inputs (dots in triangular network). Across the whole frequency band of the echoes, the pattern of frequency nulls is entered into the triangular network as a corresponding pattern of responses arranged along vertical rows corresponding to different intersections between pairs of null frequencies. In the middle panel of Figure 2.7(b), the pattern for the 2-glint echo with the short delay separation between reflections has just two widely spaced nulls that converge onto one or two of the vertical rows. The activity in these rows is transposed onto the delay axis of the target's image in tandem with the delay derived from the frequencies that are unaffected by the nulls (additional peak to the right of the peak that represents overall perceived echo delay). The image contains an estimate for the delay separation between the two glint reflections added to the estimate for overall echo delay for target range to create a composite image of both target location and structure on the same spatially dimensioned scale. In the right panel of Figure 2.7(b), the 2-glint echo with the long delay separation between reflections contains five more closely spaced nulls. The responses representing their frequencies and neural frequency separations converge onto a different vertical row in the deconvolution network. Activation of this row is transposed onto the delay axis of the image as an additional peak at a longer delay after the overall delay peak for target range. Comparing the delay images for both of the 2-glint echoes, the target's distance from the bat and the distance from the first glint to the second glint are depicted with sharp peaks. The image of either 2-glint targets is in focus because all of the frequencies in the echo contribute to forming a sharp representation of either the first glint or the second glint. Note that the converging lines from null responses ought to converge on multiple vertical rows further to the right in the triangular network, too. We believe that the bat might limit this spread of converging responses to a "lowest-common-denominator" estimate for the frequency spacing between the nulls.

The final point to make about the deconvolution network is that the order of decreasing to increasing glint-delay separations represented by the vertical rows is reversed in this diagram (right to left) relative to the overall delay axis (left to right). The overall perceived delay of the echo is scaled from short at the left to long at the right, corresponding to short target ranges at the left to long target ranges at the right. In contrast, the leftmost vertical rows in the triangular convolution network register long delay separations between the reflections, while the rightmost vertical rows represent short delay separations. This feature reflected the inverse relation between the time domain and the frequency domain. It is not an arbitrary feature of the diagram.

2.2.4 Defocusing of images for suppressing clutter

Clutter interference is a serious problem for radar and sonar systems that have to operate in complex surroundings with potential targets of interest mixed with a variety of extraneous objects distributed around the scene. Echoes from clutter mask the presence of echoes from any targets, particularly if the clutter and the target are at similar distances but in different directions. For bats, clutter typically is discrete, consisting of leaves and vegetation or the ground at relatively long ranges plus patches of vegetation closer in range. The kind of dense, low-level backscatter common in underwater surroundings is absent in air. The bat's very accurate tracking of the target provides a clue as to how clutter interference is suppressed in biosonar. The target is ensonified by the full spectrum of the broadcast and returns echoes that have spectral modifications almost entirely due to the target's reflectivity, which depends on its shape. The bat keeps its broadcast and receiving beam axes locked onto the target so direction-related spectral coloration does not occur, and at distances of no more than several meters spectral coloration due to atmospheric absorption is negligible. In contrast, for clutter, ensonification is virtually always lowpass in nature. Even though the bat's broadcast beam is quite wide, higher frequencies are less widely spread than lower frequencies [35]. Cluttering objects usually are located off to the sides of the beam due to the target's occupation of the beam's axis, so they are mostly ensonified by lowpass versions of broadcasts and necessarily return echoes that have lowpass spectra as their dominant feature. That is, the prevalence of lowpass filtering in clutter echoes is due to the frequency dependence of the beamwidth. Figure 2.8 illustrates how the lowpass nature of echoes from offside clutter impacts the imaging process that normally produces focused images of a target's glints. Two degrees of lowpass filtering are depicted in the spectrograms in Figure 2.8(a) – slight (left) and strong (right). For such lowpass filtered echoes, the progressive weakening of higher frequencies, especially in the second harmonic, FM2, relative to the first harmonic, FM1, causes the neural responses evoked by these higher frequencies to shift to longer latencies as a body relative to responses for lower frequencies. This pattern is distinct from the longer latencies at local frequencies corresponding to nulls, as in 2-glint echoes (Figure 2.7(a)). The resulting amplitude–latency trading affect tilts the slope of the entire FM2 sweep in the neural spectrograms: the slope changes to become steeper in the neural representation of the spectrogram (dots tracing FM2), even though both FM sweeps in the acoustic spectrogram are the same as in the broadcast (sloping black lines for FM2 compared to row of dots for FM2). Longer response

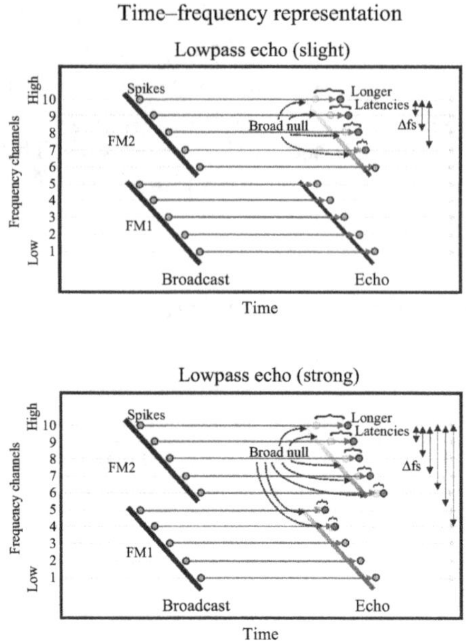

(a)

Figure 2.8. Diagram of the SCAT model III: defocusing of delay images for lowpass-filtered echoes from clutter. The off-axis location of clutter virtually guarantees that these echoes will be weaker at high frequencies because the emitted beam (Figure 2.3) does not ensonify the clutter with the full spectrum of each broadcast.
(a) Time–frequency plots for a slightly lowpass-filtered echo (top) and a strongly lowpass-filtered echo (bottom). For the lower frequencies, where attenuation due to lowpass filtering is minimal, responses (dots) are used to estimate echo delay for target ranging. However, for the higher frequencies, chiefly in FM2, the same amplitude–latency trading effect that isolated the frequencies of individual spectral interference nulls for separate processing in Figure 2.6(a), now segregates the higher frequencies as a block for more global separate processing that causes the images to become defocused. All of the responses to higher frequencies occur at longer latencies (dots), which routes them all into the deconvolution network previously described for extracting glint delays.

latencies at higher frequencies trigger the same neural deconvolution mechanism that detects discrete spectral nulls and serves to render individual target glints into focused images for portraying target shape (Figure 2.7). However, now there are no discrete, local interference nulls surrounded by relatively flat segments of the

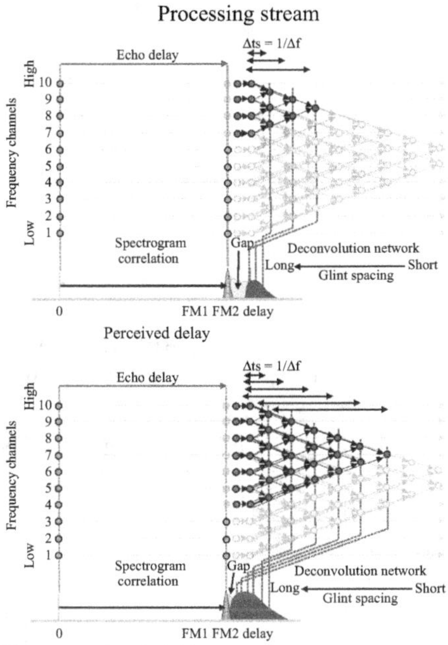

(b)

Figure 2.8. (Continued)

(b) The deconvolution network registers these many delayed responses as equally many cross-frequency coincidences that correspond to numerous putative glint separations. The resulting images contain well-defined estimates of echo delay followed by wide swaths of glint delay estimates that blur, or defocus, the perceived range and shape of the target. The blurring deprives the clutter of sharply focused images and prevents clutter from interfering with perception of the target's focused images

spectrum to unambiguously reconstruct the spacing of a target's glints. Instead, a wide swath of frequencies is lower in amplitude, causing displacement of numerous neural responses to longer latencies. The wide spread of latency retardation caused by lowpass filtering defeats the ability of the converging neural circuits in the deconvolution network to identify individual, discrete nulls [11,12]. Instead, the network interprets all of these longer latencies as coming from numerous individual interference nulls spread over much of the high-frequency (FM2) half of the echo spectrogram. The neural inhibition that finds the misaligned neural response becomes, in effect, a mechanism for detecting the global change in the slope of the FM2 sweep. As is the case for the frequencies that have responses at normal latencies in 1-glint and 2-glint echoes, only neural responses to the lower frequencies in lowpass echoes contribute to the overall perceived delay.

The uniqueness of the bat's solution to clutter is derived from the nature of activation induced by lowpass echoes in the hypothesized triangular frequency-comparing deconvolution network by the wider swath of responses with anomalously longer latencies. The network cannot converge on any single value for the delay of a putative second glint (i.e., there is no salient vertical row of activated nodes in the deconvolution network), so, as illustrated in Figure 2.8(b) it reports all the possible values of glint spacing that might be present over the higher frequencies as a correspondingly large number of glints in the "target". The various pairwise combinations of frequencies that can be extracted from the retarded responses are all registered as valid glint-related responses, which are then transposed into a broad image peak (in Figure 2.8(b)) representing all possible second glints at different delay separations from the first glint [59]. The overall image still registers the target's range, but the part of the image representing shape is blurred by the failure of the network to find a well-defined glint delay. The overall image is defocused to an extent that no part of the image stands out as salient at a specific delay. The defocusing of lowpass delay images illustrated in Figure 2.8(b) has a significant role in biosonar. When the defocused image of a lowpass echo is in competition with the focused image of a flat-spectrum or a null-patterned echo, it does not cause masking. Instead, the lack of a salient peak at a well-defined delay in a defocused image which characterizes clutter echoes, prevents interference with a focused image that has well-defined peaks for discrete glints, which characterizes target echoes. Behavioural tests reveal the bat's "trick" of transforming the individual spectral levels across frequencies into a corresponding profile of neural response-latencies as a novel aspect of signal processing that bats have evolved, one that has no counterpart in contemporary signal processing [12]. It goes far towards explaining the effectiveness of target classification and clutter rejection in real time.

2.3 Principles of biosonar imaging by SCAT

The processing scheme illustrated in Figures 2.6–2.8 summarizes what we know or suspect of how bats create wideband biosonar images. This scheme is based on a time–frequency, instantaneous-frequency representation of the FM sweeps in both the transmitted sounds and the returning echoes. The use of numerous parallel bandpass filters based on mechanical tuning in the cochlea to form the frequency axis of this representation is a big step towards the bat's real-time capability. All of the frequencies in the time–frequency representation are extracted in parallel, not serially as would be the case if a short-term Fourier transform were used to determine spectral history of the sounds for spectrograms. The bat's time–frequency representation includes an odd distorting effect discovered from experiments with the big brown bat [8,12,17,26,42,59] that converts the amplitude of sounds at different frequencies into changes in the latency of neural responses registering the time-of-occurrence of these frequencies. Having the perceived delay of echoes depend on echo amplitude introduces a quantitative misrepresentation of target range that ought to be troublesome for an engineer. The size of the distortion – about 150–200 μs for a 10-dB attenuation of echoes – corresponds to a target-ranging error of about 2.5–3.3 cm. But, when the bat captures a flying insect in its wing or tail membrane, the accuracy of its reaching out to seize the target allows for errors of this size, so in fact the distortion is acceptable in terms of capture success.

The real surprise comes when the consequences of latency shifts are considered in relation to the bat's use of a time–frequency representation for the FM sweeps in its broadcasts and echoes: What the bat gains from amplitude–latency trading is all out of proportion to the nature of the effect itself. The significant outcome is that the profile of amplitude as a function of frequency becomes a profile of latency as a function of frequency [12], which brings the shape of the spectrum into a new computational regime based on converging spikes at coincidence-detecting neurons executing AND operations, which favours real-time action.

Conventional digital signal processing methods rely on transforms, such as the fast Fourier transform, to move information back and forth between the domains of time and frequency. These transforms require executing large numbers of multiplication instructions on arrays of data, which is computationally intensive and time-consuming due to the serial nature of digital computing routines. The operations are made to go faster by combining multiple digital processors in parallel, but many arithmetic steps still have to be executed to form images. We hypothesize that the bat's scheme is, first, to distribute its computations totally in parallel over many single-bit AND operations for detection of coincidences between several input spikes arriving at the same moment at target neurons. These target cells use coincidence detections in conjunction with delay-lines to carry out SC for determining echo delay, and they use coincidence detections in conjunction with frequency tuning of neural channels to carry out ST for determining the frequency separation of spectral interference nulls. Information about the frequency separation of nulls is translated into a representation of the target's glints in the perceived echo-delay images. Even more, the lowpass effect common to virtually all echoes from clutter engages the same amplitude–latency effect over a broader span of frequencies than just the specific frequencies of interference nulls [59]. The outcome is a splatter of putative glints in the images of clutter that cause the bat's images to be defocused for the clutter, which renders them harmless for interfering with focused images of targets [12].

Acknowledgements

This chapter describes the results of work supported by ONR Grant N00014-14-1-05880 to JAS, by a grant from the Capita Foundation to JAS, by internal investments by NUWC Division Newport to JEG, and by a fellowship from the Studienstiftung Deutschland to MW.

References

[1] Griffin, D. R. (1958). *Listening in the Dark* (Yale University Press, New Haven, CT; reprinted by Cornell University Press, Ithaca, NY, 1986).
[2] Neuweiler, G. (2000). *The Biology of Bats* (Oxford University Press, New York).
[3] Simmons, J. A., Ferragamo, M. J., Saillant, P. A., *et al.* (1995). Auditory dimensions of acoustic images in echolocation. In *Hearing by Bats*, edited by Popper, A. N., and Fay, R. R. (Springer-Verlag, New York), pp. 146–190.
[4] Wohlgemuth, M. J., Luo, J., and Moss, C. F. (2016). "Three-dimensional auditory localization in the echolocating bat". *Current Opinion in Neurobiology* 41, 78–86.
[5] Simmons, J. A. and Gaudette, J. E. (2012). "Biosonar echo processing". *IET Radar, Sonar and Navigation* 22, 311–319.
[6] Simmons, J. A. and Stein, R. A. (1980). "Acoustic imaging in bat sonar: echolocation signals and the evolution of echolocation". *Journal of Comparative Physiology* 135, 61–84.
[7] Moss, C. F. and Surlykke, A. (2010). "Probing the natural scene by echolocation in bats". *Frontiers in Behavioral Neuroscience* 4, 33, doi:10.3389/fnbeh.2010.00033.
[8] Simmons, J. A. (2012). "Bats use a neuronally implemented computational acoustic model to form sonar images". *Current Opinion in Neurobiology* 22, 311–319.
[9] Skolnik, M. I. (1980). *Introduction to Radar Systems*, 2nd ed. (McGraw-Hill, New York).
[10] Stimson, G. W. (1998). *Introduction to Airborne Radar*, 2nd ed. (Institution of Engineering and Technology, Mendham, NJ, 584 pages).
[11] Simmons, J. A. (2014). "Temporal binding of neural responses for focused attention in biosonar". *Journal of Experimental Biology* 217, 2834–2843.
[12] Bates, M. E., Simmons, J. A., and Zorikov, T. V. (2011). "Clutter rejection in bat sonar is achieved by image defocusing from harmonic beamforming". *Science* 333, 627–630.
[13] Warnecke, M., Bates, M. E., Flores, V., and Simmons, J. A. (2014). "Spatial release from simultaneous echo masking in bat sonar". *Journal of the Acoustical Society of America* 135, 3077–3085.
[14] Moss, C. F., Bohn, K., Gilkenson, H., and Surlykke, A. (2006). "Active listening for spatial orientation in a complex auditory scene". *PLoS Biology* 4, 615–626.
[15] Moss, C. F., Chiu, C., and Surlykke, A. (2011). "Adaptive vocal behavior drives perception by echolocation in bats". *Current Opinion in Neurobiology* 21, 645–652.
[16] Kothari, N. B., Wohlgemuth, M. J., Hulgard, K., Surlykke, A., and Moss, C. F. (2014). "Timing matters: sonar call groups facilitate target localization in bats". *Frontiers in Physiology* 5, 1–13, doi:10.3389/fphys.2014.00168.

[17] Petrites, A. E., Eng, O. S., Mowlds, D. S., Simmons, J. A., and DeLong, C. M. (2009). "Interpulse interval modulation by echolocating big brown bats (*Eptesicus fuscus*) in different densities of obstacle clutter". *Journal of Comparative Physiology A* 195, 603–617.

[18] Barchi, J. R., Knowles, J. M., and Simmons, J. A. (2013). "Spatial memory and stereotypy of flight paths by big brown bats in cluttered surroundings". *Journal of Experimental Biology* 216, 1053–1063.

[19] Falk, B., Jakobsen, L., Surlykke, A., and Moss, C. F. (2014). "Bats coordinate sonar and flight behavior as they forage in open and cluttered environments". *Journal of Experimental Biology* 217, 4356–4364.

[20] Knowles, J., Barchi, J., Gaudette, J., and Simmons, J. A. (2015). "Effective biosonar echo-to-clutter rejection ratio in a complex dynamic scene". *Journal of the Acoustical Society of America* 138, 1090.

[21] Warnecke, M., Lee, W.-J., Krishnan, A., and Moss, C. F. (2016). "Dynamic echo information guides flight in the big brown bat". *Frontiers in Behavioral Neuroscience* 10, 81. doi:10.3389/fnbeh.2016.00081.

[22] Wheeler, A. R., Fulton, K. M., Gaudette, J. E., Simmons, R. A., Matsuo, M., and Simmons, J. A. (2016). "Echolocating big brown bats, *Eptesicus fuscus*, modulate pulse intervals to overcome range ambiguity in cluttered surroundings". *Frontiers in Behavioral Neuroscience* 10, 125. doi:10.3389/fnbeh.2016.00125.

[23] Moss, C. F. and Zagaeski, M. (1994). "Acoustic information available to bats using frequency modulated sonar sounds for the perception of insect prey". *Journal of the Acoustical Society of America* 95, 2745–2756.

[24] Kober, R. and Schnitzler, H.-U. (1990). "Information in sonar echoes of fluttering insects available for echolocating bats". *Journal of the Acoustical Society of America* 87, 883–896.

[25] Schnitzler, H.-U. and Denzinger, A. (2011). "Auditory fovea and Doppler shift compensation: adaptations for flutter detection in echolocating bats using CF/FM signals". *Journal of Comparative Physiology A* 197, 541–559.

[26] Mantani, S., Hiryu, S., Fujioka, E., Matsuta, N., Riquimaroux, H., and Watanabe, Y. (2012). "Echolocation behavior of the Japanese horseshoe bat in pursuit of fluttering prey". *Journal of Comparative Physiology A* 198, 741–751.

[27] Simmons, J. A. and Chen, L. (1989). "Acoustic basis for target discrimination by FM echolocating bats". *Journal of the Acoustical Society of America* 86, 1333–1350.

[28] Kurta, A. and Baker, R. H. (1990). "*Eptesicus fuscus*". *Mammalian Species* 356, 1–10.

[29] Clare, E. L., Symondson, W. O. C., and Fenton, M. B. (2015). "An inordinate fondness for beetles? Variation in seasonal dietary preferences of night-roosting big brown bats (*Eptesicus fuscus*)". *Molecular Ecology*, doi:10.1111/mec.12519.

[30] Simmons, J. A. (2005). "Big brown bats and June beetles: multiple pursuit strategies in a seasonal acoustic predator–prey system". *Acoustics Research Letters Online* 6, 238–242.

[31] Surlykke, A. and Moss, C. F. (2000). "Echolocation behavior of big brown bats, *Eptesicus fuscus*, in the field and the laboratory". *Journal of the Acoustical Society of America* 108, 2419–2429.

[32] Saillant, P. A., Simmons, J. A., Bouffard, F. H., Lee, D. N., and Dear, D. S. (2007). "Biosonar signals impinging on the target during interception by big brown bats, *Eptesicus fuscus*". *Journal of the Acoustical Society of America* 121, 3001–3010.

[33] Suthers, R. A. and Fattu, J. M. (1973). "Mechanisms of sound production by echolocating bats". *American Zoologist* 13, 1215–1226.

[34] Kloepper, L. N., Gaudette, J. E., Simmons, J. A., and Buck, J. R. (2014). "Mouth gape angle has little effect on the transmitted signals of big brown bats". *Journal of the Acoustical Society of America* 136, 1964–1971.

[35] Hartley, D. J. and Suthers, R. A. (1989). "The sound emission pattern of the echolocating bat, *Eptesicus fuscus*". *Journal of the Acoustical Society of America* 85, 1348–1351.

[36] Wotton, J. M., Haresign, T., and Simmons, J. A. (1995). "Spatially dependent acoustical cues generated by the external ear of the big brown bat, *Eptesicus fuscus*". *Journal of the Acoustical Society of America* 98, 1423–1445.

[37] Simmons, J. A., Kick, S. A., Hale, C., Lawrence, B. D., Bard, C., and Escudié, B. (1983). "Acuity of horizontal-angle discrimination by the echolocating bat, *Eptesicus fuscus*". *Journal of Comparative Physiology* 153, 321–330.

[38] Stilz, W.-P. and Schnitzler, H.-U. (2012). "Estimation of the acoustic range of bat echolocation for extended targets". *Journal of the Acoustical Society of America* 132, 1765–1775.

[39] Popper, A. N. and Fay, R. R., eds. (1995). *Hearing by Bats* (Springer-Verlag, New York).

[40] Sharma, N., Buck, J. R., and Simmons, J. A. (2011). "Trading detection for resolution in active sonar receivers". *Journal of the Acoustical Society of America* 130, 1272–1281.

[41] Sanderson, M. I., Neretti, N., Intrator, N., and Simmons, J. A. (2003). "Evaluation of an auditory model for echo delay accuracy in wideband biosonar". *Journal of the Acoustical Society of America* 114, 1648–1659.

[42] Simmons, J. A., Ferragamo, M. J., and Moss, C. F. (1998). "Echo-delay resolution in sonar images of the big brown bat, *Eptesicus fuscus*". *Proceedings of the National Academy of Sciences* 95, 12647–12652.

[43] Neretti, N., Sanderson, N. I., Intrator, N., and Simmons, J. A. (2003). "Time-frequency computational model for echo-delay resolution in sonar images of the big brown bat, *Eptesicus fuscus*". *Journal of the Acoustical Society of America* 113, 2137–2145.

[44] Ghose, K. and Moss, C. F. (2003). "The sonar beam pattern of a flying bat as it tracks tethered insects". *Journal of the Acoustical Society of America* 114, 1120–1131.

[45] Ghose, K. and Moss, C. F. (2006). "Steering by hearing: a bat's acoustic gaze is linked to its flight motor output by a delayed, adaptive filter law". *Journal of Neuroscience* 26, 1704–1710.

[46] Ghose, K., Horiuchi, T. K., Krishnaprasad, P. S., and Moss, C. F. (2006). "Echolocating bats use a nearly time-optimal strategy to intercept prey". *PLoS Biology* 4, e108, 0865–0873.

[47] Fujioka, E., Aihara, I., and Watanabe, S. (2014). "Rapid shifts of sonar attention by *Pipistrellus abramus* during natural hunting for multiple prey". *Journal of the Acoustical Society of America* 136, 3389–3400.

[48] Hulgard, K., Moss, C. F., Jakobsen, L., and Surlykke, A. (2016). "Big brown bats (*Eptesicus fuscus*) emit intense search calls and fly in stereotyped flight paths as they forage in the wild". *Journal of Experimental Biology* 219, 334–340.

[49] Masters W. M., Moffat, A. J. M., and Simmons, J. A. (1985). "Sonar tracking of horizontally moving targets by the big brown bat, *Eptesicus fuscus*". *Science* 228, 1331–1333.

[50] Hiryu, S., Bates, M. E., Simmons, J. A., and Riquimaroux, H. (2010). "FM echolocating bats shift frequencies to avoid broadcast-echo ambiguity in clutter". *Proceedings of the National Academy of Sciences* 107, 7048–7053.

[51] Müller, R. and Kuc, R. (2000). "Foliage echoes: a probe into the ecological acoustics of bat echolocation". *Journal of the Acoustical Society of America* 108, 836–845.

[52] Yovel, Y., Franz, M. O., Stilz, P., and Schnitzler, H.-U. (2008). "Plant classification from bat-like echolocation signals". *PLoS Computational Biology* 4, e1000032. doi:10.1371/journal.pcbi.1000032.

[53] Simmons, J. A., Saillant, P. A., Ferragamo, M. J., *et al.* (1996). Auditory computations for acoustic imaging in bat sonar. In: H. L. Hawkins, T. A. McMullen, A. N. Popper, and R. R. Fay (Eds.) Auditory Computation, *Springer Handbook on Auditory Research*, Springer-Verlag, New York, pp. 401–468.

[54] Saillant, P. A., Simmons, J. A., Dear, S. P., and McMullen, T. A. (1993). "A computational model of echo processing and acoustic imaging in FM echolocating bats: the SCAT receiver". *Journal of the Acoustical Society of America* 94, 2691–2712.

[55] Simmons, J. A., Neretti, N., Intrator, N., Altes, R. A., Ferragamo, M. J., and Sanderson, M. I. (2004). "Delay accuracy in bat sonar is related to the reciprocal of normalized echo bandwidth, or *Q*". *Proceedings of the National Academy of Sciences* 101, 3638–3643.

[56] Kick, S. A. and Simmons, J. A. (1984). "Automatic gain control in the bat's sonar receiver and the neuroethology of echolocation". *Journal of Neuroscience* 4(11), 2725–2737.

[57] Burkard, R. and Moss, C. F. (1994). "The brain-stem auditory-evoked response in the big brown bat (*Eptesicus fuscus*) to clicks and frequency-modulated sweeps". *Journal of the Acoustical Society of America* 96(2), 801–810.

[58] Ma, X. and Suga, N. (2008). "Corticofugal modulation of the paradoxical latency shifts of inferior collicular neurons". *Journal of Neurophysiology* 100(2), 1127–1134.

[59] Warnecke, M. and Simmons, J. A. (2016). "Target shape perception and clutter rejection use the same mechanism in bat sonar". *Journal of Comparative Physiology A* 202(5), 371–379.

Chapter 3

Enhanced range resolution: comparison with the matched filter

Krasin Georgiev[1], Alessio Balleri[1], Andy Stove[2] and Marc W. Holderied[3]

3.1 Introduction

Detection and recognition of targets amongst other interfering objects has been both a theoretical and a practical problem since the early ages of radar systems [1]. Research on this topic started with investigations looking at resolving individual scatterers within a single target in complex clutter environments and, in the course of the years, has advanced towards much more sophisticated challenges that relate to modern cognitive sensing capabilities [2,3].

Echolocating bats have evolved the ability to detect, resolve and discriminate targets in highly challenging environments using biological sonar. Fifty million years of evolution have resulted in an echolocation system that can achieve remarkable perceptual and navigational performance even in the presence of dense clutter. The way we believe bats process acoustic echoes in their auditory system presents key differences with respect to typical radar and sonar receivers, and hence investigating these differences as well as similarities may result in useful lessons that can be learnt and ultimately applied to synthetic sensors.

Bats can use echolocation alone to detect, localise and discriminate targets [4,5] and, behavioural experiments have demonstrated they can deal successfully with many of the challenges faced by synthetic sensors. They exploit waveform, spatial and temporal adaptations in the process of target detection, resolution and recognition. At the same time, adaptive waveform diversity techniques [6] and cognitive guidance and control [6,7] and memory [7,8] are all characteristics that are studied in the context of cognitive synthetic sensing.

One of the very attractive features of bat echolocation is the ability of bats to resolve very closely located scatterers and various experiments have been carried

[1]Centre for Electronic Warfare, Information and Cyber, Cranfield University, Defence Academy of the UK, Shrivenham SN6 8LA, UK
[2]Stove Specialties, Hove, E. Sussex BN3 5NL, UK
[3]School of Biological Sciences, Bristol University, Bristol BS8 1TQ, UK

out to assess and better understand this capability. In [9,10], for example, Simmons showed that the big brown bat (*Eptesicus fuscus*) could discriminate a single-point target from a two-point target for spacings generating time delays above as little as 2 μs with chirp-like echolocation calls of 2−3 ms duration and a bandwidth of 85 kHz. For this type of echolocation call, the corresponding conventional (or nominal) range resolution of a radar or sonar system, defined as the half-power width of a point target response [11], would be about 12 μs and hence six times worse than the resolution of the bat. Similarly, Schmidt showed that the bat *Megaderma lyra* could discriminate between two phantom target echoes separated by a time-delay of about 1 μs with 0.5 ms long waveforms and a bandwidth of up to 100 kHz [12]. The corresponding nominal resolution is 10 μs. Discrimination performance in the presence of surface structures was demonstrated in [13] where *Glossophaga soricina* were trained to distinguish a smooth surface from a coarse structure with a depth bigger than 0.38 mm. The emitted signal in these experiments was a frequency modulated down-chirp consisting of three harmonics spanning the bandwidths between 95–55, 150–86 and 190–140 kHz. The predicted range resolution is 1.26 mm. There might be little differences in the different reportings but overall the literature presents a consistent story of the bat resolution being better than expected.

The matched filter is a linear filter designed to maximise the Signal-to-Noise Ratio (SNR) when the target echo is a delayed replica of the transmitted signal buried in white noise [14,15]. Work by radar and sonar engineers on bat-inspired sensing has generally assumed that the receiver includes a matched filter (e.g., see [5,6]) and, in the literature, the target range profiles are implicitly associated with the target echo matched filtered to the transmitted signal.

The Spectrogram Correlation And Transformation (SCAT) receiver is a model of the auditory system proposed by Saillant *et al.* in [16] that takes into account the physiology and neural organisation of the bat *Eptesicus fuscus*, that is a bat that emits broadband signals. The SCAT model uses only range information, a single receiver, and does not account for directional sensitivity [16–18]. In the SCAT, the emitted call and received echoes are frequency-modulated signals which are passed through a filter bank to generate an auditory spectrogram which is a model for the bat cochlea. This is followed by a temporal processing block, that models the delay tuned neurons in the brain responsible for target ranging, and by a spectral processing block whose function is to resolve close targets by exploiting echo interferences [16,18]. While the aim of the bat auditory system models is to reproduce the acoustic images perceived by bats, there is no explanation of what exactly brings the range resolution performance improvement in a way that can be exploited in a technological system.

In this chapter, we describe a baseband receiver equivalent to the SCAT that can be applied to both radar and sonar signals. Radar systems, in fact, operate in the microwave band and use carrier frequencies (of the order of a few GHz) and typical bandwidths (of a few MHz) which are much larger than those of sonar waveforms accepted by the original SCAT receiver. The model also allows a theoretical treatment of the output signals of the SCAT and is used to mathematically analyse the resolution capabilities of the spectrogram transformation block. The performance of the model is compared with that of the matched filter with simulated and experimental data from two closely spaced targets. Results show that the Baseband Spectrogram Correlation and Transformation receiver (BSCT) can provide improved range resolution performance with respect to the matched filter.

3.2 Description of the spectrogram correlation and transformation model

The SCAT receiver is a model of the auditory system that accounts for the underlying neural organisation in bats using frequency modulated signals by special emphasis on the physiological correlation of the model. The main modules of the SCAT are the cochlear block, the temporal (spectrogram correlation) block and the spectral (spectrogram transformation) block. The bat auditory system processes both the emitted call and the received echo through an auditory periphery and then through some higher level brain structures. The auditory periphery includes the outer, the middle and the inner ear and provides a time-frequency representation of the input signal. The frequency content is sampled in a non-linear scale (hyperbolic or logarithmic). The consequent central nervous system processing provides an estimate of the time delay between the call and the echo and provides cues about the structure of each echo. A short summary of the SCAT model building blocks (Figure 3.1) based on [16] follows.

Different aspects of the way bats process echoes from stationary targets and the surrounding environment have been studied in the literature and variations of the SCAT have been also presented. The interested reader can refer to [17,19,20] for additional information.

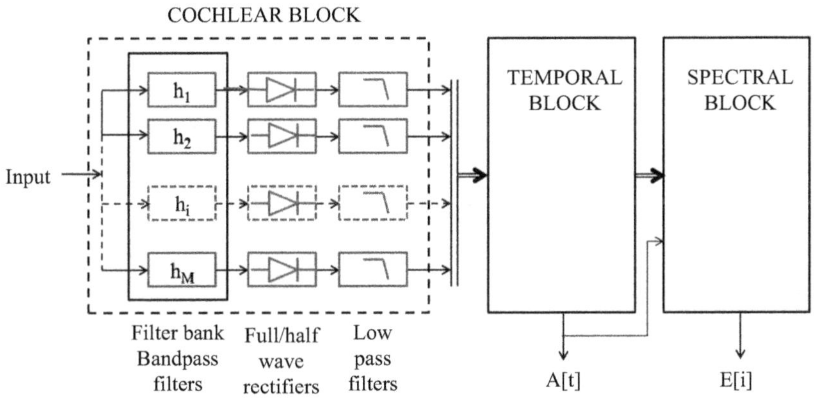

Figure 3.1 Cochlear block of the SCAT receiver. The filter bank consists of $M = 81$ bandpass filters h_i with central frequencies from 20 to 100 kHz. A rectifier and a Butterworth low-pass filter follow after each bandpass filter. Reprinted with permission from [22]

3.2.1 Cochlear block

The cochlear block is modelled with a bank of 81 Butterworth band-pass filters of order 10 and bandwidth $B = 4$ kHz. Each filter is followed by a signal rectifier and a 3 kHz bandwidth low-pass filter (Figure 3.1) in order to extract the envelope of the signal.

The central frequencies f_i of the band-pass filters span the bandwidth between 20 to 100 kHz and are arranged in a hyperbolic scale as $f_i = 1/T_i$, where the central period T_i changes linearly from 10 to 50 μs with increments of 0.5 μs. With a 4 kHz bandwidth at the lower-frequency end of the spectrum the filters overlap (channel spacing about 2.5 kHz) but at the upper end the spectrum is slightly undersampled (channel spacing about 5 kHz).

Two levels of smoothing of the envelope have been proposed with the SCAT, namely a high smoothing and a low smoothing with the high smoothing consisting of a full wave rectifier followed by a second-order Butterworth low-pass filter, whilst the low smoothing consists of a half-wave rectifier followed by a first-order low-pass filter. In both cases the low-pass filter has a bandwidth of 3 kHz. The output of the cochlea is called the auditory spectrogram or cochleogram (Figure 3.2).

In the literature, some modified versions of the original SCAT have been proposed, which differ in how the initial splitting of the signal into frequency channels is carried out. For example, the bank of constant bandwidth Butterworth filters may be replaced by gammatone filters [18] with frequency-dependent bandwidths after [21] or Gaussian chirplets with carrier frequencies compatible with the emission sweep rate [17]. Some models use additional non-linear transformations, e.g., raising the amplitude of the detected signal to a power of 0.4 to account for the non-linear interactions within the organ of corti [18].

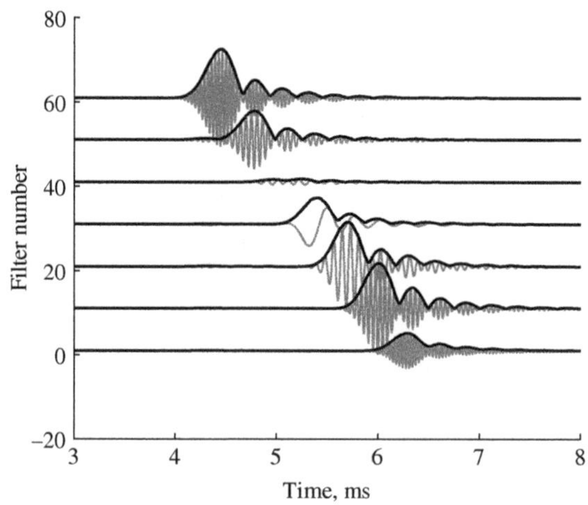

Figure 3.2 Bandpass filter bank output, in-phase (real) component and envelope. Reprinted with permission from [22]

3.2.2 Temporal block

The purpose of the temporal block is to estimate the time delay between the call and the echo based on the output of the cochlear block. It consists of a set of tapped delay lines that implement a cross-correlation function between the call and the echo.

The temporal block processes the output of the cochlear block and converts it into neural spikes. The neural coding can either be considered as part of the cochlear block or not. The spike decay period is 4 μs [16]. A spike rise results in an increase of the threshold and inhibits the same neuron from producing another spike. The spike activation threshold decays to its initial value over a period of 1 ms.

The temporal block carries out a "dechirping" of the signal by adding appropriate delays to each frequency channel (Figure 3.3). Delays are calculated using the emitted signal as a trigger. Simultaneous activity in multiple channels is detected by a set of coincidence detection neurons and is a sign of the target presence. A different approach for signal "dechirping" by channel-wise autocorrelation is proposed in [18]. Target detection is implemented by summing the output over all channels and the target is declared with a peak-detection algorithm.

The mechanism described does not require coherence between the transmitted and received waveform and does not manipulate the phase information. It provides compression of the input chirp from several milliseconds (0.5–15 ms) to several hundred microseconds (200–300 μs) as the impulse response of the 4 kHz filters is relatively wide (Figure 3.3). Therefore it can give the location of the targets with relatively low resolution, compared to the matched filter nominal resolution of 15 μs for 65 kHz bandwidth.

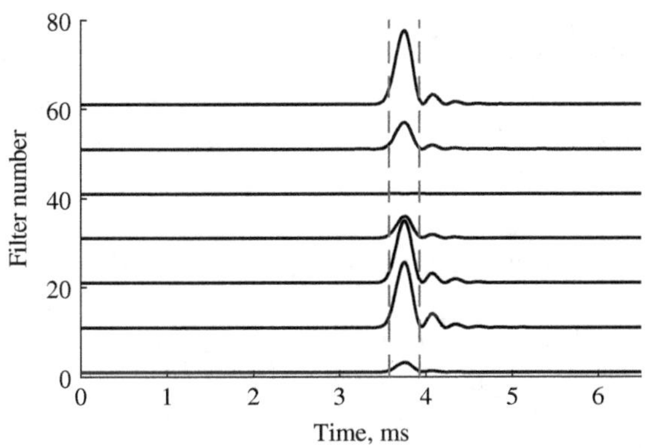

Figure 3.3 Example of the de-chirped squared envelopes for seven filters and integration interval (dashed lines) over which the total energy for each filter is calculated. Reprinted with permission from [22]

3.2.3 Spectral block

The spectral block is responsible for extracting the fine structure of the target. It is used to detect and measure the delay between highly overlapping echoes that cannot be resolved by the temporal block. It is important because it provides the fine range resolution needed to resolve targets [16].

The spectral block exploits the interference pattern between overlapping echoes, which results in the suppression or amplification of the power of the output of some of the filters of the cochlear block. It integrates the output signal of each frequency channel for a specific time interval (about 350 μs [16]). De-chirp delays are also compensated so that only the signals coming from the same target are integrated. Each target is processed by the spectral block separately. Its interference pattern is next normalised by the interference pattern of the emitted call to get a target signature in the frequency domain.

The output of the integration represents the frequency spectrum of the signal. A modified inverse cosine transform is then used to convert the frequency spectrum into the time domain. Some studies concentrate on the time domain reconstruction of the target image [16,17,19] and others emphasise the spectral based perception of the target [13,18]. A "pattern-matching" interpretation of the transformation of spectral interference patterns into fine delays is proposed in [20].

3.2.4 Model output

The outputs of the temporal and spectral blocks could be considered separately. For a group of closely positioned targets the former gives information about location on the time (range) axis for the group as a whole, and the latter describes the intra-group behaviour either as a spectrum which contains the interference between the targets or as fine delays (relative distances).

It is interesting to note that the idea that the target is perceived as a set of groups of echoes is supported by an unrelated experiment [23]. In this experiment the bat should recognise a known target (two scatterers at predefined distance) in the presence of clutter consisting of a single scatterer with variable position. The interference between the clutter and the target was studied. It was observed that delay of less than 50 μs from the clutter to a scatterer of the target leads to increase of the discrimination error. If the delay is higher the clutter does not mask the target even if it is positioned between the target scatterers.

3.3 The baseband spectrogram transformation receiver

In this section, a baseband equivalent of the spectrogram transformation receiver is proposed that can treat analytical input signals at baseband. It is also applicable to RF signals at very high frequencies (bandpass signals) which are acquired as low frequency complex signals (baseband signals). Down-conversion to baseband is essential to digitise bandpass signals meeting the Nyquist criterion and, it also provides analytical advantages as signals can be represented in a much simpler complex form.

Let us consider a filter bank of M filters with central frequencies f_i, $i = 0, \ldots, M-1$ and bandwidth B. If the filters have all the same design, the Fourier transform $H_i(f)$ of the impulse response $h_i(t)$ for the ith filter can be obtained by shifting in frequency a baseband (low-pass) filter response $h(t)$ as

$$h_i(t) = h(t)e^{j2\pi f_i t} \tag{3.1}$$

$$H_i(f) = H(f - f_i) \tag{3.2}$$

where $H(f)$ is the Fourier transform of $h(t)$. The response of the ith filter is the convolution between the filter impulse response and the input signal $x(t)$

$$y_i(t) = x(t) * h_i(t) \tag{3.3}$$

The rectifiers and the low-pass filters that follow each bandpass filter of the original cochlear block (Figure 3.1) are replaced with an ideal amplitude extractor, so that the envelope of the bandpass filtered signal can be modelled with the amplitude of the signal $y_i(t)$ (Figure 3.2). Then the envelopes are converted to an auditory spectrogram by squaring the amplitude of $y_i(t)$ (Figure 3.3).

In the spectral block, the total energy of the output of each filter is computed by integration (Figure 3.4). In order to separate the spectral signature of multiple groups of scatterers, the integration is limited to an interval around the location of the group of echoes under consideration (Figure 3.3) so that each group can be analysed separately. The group locations for each frequency are extracted by the temporal block.

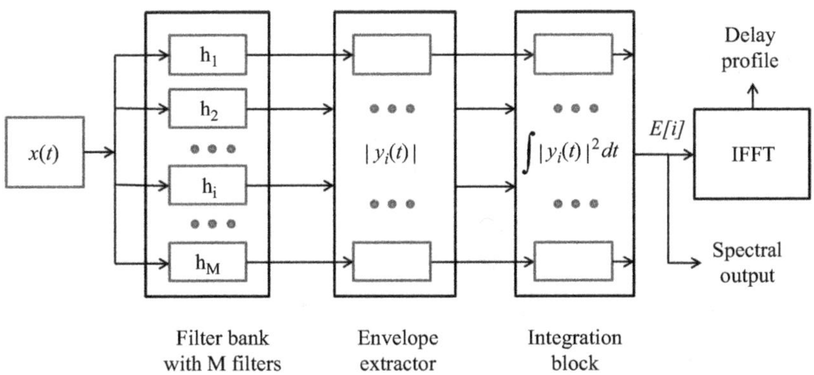

Figure 3.4 BSCT spectral processing diagram. The model input $x(t)$ is an analytical signal. It is passed through a filter bank of M complex bandpass filters h_i. The absolute value of each filter output y_i is squared and integrated over time to get the energy $E[i]$ of the corresponding frequency

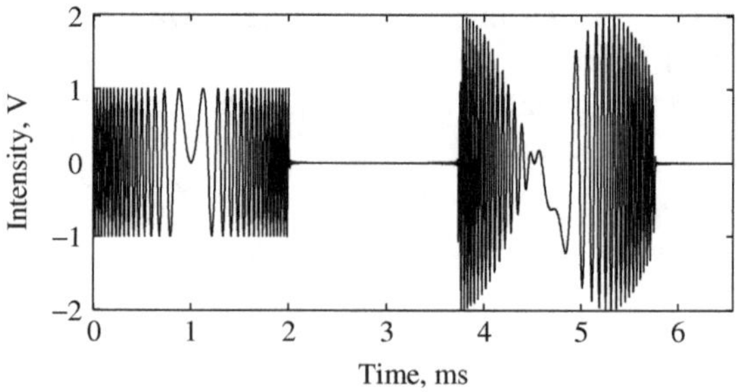

Figure 3.5 Baseband model input, real component. The signal frequency is shifted to baseband. The emitted call has a duration of 2 ms and spans linearly from −32.5 to +32.5 kHz. The received echo comes from two scatterers with a relative delay of 20 μs and equal amplitude. Reprinted with permission from [22]

When only a single group of scatterers is present, the temporal block can be ignored and the output of the spectral block can be obtained by integrating the whole filter output following equation (3.4):

$$E[i] = \int_{-\infty}^{\infty} |y_i(t)|^2 dt = \int_{-\infty}^{\infty} |Y_i(f)|^2 df \tag{3.4}$$

$$E[i] = \int_{-\infty}^{\infty} |X(f)|^2 |H(f - f_i)|^2 df \tag{3.5}$$

When all narrow-band filters in the filter bank have a perfectly flat frequency response over a bandwidth B, i.e.,

$$H_i(f) = \text{rect}\left(\frac{f - f_i}{B}\right) \tag{3.6}$$

where

$$\text{rect}\left(\frac{f}{B}\right) = \begin{cases} 1, & |f| \leq \dfrac{B}{2} \\ 0, & \text{otherwise} \end{cases} \tag{3.7}$$

The filter bank spectral output is

$$\begin{aligned} E[i] &= \int_{-\infty}^{\infty} |X(f)|^2 |H(f - f_i)|^2 df \\ &= \int_{f_i - B/2}^{f_i + B/2} |X(f)|^2 df \end{aligned} \tag{3.8}$$

Equation (3.8) shows the spectrogram transformation provides the signal spectral energy of the input signal integrated over a bandwidth B around the central frequencies f_i, that is equivalent to digitising the energy spectral density by averaging across the filter bandwidth. Finally, the sequence $E[i]$ is transformed into the time domain to obtain the output signal of the spectral block. Assuming the central frequencies f_i are linearly spaced this transformation can be obtained with an Inverse Fast Fourier Transform (IFFT).

Some preliminary experimental results showing the equivalence between the BSCT and original SCAT were presented in [22].

3.4 Response of the BSCT to two closely spaced ideal reflectors

We study the response of the BSCT to the echo from two closely spaced ideal point reflectors in order to investigate the range resolution properties of the model and to allow a comparison with the matched-filter. When two ideal targets are present, the complex envelope of the input signal $x(t)$ is the sum of two delayed replicas of the complex envelope of the transmitted signal $x_C(t)$

$$x(t) = x_C(t - t_1)e^{-j2\pi f_0 t_1} + x_C(t - t_2)e^{-j2\pi f_0 t_2} \tag{3.9}$$

where t_1 and t_2 are the time-delays of the echo from the first and second targets, respectively, and f_0 is the carrier frequency. For simplicity of notation, the difference between the two delays will be denoted as $\tau = t_2 - t_1$.

The Fourier transform of $x(t)$ can be written as a function of the Fourier transform $X_C(f)$ of the signal $x_C(t)$ as

$$\begin{aligned} X(f) &= X_C(f)e^{-j2\pi(f+f_0)t_1} + X_C(f)e^{-j2\pi(f+f_0)t_2} \\ &= X_C(f)e^{-j2\pi f^I t_1}e^{-j\pi f^I \tau}2\cos\left(\pi f\tau + \frac{\psi_\tau}{2}\right) \end{aligned} \tag{3.10}$$

and its energy spectral density is

$$\begin{aligned} |X(f)|^2 &= |X_C(f)|^2 4\cos^2\left(\pi f\tau + \frac{\psi_\tau}{2}\right) \\ &= 2|X_C(f)|^2[\cos(2\pi f\tau + \psi_\tau) + 1] \end{aligned} \tag{3.11}$$

where $f^I = f + f_0$ is the frequency before down-conversion and $\psi_\tau = 2\pi\tau f_0$ is the phase of the cosine wave after down-conversion.

Let us assume that the transmitted signal $x_C(t)$ is a Linearly Frequency Modulated chirp (LFM) with a bandwidth B_C and duration T of the form

$$x_C(t) = e^{j\pi\gamma t^2}\operatorname{rect}\left(\frac{t}{T}\right) \tag{3.12}$$

where $\gamma = B_C/T$ is the chirp rate. The Fourier transform of the chirp

$$X_C(f) = \int_{-\infty}^{\infty} x_C(t) e^{-j2\pi ft} dt$$
$$= \int_{-T/2}^{T/2} e^{-j2\pi\left(ft - \frac{\gamma t^2}{2}\right)} dt \tag{3.13}$$

for large time-bandwidth products ($B_C T \gg 1$), after the stationary phase approximation [24], can be written as

$$X_C(f) = \mathrm{rect}\left(\frac{f}{B_C}\right)\sqrt{\frac{1}{\gamma}} \cdot e^{-j\pi\frac{1}{\gamma}f^2} e^{j\frac{\pi}{4}} \tag{3.14}$$

The result consists of a constant magnitude spectrum with a square law phase modulation and a constant phase residual of $\pi/4$. From (3.11), the energy spectral density of the return $x(t)$ from two scatterers can be written as

$$|X(f)|^2 = \frac{2}{\gamma} \cdot [\cos(2\pi f\tau + \psi_\tau) + 1] \cdot \mathrm{rect}\left(\frac{f}{B_C}\right) \tag{3.15}$$

and the output is

$$E_E[i] = \frac{2}{\gamma}\int_{f_i - B/2}^{f_i + B/2} [\cos(2\pi f\tau + \psi_\tau) + 1] df$$
$$= \frac{2B}{\gamma}[\mathrm{sinc}\,(\tau B)\cos(2\pi\tau f_i + \psi_\tau) + 1] \tag{3.16}$$

for $-B_C/2 \le f_i \le B_C/2$ and $E_E[i] = 0$ otherwise, where $\mathrm{sinc}\,(x) = \sin(\pi x)/(\pi x)$.

Equation (3.16) shows that the spectral output of the SCAT is the sum of a weighted sinusoid and a constant multiplied by a rectangular window. The frequency of the sinusoid is actually the time spacing between the targets τ.

Figure 3.6 shows an example of the energy output for a filter bank of 65 filters with central frequencies distributed linearly between −32 and 32 kHz and with 4 kHz bandwidth. The results are for an input signal consisting of the sum of two LFM chirps with a bandwidth of 65 kHz, centre frequency of 67.5 kHz, a duration of 2 ms and a relative time delay of 36 μs. Setting $\tau = 36$ μs, $B = 4$ kHz and $\psi_\tau = 2\pi\tau f_0$ in (3.16) gives the theoretical curve. Applying the FFT on a simulated chirp sequence with the above parameters and integrating as in (3.8) gives the simulated discrete points on the graph. The simulated curve is a good approximation of the theoretical output of the BSCT. The deviations near the boundaries appear because of those filters at the two ends of the spectrum that cover frequencies with no signal and because the chirp magnitude is not a perfect rectangular function in the frequency domain.

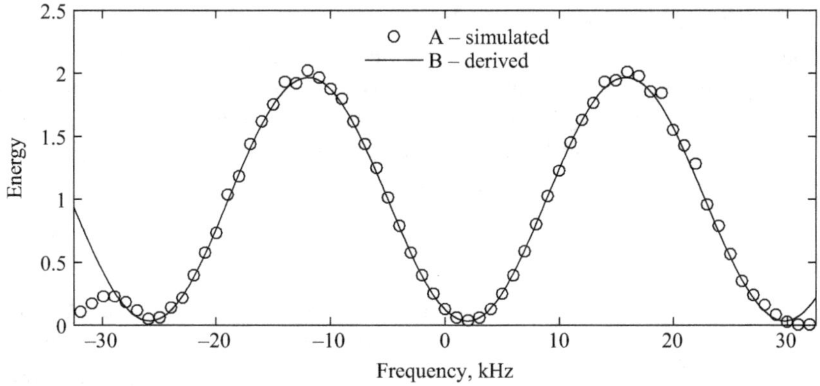

Figure 3.6 Model output, energy E_E by filter i, i = 1, ..., 65, B = 4 kHz, for two overlapping chirps with duration 2 ms, f_l = 35 kHz, f_n = 100 kHz and a relative delay τ = 36 μm. Based on (a), simulated input and (b), the derived theoretical solution in (3.16)

We proceed by considering $E_E(k)$ as the Discrete Fourier Transform of a sequence with a limited bandwidth, B_C, and digitised with a sampling rate $f_s > 2B_C$. The resulting sequence in the time domain is

$$\begin{aligned}\mathscr{F}_f^{-1}[E_E[k]](t) &= 2\frac{B}{\gamma}\mathscr{F}_f^{-1}[\operatorname{sinc}(B\tau)\cos(2\pi f_k\tau + \psi_\tau) + 1](t) * K\operatorname{sinc}(B_C\tau) \\ &= 2\frac{BK}{\gamma}\left[\frac{b\cdot e^{j\psi_\tau}}{2}\operatorname{sinc}(B_C(t-\tau))\right. \\ &\quad \left. + \frac{b\cdot e^{-j\psi_\tau}}{2}\operatorname{sinc}(B_C(t+\tau)) + \operatorname{sinc}(B_C t)\right] \qquad (3.17)\end{aligned}$$

where $b = \operatorname{sinc}(B\tau)$. The inverse Fourier transform of (3.16) is the sum of three sinc functions one of which does not depend on the relative delay between the two targets. The two sinc functions centred at $\pm\tau$, instead, allow an estimation of the delay between the two echoes. It is important to highlight that the two peaks centred at $\pm\tau$ are scaled by a constant $b = \operatorname{sinc}(B\tau)$, which depends on the filter bandwidth B and the delay τ between the scatterers, and this indicates that distant scatterers are attenuated (or "low-pass filtered").

In the theoretical developments above we assumed that the targets are ideal reflectors and the potential phase shifts from the reflectors were not taken into account. Such shifts will introduce an unknown or random element in the phase ψ_τ, e.g., in (3.11) and (3.15)–(3.17).

3.4.1 Central lobe suppression

Without further processing the time resolution of two closely located scatterers of the bioinspired spectrogram transformation is the same as the resolution achieved by the matched filter. The distance between the sinc functions is τ as for the matched filter and the width of the lobes in the time domain at -3 dB is the reciprocal of the bandwidth of the signal (or that of the filter bank if it is less):

$$\Delta\tau = \frac{1}{B_C} \tag{3.18}$$

However, the central lobe does not depend on the position of the targets and can be suppressed to double the distance between the two lobes at $\pm\tau$. If the suppression of the central lobe is achieved then the distance between the two sinc functions is increased from τ to 2τ and the functional range resolution is improved so that two targets at half the range separation can be resolved (Figure 3.7). Furthermore, because of the resulting representation is symmetrical, we can imagine folding it back on itself so there is then essentially only a single peak. This means that measuring the relative distance between two targets has been made as simple as measuring the range to a single target [1, Section 3.1]. There are a few approaches that can be considered to remove the central lobe. One is to use the knowledge of the transmitted signal energy and take a suitably scaled difference and another is to remove the average value of the spectral energy over the bandwidth before taking the inverse Fourier transform [20].

Removing the mean of (3.16) leads to

$$\begin{aligned} E_{E'}(f) &= E_E(f) - \frac{1}{B_C}\int_{(-B_C/2)}^{(B_C/2)} E_E(f)df \cdot \mathrm{rect}\left(\frac{f}{B_C}\right) \\ &= 2\frac{1}{\gamma} \cdot [\mathrm{sinc}(\tau B)\cos(2\pi f\tau + \psi_\tau) - \mathrm{sinc}(\tau B)\,\mathrm{sinc}(\tau B_C)\cos\psi_\tau] \end{aligned} \tag{3.19}$$

Results show that, with this method, the inverse Fourier transform of (3.19) will still have a residual zero lobe component but it will be reduced from one to less than sinc $(\tau B)\,\mathrm{sinc}(\tau B_C)$. The zero lobe suppression mechanism can be interpreted in the time domain also. Simulations show that subtracting the normalised time domain representation of the BSCT for a single scatterer from the corresponding representation for two scatterers have the same effect.

The resolution of two closely located targets is illustrated in Figure 3.8 where the lobe interference is presented for different spacings between the targets and different types of processing. It shows the output of the matched-filter and the inverse FFT of (3.16) and (3.19) which represents the output of the matched filter, the inverse of the spectrogram transformation and the effect of the zero lobe removal correspondingly. The magnitudes for both spectrogram related approaches are normalised by the magnitude of the central lobe before the suppression.

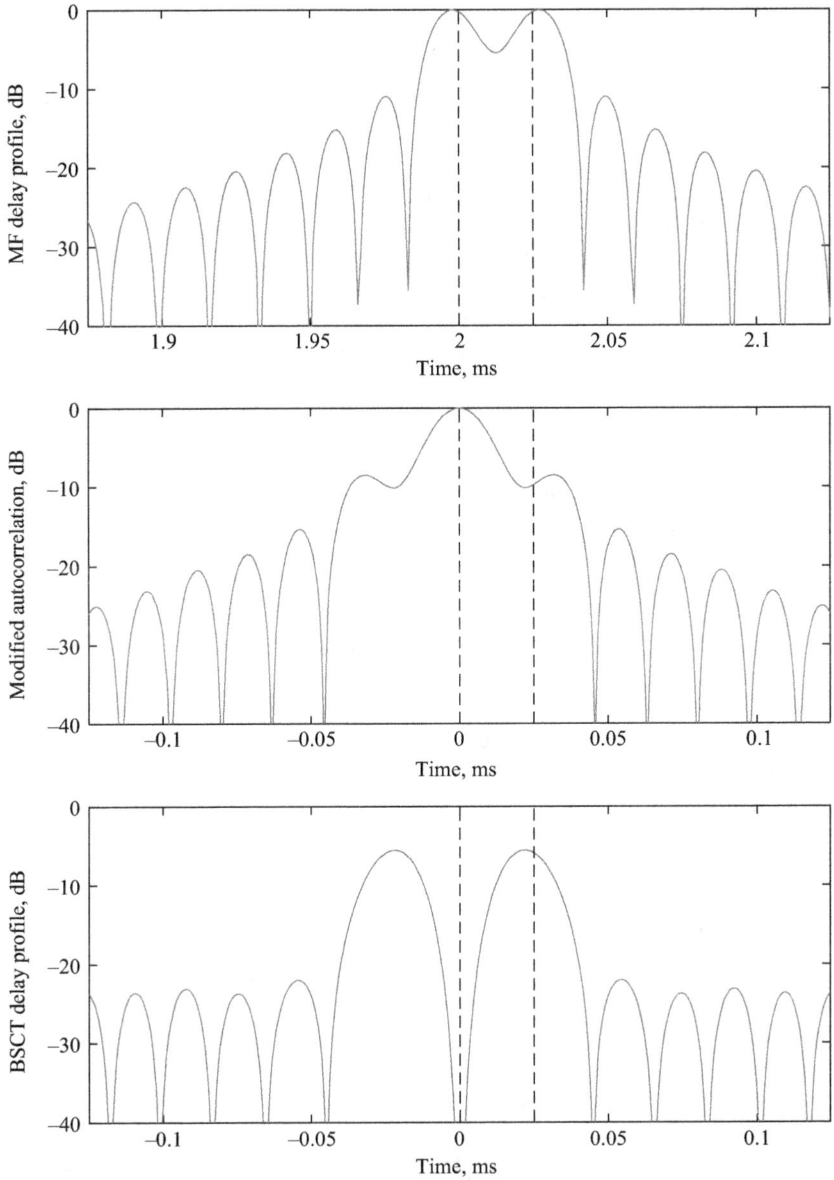

Figure 3.7 Matched filter compared to BSCT and the effect of the zero lobe suppression, $B_C = 65$ kHz, $t_1 = 2$ ms, $\tau = 1.6(1/B_C) = 25$ μs. While the lobe width is the same the functional resolution is improved by doubling the spacing between the lobes

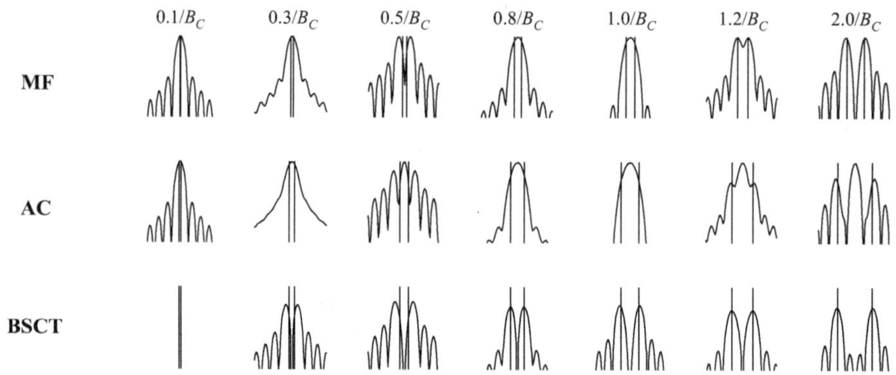

Figure 3.8 Fine delay profiles for different delays between two close targets near the resolution limit $1/B_C$. The two sinc *interference pattern typical of the matched filtering [11] is presented in the top row. The simple inverse transform of the spectrogram transformation is denoted as "AC" as a form of autocorrelation. The bottom row shows the zero lobe removal effect. The true locations are indicated by the vertical bars*

3.5 Experimental setup and data collection

This section describes experiments employing real and phantom targets at ultrasound. Simulated data that mirror the experiments is generated to allow a comparison between experiments and simulations. Phantom target experiments are useful to test the robustness of the algorithm with respect to the hardware imperfections, such as non-linearities and a non-flat frequency response and propagation effects. Real target experiments, in addition, test the robustness of the algorithm for non-ideal reflectors.

3.5.1 General settings and equipment

For both the simulations and the experiments, 10 ms long recordings were sampled at a rate of 10 MHz to collect 100,000 samples per realisation. Echoes, in the experiments, were recorded with an ultrasound microphone (type CM16, Avisoft Bioacoustics, Berlin, Germany) and waveforms were transmitted with an ultrasound loudspeaker (type S55/6, Ultra Sound Advice, London, UK). Digital-to-analogue and analogue-to-digital conversions were achieved with a TiePie Handyscope dual channel oscilloscope carrying an arbitrary signal generator (type HS5-540, TiePie Engineering, Sneek, the Netherlands). The transmitted signal was a linear down-chirp spanning the frequencies between 100 to 35 kHz with a duration of 2 ms. The target under test consisted of two discrete point scatterers delayed of $\tau = 2d/c$, with d being the spacing between the two scatterers and c the speed of propagation. All experimental measurements were preprocessed by removing the mean value and converted to baseband (Figure 3.5).

Equipment control and data processing were performed in MATLAB® (release 2015a, MathWorks, Natick, MA, USA).

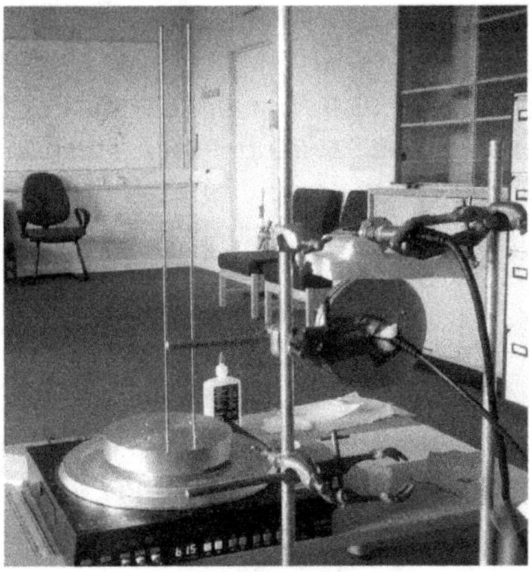

Figure 3.9 Experimental setup with the two vertical rods on the turntable

3.5.2 Simulations

In the simulations, echoes were simply generated by summing two time-delayed replicas of the transmitted waveform for different delays τ.

3.5.3 Phantom targets

Echoes from phantom targets were generated with the loudspeaker located at about 2 m from the microphone. Two scatterers were reproduced by emitting the call and a replica of the call delayed of τ. The relative position between the two scatterers was varied by using different values of the delay in the interval $\tau = [0, 2, \ldots, 50]$ μs, corresponding to a separation $d = [0, 0.34, \ldots, 8.5]$ mm. Phantom targets were used to study the response to ideal point targets in a real environment, accounting for the effects due to air attenuation, noise and transmitter/receiver imperfections.

3.5.4 Physical targets

The two physical targets used in the experiments were two vertical rods with a diameter of 6 mm and length of 0.6 m. These were placed on a turntable (LT360EX, LinearX Systems, Battle Ground, WA, USA) so that their relative distance with respect to the ultrasound radar could be changed by rotating the turntable with a step of 0.5°. One of the rods was placed at the centre of the turntable and the other at a distance of 25 mm from the centre of rotation. The speaker and the microphone were deployed at a height of approximately 30 cm above the level of the turntable surface. The distance between the centre of rotation of the turntable and the sensor was about 1 m. A photo of the experimental setup is shown in Figure 3.9.

3.6 Results

In this section, the outputs of the BSCT and of the matched filter are compared in order to assess their relative range resolution performance. The BSCT frequency profile, describing the energy of the output of each filter of the BSCT, was calculated by averaging the spectrum over a bandwidth $B = 4$ kHz for $K = 65$ filters with linearly spaced central frequencies between -32 and 32 kHz (step of 1 kHz). An example of the BSCT frequency profiles as a function of the time delay τ is shown in Figure 3.10 for simulated and real targets.

The output of the BSCT representing the delay profile was calculated by removing the average of the spectrum (as in (3.19)) and by applying the IFFT to the resulting zero-mean spectrum. Zero-padding was applied to the spectrum before the IFFT by adding a sequence of N zeros to achieve a suitable sampling rate

$$f_s = \frac{NB_C}{K} \tag{3.20}$$

The output was then normalised to unit energy and displayed in a dB scale.

Figure 3.11(a) displays the fine delay profiles generated by the model for different spacings between the simulated scatterers. Unlike the matched filter output, these profiles encode directly the spacing between the scatterers and not the absolute range of the target. As the representation is symmetrical, there is a second

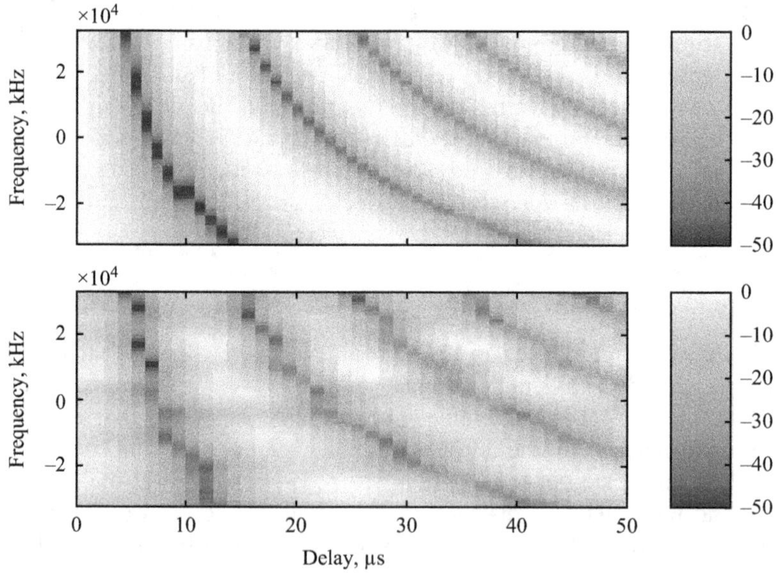

Figure 3.10 BSCT frequency profiles for simulated data (top) and real target measurements (bottom). The grey colour scale encodes the energy in dB

peak at negative delays but this peak interferes less than the second peak produced by the matched filter because the separation is double. Because the bandwidth of the chirp was 65 kHz the nominal resolution was 15 μs. We can see that the second scatterer can be resolved at delays below 10 μs. The delay profiles based on the

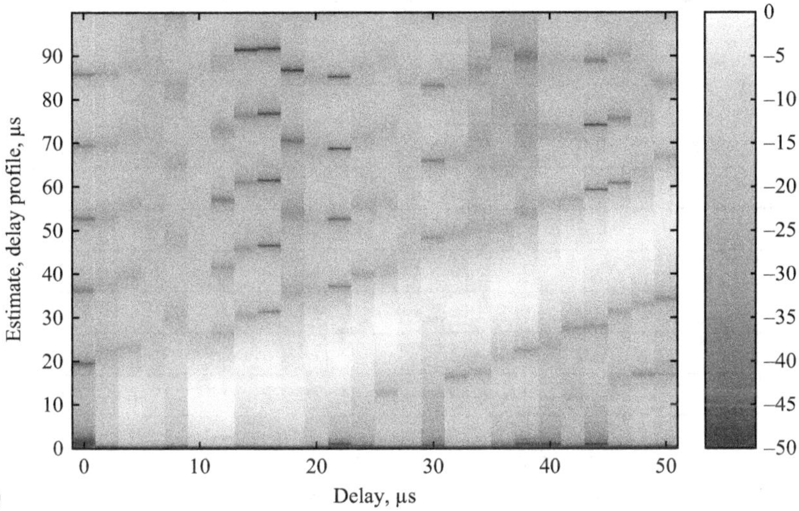

(a)

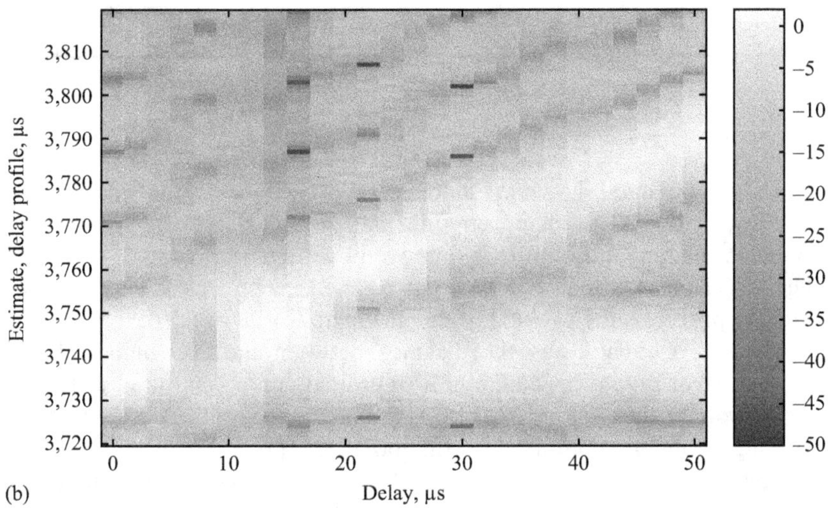

(b)

Figure 3.11 Fine delay profiles as a function of the delay between the scatterers for the simulated data. The signal bandwidth is 65 kHz. In the spectrogram transformation-based approach the profile encodes the spacing between the scatterers. From the matched filter we get the absolute position of the scatterers. (a) Baseband spectrogram transformation output. (b) Matched filter output

Table 3.1 Fine delay estimates on simulated targets

Delay (μs)	Delay estimation approach				
	MF	**BST**	**NLS**	**BSTb**	**NLSb**
0.0	0.0	—	93.7	—	7.8
2.0	0.0	—	0.0	10.3	3.0
4.0	0.0	10.6	8.8	10.3	4.4
6.0	0.0	11.2	10.1	10.9	6.3
8.0	20.0	16.0	13.3	15.2	8.6
10.0	0.0	11.2	11.5	11.0	9.9
12.0	0.0	11.7	12.4	11.4	11.8
14.0	0.0	16.5	13.7	16.6	14.0
16.0	0.0	16.7	15.6	16.8	15.9
18.0	0.0	15.7	17.4	16.5	17.8
20.0	22.0	20.3	20.3	19.7	19.8
22.0	22.0	21.0	22.7	19.8	21.6
30.0	34.0	32.5	31.3	31.2	29.9
b – denotes equalised spectrum.					
— the algorithm failed to work.					

matched filter output are shown in Figure 3.11(b). The lobes of the scatterers merge for spacings below 20 μs demonstrating that, for ideal reflectors and an ideal environment, the range resolution is improved by a factor of two.

The BSCT point estimate of the relative time spacing was calculated from the BSCT delay profile as the position of the maximal profile peak (denoted further as BST). The MF estimate was the distance between the two highest peaks in the matched filter output. In addition, a sinusoidal curve was fitted to the BSCT frequency profile using a Non-linear Least Square method (NLS) with the "Thrust-Region" algorithm and optimised starting point [25, pp. 4-46–4-56].

The results are summarised in Table 3.1 and Figure 3.12(a) where the estimates of the delay between the scatterers based on the BSCT peaks (BST), matched filter (MF) and non-linear curve fitting (NLS) approaches are presented together with the true delays. The resolution capability of the matched filter for the bandwidth available is 20 μs as already observed. The spectrogram transformation delay profile-based approach gives a range resolution down to 10 μs and can indicate the presence of the second scatterer down to 4 μs delay separation. The NLS results are comparable to those of the BSCT.

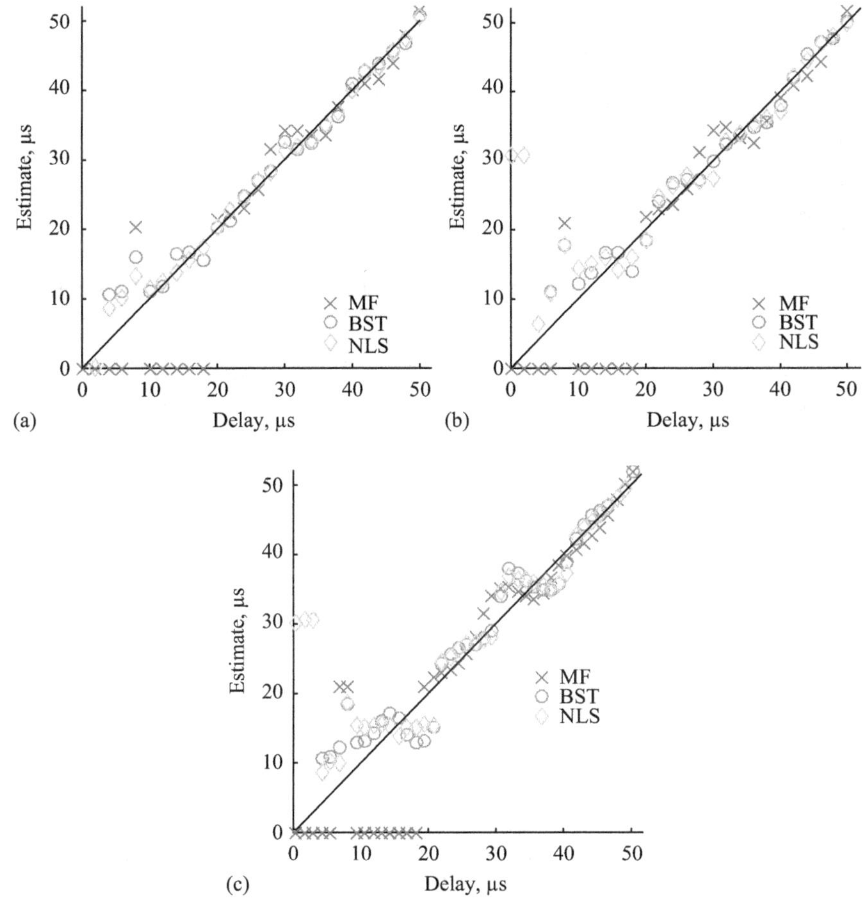

Figure 3.12 Estimated delay as a function of the delay between the scatterers. The signal is a linear chirp with a bandwidth of 65 kHz and a duration of 2 ms. The delay between the scatterers is estimated using the matched filter (MF), the baseband spectrogram transformation (BST) and non-linear least square fitting (NLS) techniques. The phantom target is generated by a speaker emitting two delayed chirps towards a microphone located at 1.886 m. The physical target consists of two vertical rods on a turntable. (a) Simulated target. (b) Phantom target. (c) Physical target

If the frequency response of a single scatterer is known (e.g., can be measured or calculated from the transmitter and receiver frequency responses and the air attenuation at the frequencies and the range of interest), then the spectrum can be equalised. This compensation of the expected chirp spectrum improves the resolution accuracy. The improvements are significant for the NLS estimation

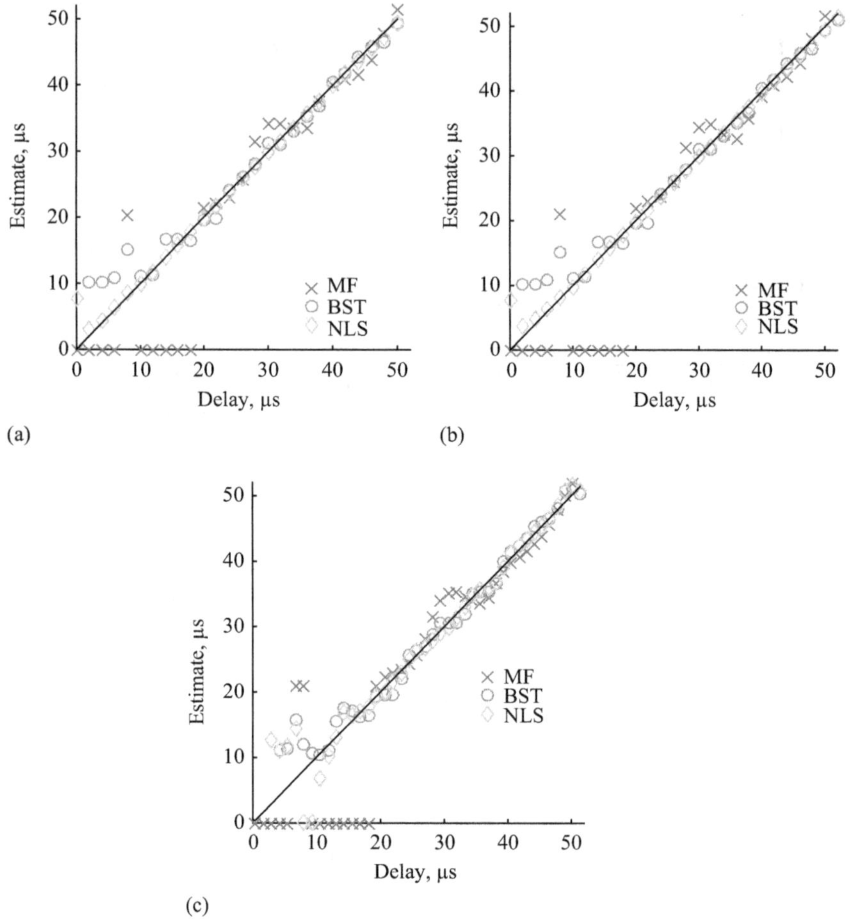

Figure 3.13 Estimated delays as a function of the delay between the scatterers when the signal spectrum is equalised by a single scatterer spectrum. Equalising the spectrum improves mostly the NLS estimates for the simulations and the phantom targets but has no big effect on the real target estimates (compare with Figure 3.12). (a) Simulated target. (b) Phantom target. (c) Physical target

down to 2 μs, e.g., from Table 3.1 we can read that for a true delay of 4 μs the estimate without equalisation is 8.8 μs (column NLS) and with equalisation is 4.4 μs (column NLSb). The changes in the BST performance are minor (compare columns BST and BSTb from Table 3.1 or Figures 3.12(a) and 3.13(a)). The results relative to the phantom echoes in Figures 3.12(b) and 3.13(b) show that the range resolution performance is compatible with that of the simulations.

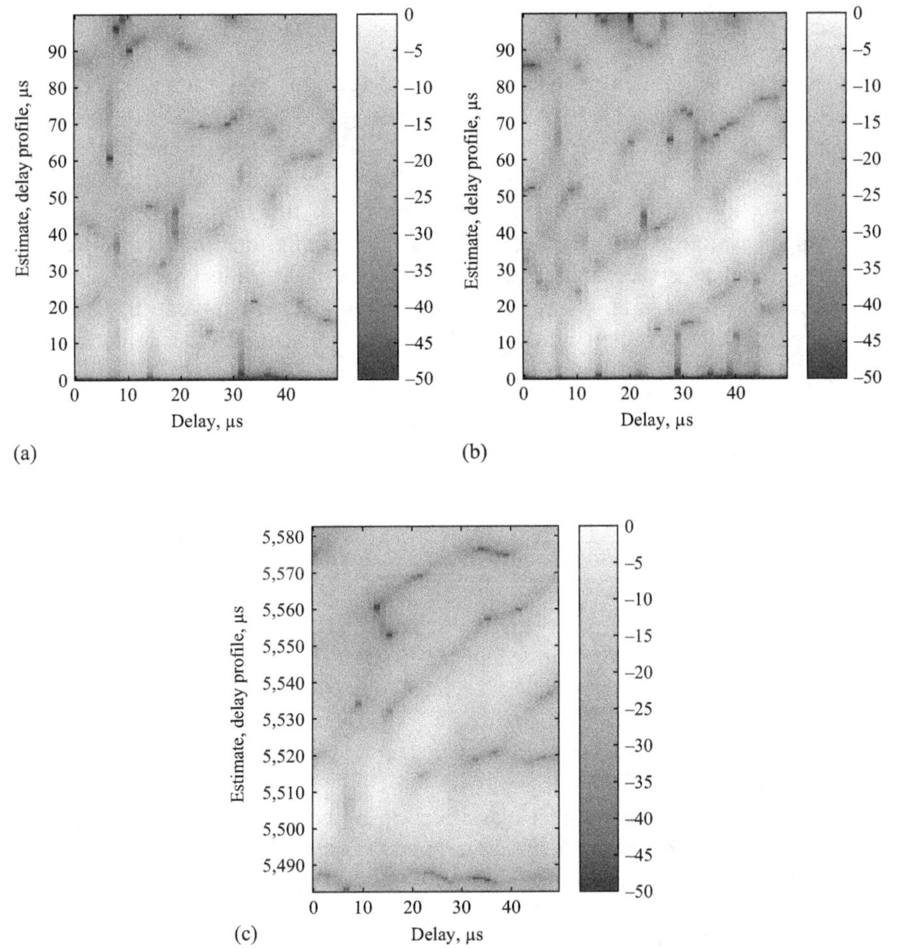

Figure 3.14 Fine delay profiles as a function of the delay between the scatterers for the real data. The signal bandwidth is 65 kHz. (a) Baseband spectrogram transformation output. (b) Baseband spectrogram transformation equalised output. (c) Matched filter output

The results relative to the experiments with the two vertical rods on the turntable are shown in Figure 3.14. These confirm that the BSCT provides reliable range resolution down to 10 μs, that is two times better than the matched filter resolution of 20 μs (Figures 3.12(c) and 3.13(c)). For the real data, the NLS approach does not perform well and, although it improves with equalisation, the improvements are minor. The BSCT is capable of handling two non-perfect point scatterers as well as the more computationally expensive non-linear least-square parameter estimation approach.

3.7 Conclusion

In this chapter, we have presented a baseband receiver that allows an analytical treatment of the output of a SCAT-like processing and that can also be applied to RF signals. The output of the Baseband SCAT (BSCT) has been compared analytically with that of a conventional matched filter for two closely spaced scatterers. Results have shown that a bat-inspired spectrogram transformation can provide better range resolution performance than that of a conventional matched filter. The results have been derived analytically and then verified with a set of laboratory experiments at ultrasound, using phantom echoes and real targets, to assess the robustness of the algorithm with respect to operational hardware limitations and propagation effects.

References

[1] A. W. Rihaczek, *Principles of High-Resolution Radar*. New York: McGraw-Hill, 1969.

[2] D. Blacknell and H. Griffiths (Eds.), *Radar Automatic Target Recognition (ATR) and Non-Cooperative Target Recognition (NCTR)*. Stevenage, UK: Institution of Engineering and Technology, 2013.

[3] G. W. Stimson, H. D. Griffiths, C. J. Baker, and D. Adamy, *Stimson's Introduction to Airborne Radar* (3rd ed.). Stevenage, UK: Institution of Engineering and Technology, 2014.

[4] D. von Helversen, "Object classification by echolocation in nectar feeding bats: Size-independent generalization of shape," *Journal of Comparative Physiology A, Neuroethology, Sensory, Neural, and Behavioral Physiology*, vol. 190, no. 7, pp. 515–521, Jul.2004.

[5] A. Balleri, H. D. Griffiths, C. J. Baker, K. Woodbridge, and M. W. Holderied, "Analysis of acoustic echoes from a bat-pollinated plant species: Insight into strategies for radar and sonar target classification," *IET Radar, Sonar and Navigation*, vol. 6, no. 6, pp. 536–544, 2012.

[6] C. J. Baker, H. D. Griffiths, and A. Balleri, "Biologically inspired waveform diversity," in *Waveform Design and Diversity for Advanced Radar Systems*, ser. Radar, Sonar, Navigation and Avionics, F. Gini, A. De Maio, and L. Patton, Eds. Stevenage, UK: Institution of Engineering and Technology, 2012, vol. 22, ch. 6, pp. 149–172.

[7] C. J. Baker, G. E. Smith, A. Balleri, M. Holderied, and H. D. Griffiths, "Biomimetic echolocation with application to radar and sonar sensing," *Proceedings of the IEEE*, vol. 102, no. 4, pp. 447–458, 2014.

[8] J. R. Barchi, J. M. Knowles, and J. A. Simmons, "Spatial memory and stereotypy of flight paths by big brown bats in cluttered surroundings," *Journal of Experimental Biology*, vol. 216, no. 6, pp. 1053–1063, 2013.

[9] J. A. Simmons, M. Ferragamo, C. F. Moss, S. B. Stevenson, and R. A. Altes, "Discrimination of jittered sonar echoes by the echolocating bat, *Eptesicus fuscus*: The shape of target images in echolocation," *Journal of Comparative Physiology A*, vol. 167, no. 5, pp. 589–616, Nov. 1990.

[10] J. A. Simmons, P. A. Saillant, J. M. Wotton, T. Haresign, M. J. Ferragamo, and C. F. Moss, "Composition of biosonar images for target recognition by echolocating bats," *Neural Networks*, vol. 8, no. 7–8, pp. 1239–1261, 1995.

[11] A. W. Rihaczek, "Radar resolution of ideal point scatterers," *IEEE Transactions on Aerospace and Electronic Systems*, vol. 32, no. 2, pp. 842–845, Apr. 1996.

[12] S. Schmidt, "Perception of structured phantom targets in the echolocating bat, *Megaderma lyra*," *Journal of the Acoustical Society of America*, vol. 91, no. 4, pp. 2203–2223, 1992.

[13] R. Simon, M. Knörnschild, M. Tschapka, *et al.*, "Biosonar resolving power: Echo-acoustic perception of surface structures in the submillimeter range," *Frontiers in Physiology*, vol. 5, p. 64, Jan. 2014.

[14] N. Levanon and E. Mozeson, *Radar Signals*. Hoboken, NJ: IEEE Press; Wiley-Interscience, 2004.

[15] C. E. Cook and M. Bernfeld, *Radar Signals: An Introduction to Theory and Application*. New York: Academic Press; 1967.

[16] P. A. Saillant, J. A. Simmons, S. P. Dear, and T. A. McMullen, "A computational model of echo processing and acoustic imaging in frequency-modulated echolocating bats: The spectrogram correlation and transformation receiver," *Journal of the Acoustical Society of America*, vol. 94, no. 5, pp. 2691–2712, Nov. 1993.

[17] I. Matsuo, K. Kunugiyama, and M. Yano, "An echolocation model for range discrimination of multiple closely spaced objects: Transformation of spectrogram into the reflected intensity distribution," *Journal of the Acoustical Society of America*, vol. 115, no. 2, pp. 920–928, Feb. 2004.

[18] L. Wiegrebe, "An autocorrelation model of bat sonar," *Biological Cybernetics*, vol. 98, no. 6, pp. 587–595, 2008.

[19] H. Peremans and J. Hallam, "The spectrogram correlation and transformation receiver, revisited," *Journal of the Acoustical Society of America*, vol. 104, no. 2, pp. 1101–1110, Aug. 1998.

[20] M. Park and R. Allen, "Pattern-matching analysis of fine echo delays by the spectrogram correlation and transformation receiver," *Journal of the Acoustical Society of America*, vol. 128, no. 3, pp. 1490–1500, 2010.

[21] R. D. Patterson, M. H. Allerhand, and C. Giguère, "Time-domain modeling of peripheral auditory processing: A modular architecture and a software platform," *Journal of the Acoustical Society of America*, vol. 98, no. 4, pp. 1890–4, Oct. 1995.

[22] K. Georgiev, A. Balleri, A. Stove, and M. Holderied, "Baseband version of the bat-inspired spectrogram correlation and transformation receiver," in *2016 IEEE Radar Conference*, Philadelphia, PA, USA, May 2016.

[23] M. E. Bates, J. A. Simmons, and T. V. Zorikov, "Bats use echo harmonic structure to distinguish their targets from background clutter." *Science (New York, NY)*, vol. 333, no. 6042, pp. 627–30, Jul. 2011.

[24] R. Wong, *Asymptotic Approximations of Integrals*, ser. Classics in Applied Mathematics. Society for Industrial and Applied Mathematics, 2001.

[25] *MATLAB and Curve Fitting Toolbox Release 2015a, User's Guide*. Natick, MA: The MathWorks, Inc., 2015.

Chapter 4

Air-coupled sonar systems inspired by bat echolocation

James F.C. Windmill[1] *and Francesco Guarato*[1]

4.1 Introduction

This chapter is a review of work on biomimetic sonar systems developed in the last 15 years which are inspired by the bat echolocation process. Inspiration from bat biosonar is not intended as a bare transfer of features from the bat biosonar into an engineering sonar system. The various contributions considered in this chapter aim at developing one specific technology or localisation method using cues from the bat biosonar. In some cases, for example, the reproduction of one or more functionalities of the bat echolocation process is targeted to understand the functionalities themselves by implementing them into mathematical formulations or attempting the same tasks through empirical investigation.

Bat biosonar is optimised by nature through an evolutionary process. And "biological systems tend to be non-linear, special-purpose designs" so that they are optimised for a "particular set of tasks" [1]. On the engineering side, sonar systems have applications in many different areas, with tasks other than those accomplished by the biosonar of some bat species. One structural problem in engineering bat-inspired sonar systems is therefore the different usability of the sonar system once it is implemented [1].

Whoever intends to design and implement a sonar system inspired by bat echolocation therefore has to be aware that the biosonar of all species of bats are optimum in natural contexts, and for very specific tasks typically related to the particular diet of one bat specie. This of course is a limit in the applicability of bat biosonars in an engineering context, as tasks are different and physical dimensions are much bigger than the natural environment where bat species operate. Hence, a bare transfer of bat biosonar, once understood, into a technical sonar system is rarely feasible. With this in mind, the contributions included in this chapter generally focus on one feature in the biosonar of one bat specie, and reproduce it to investigate its reliability and accuracy in

[1]Centre for Ultrasonic Engineering, Department of Electronic and Electrical Engineering, University of Strathclyde, Royal College Building, 204 George Street, Glasgow G1 1XW, UK

accomplishing a task within an engineering context. Reproducing a capability specific of bat biosonar into a man-made sonar system is also a way of understanding the functionality of the sonar itself. This also makes it possible to analyse the potential of the engineered sonar system in different applications by performing measurements and conducting tasks in a real environment using a prototype.

In addition, the sensory basis that bat species use to perform their tasks has not been completely established yet. Hence, even if their biosonar has been acknowledged as being the main information source during, for instance, hunting tasks, a certain degree of *a priori* knowledge has to be taken into account. This is of course a limitation in taking bat biosonar as a sonar-based solution to a wide range of problems.

Because of this difficulty in transferring knowledge from bat biosonar into an engineered application, researchers in the last 15 years have sought for the unique advantage of bat biosonar in a particular application in the natural world, why it is used by bats and hence where and with what technology can it be best reproduced in an engineering context or what technology does it suggest. The bioinspired sonar systems, localisation techniques and technology designs considered in this chapter take into account single distinctive elements of the bat biosonar and attempt to reproduce the same functionality in an engineered prototype or signal processing method.

Another difficulty in reproducing bat echolocation with the same efficiency is related to the technical limitation of transducers which cannot convey the same power output as bat calls.

This chapter then focuses on localisation strategies and signal processing techniques inspired by bat echolocation. These are methods to estimate the orientation of a reflecting target or to measure the distance of the target from the sonar system, or both.

4.2 What is a sonar system?

A sonar system can be defined as an assembly of transmitters and receivers, and provided with basic signal processing implementing a sonar. SONAR (SOund Navigation And Ranging) is a technique that allows navigation by means of detection and localisation of obstacles and targets using emission and reception of sound. A sonar is implemented on a system equipped with transducers used as transmitters and receivers for sound emission and reception, respectively. A transducer is a device which performs the conversion of an electrical signal into a mechanical one, and vice versa.

A sonar can be *passive* or *active*. A sonar is passive if the system implementing the sonar only receives sound and does not produce it. A sonar is active if the system produces the sound and then receives the sound reflected back from a target. Determining the distance of a sound source with a passive sonar is not straightforward as the time of the transmitted signal is not known; however, it can easily be calculated by an active sonar. The distance d between the sonar system and a target reflecting the acoustic signal emitted by an active sonar system can be calculated using

$$d = \frac{v \cdot t}{2} \tag{4.1}$$

where v is the speed of sound in the medium and t the time of flight between transmission and reception of the signal. Equation (4.1) thus takes into account half of the time for the signal to travel in the medium from the emitter to the reflecting target and then to the receiver.

Air-coupled sonar systems are implemented typically through electrostatic transducers both for emitters and receivers. Widely used transducers for the emitter in an air-coupled sonar system are the Polaroid transducer which is found in many ranging applications [2] and the S55/6 Ultrasound Loudspeaker (http://www.ultrasoundadvice.co.uk/index.php/ultrasound/usls), the latter for acoustic characterisation rather than in a fully operational sonar system. The Polaroid transducer 6500 series can produce sound up to 110 dB at 50 kHz (reference 20 uPa at 1 m), and it can operate over a wide frequency range (20–100 kHz). Electrostatic transducers are also employed as receivers (i.e., microphones) though their dimensions are much reduced compared to those operating as emitters. A microphone with good sensitivity covering the frequency ranges of most bat species (20–150 kHz) is the Brüel & Kjær 4138 microphone. Another possibility is the usage of Knowles electret microphones.

The chapter aims to encompass bioinspired sonar systems, namely bat-inspired air-coupled sonar systems, which are typically equipped with one emitter and two receivers, thus resembling the physiological features of one mouth and two ears on a bat head. The similarities between bats and such biomimetic sonar systems are not limited to only this. As many bat species perform echolocation by means of ultrasonic broadband calls, bat-inspired sonar systems also produce sound in the ultrasonic frequency range. The transmitter and the receivers are air-coupled transducers designed to operate in a broad frequency range typically spanned by bat species (i.e., 20–150 kHz).

The position of a target with respect to a sonar system can be readily defined in terms of spherical coordinates, see Figure 4.1.

Supposing each of the receivers of the sonar system can be defined as the origin, the distance and the orientation completely defining the position of the target reflector are given by spherical coordinates from which Cartesian coordinates (X, Y, Z) can be readily recovered:

$$\begin{cases} x = r \sin\theta \cos\varphi \\ y = r \sin\theta \sin\varphi \\ z = r \cos\theta \end{cases} \tag{4.2}$$

where r is the distance of the target from the sonar system while azimuth θ and elevation φ angles define the orientation of the target reflector with respect to one of the receivers in the sonar system.

While extraction of time features such as time delay between emission and reception (4.1) allows the calculation of the distance of the target, determination of orientation as a function of azimuth and elevation is more complicated. In the configuration of a sonar system with one emitter and two receivers, the difference of time of flight of the reflected acoustic signal between the two receivers can be used to determine the azimuth angle of the target. To estimate the target location in

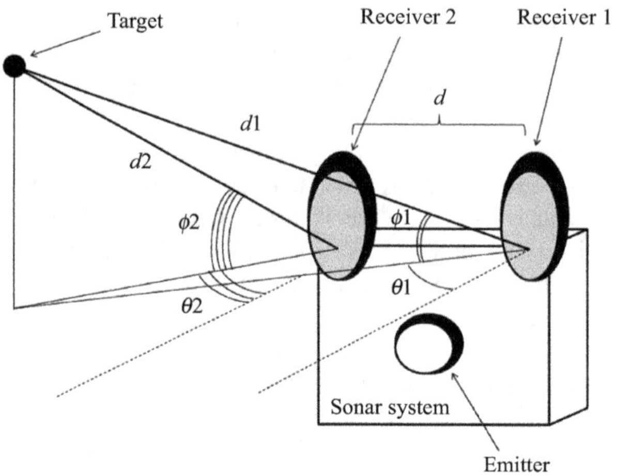

Figure 4.1 Schematics of a biomimetic sonar system whose emitter and receivers resemble their mutual displacement on a bat head. The spherical coordinates of the target with respect to each of the receivers are defined by distance, azimuth angle and elevation angle. Reprinted with permission from [3]

three dimensions more data have to be taken into account with more refined localisation strategies. The sonar localisation strategies mentioned in this chapter (and many more developed in the years before) extract both temporal and spatial features from the received signals and, to do so, knowledge on the spatial filtering provided by the receivers and/or the transmitter must be known *a priori*. The processing of the echoes from the target provides that analysis in the frequency domain performing some kind of comparison between the echoes and the binaural beam pattern associated with the receivers. Spectral information can also be combined with time using knowledge of the frequency sweep over time in the transmitted signal and thus providing further accuracy in determining time of flight, time of flight being the time between emission of the acoustic signal and reception of the echo reflected by a target.

A bat-inspired sonar system is typically characterised by one or more of the following features:

- the echoes are detected by two receivers
- the emitted signals are broadband so they are frequency modulated signals similar to those of bats
- the spatial filter of the receivers as well as of the emitter plays a fundamental role, particularly in determining azimuth or elevation direction with respect to the sonar system or both
- the analogue filtering is provided by external baffles for emitters or receivers
- the digital filtering is provided by filter banks centred at different frequencies and spanning the range of the emitted signal

In the biomimetic sonar systems developed in the last 15 years, inspiration from bat echolocation has meant innovations and novel approaches in the development of separate (but all functional to target localisation) parts of a sonar system. These parts have grown independently but are all underpinned under the general design of the sonar system according to the features listed above. In particular, in such sonar systems knowledge of the echolocation process performed by a bat has affected the following different topics:

(a) Design and size of emitter and receivers in the sonar system
(b) Frequencies covered by the emitted acoustic signal
(c) Structure of the emitted signal
(d) Signal processing of the received signals to estimate target distance and/or orientation

Indeed, the implementation of a localisation strategy inspired by bat echolocation is in most cases performed along with the design of structures for emitters and/or receivers resembling the same design observed in the external ears and nose leaves of some bat species. Thus, alongside biomimetic signal processing and localisation strategies, it is necessary to explore the variety of bat-inspired engineered solutions to modify the acoustical properties of emitters and receivers of sonar systems. The design of external structures should hopefully come along with the development of a biomimetic signal processing method.

All works considered in the following paragraphs contribute to the wide and extensive subject *Air-coupled sonar systems inspired by bat echolocation*. The division of works into groups (a)–(d) listed above is therefore functional to the clarity of exposition in this chapter. It will also allow us to better explore the individual and targeted innovations of each of these contributions.

4.3 Bioinspired (biomimetic) design of sonar systems: emitters and receivers

It is well known that spectral cues given by the outer ear play an important role in bat echolocation [4,5]. The beam pattern at several frequencies depicted in Figure 4.2 is associated with the pinna of the *Phyllostomus discolor* bat. The beam pattern is a function of frequency and orientation as it describes the attenuation of the signal depending on its frequency and direction of arrival. Similarly, biomimetic sonar systems make heavy use of spectral cues in target localisation. In particular, spatial filters can be implemented for the receivers in order to discriminate between orientations: this, along with the distance of the target calculated from time of flight, makes it possible to localise a target in three-dimensional space using spherical coordinates. To this extent, biomimetic sonar systems have taken into account the possibility of providing receivers and emitters with static as well as dynamic spatial filters, that is, beam patterns which are a function of orientation (i.e., the direction the acoustic signal is propagated or received) and frequency. The beam pattern of stationary beamforming baffles are useful in localisation tasks as they increase the signal-to-noise ratio by means of a directivity gain as well as

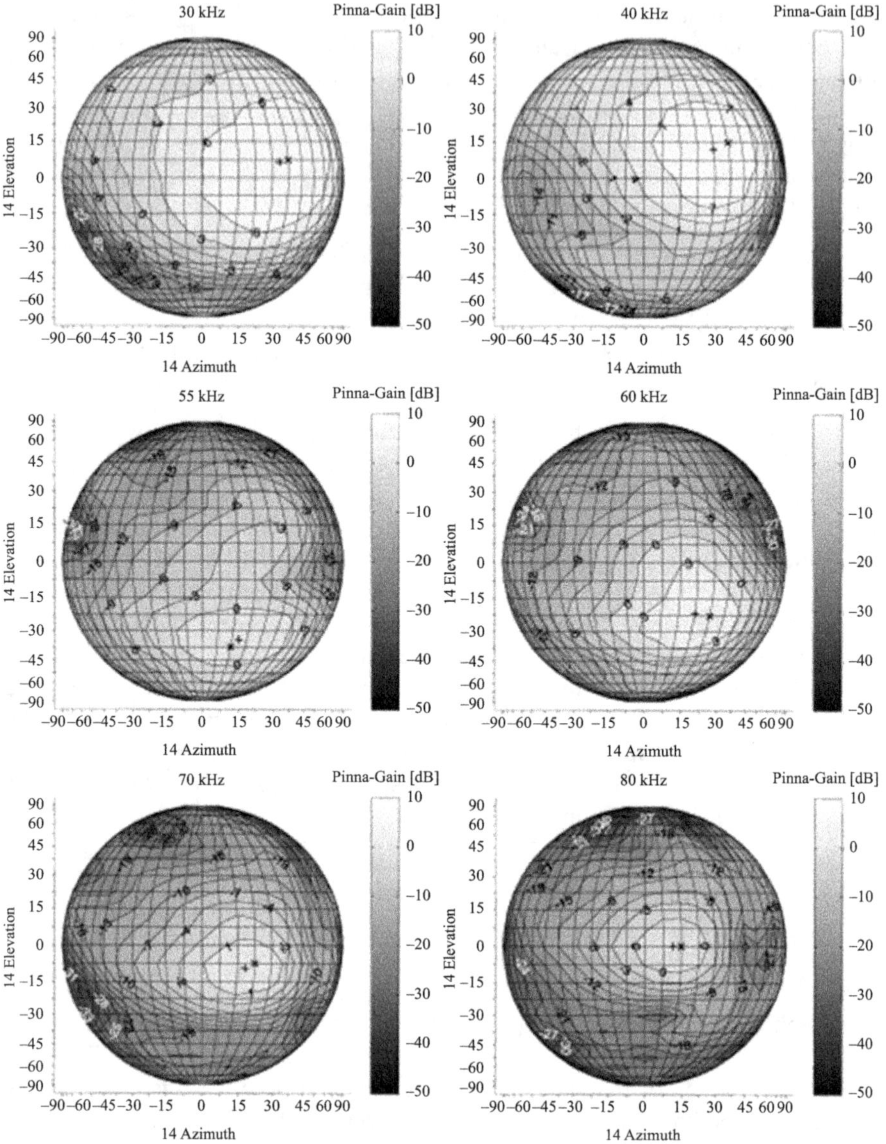

Figure 4.2 Beam profile associated with the pinna of a specimen of Phyllostomus discolor *bat specie. For different frequencies, the main beam changes its orientation and shape. Reprinted with permission from [4]*

enhancing spatial resolution with a narrow beam. Furthermore, static as well as dynamic beam patterns associated with the receivers can be used to discriminate orientations of the echoes from a target as in this case different orientations provide the sound with different attenuation values.

The beam pattern of receivers can be modified by mounting appropriate external structures on omnidirectional microphones thus providing them with a specific beam pattern to allow discrimination between orientations. Similarly, the acoustic beam pattern of an emitter can be modified using an appropriate structure in front of it.

The design of emitters and receivers mimicking the shape of bat nose leaves and ears is strictly related to sonar localisation methods and their performance: indeed, the shape of beamforming baffles for emitters and receivers and their time-variant acoustical behaviour are used in localisation strategies. For instance, in the biosonar of horseshoe bats "a baffle that is in motion while diffracting outgoing or incoming sonar signals would enhance the encoding of sensory information in some way" [6].

4.3.1 Biomimetic emitters

The design and implementation of biomimetic emitters provides an acoustic signal whose power is comparable to that of a bat on a broad frequency range [7] or provides such emitters with a beam pattern comparable to those observed in real bat species [8,9].

An air coupled transducer [7] is designed to provide 90 dB SPL at a distance of 1 m in a broad frequency range (50–150 kHz). Thus, both acoustic power and frequency range are targeted and accomplished using an ElectroMechanical film to reproduce the same output power in the emitter. The output power of transmitters is the bottle neck for currently available technologies and transducers compared to the bat's biosonar. It is inferred that this transducer can be used as a receiver too, provided that appropriate truncated conical shapes are mounted on the transducer to modify its beam pattern. The transducer covers 1 cm^2 as a receiver while it measures 2.25 cm^2 as a transmitter. The reduced size of the transducer as receiver is to improve its receiving sensitivity over 35–150 kHz as well as to fit it into an artificial pinna shape to modify its beam pattern.

External baffles to modify the directional property of commercially available transducers have been developed significantly in recent years. These baffles are mounted around the transmitting transducer thus to beam form the outgoing acoustic signal by means of directing or diffracting it through reflections on the baffles' inner surface. In particular, movable baffles to reproduce the movements of the horseshoe bat's noseleaves have been built around an air coupled emitter and tested [8]. These structures are analysed to study their usability in the future development of dynamic baffles for ultrasonic emitters. One approach is based on a finite element digital mesh that is recovered from X-ray scanned noseleaves of a horseshoe bat and stored into a 3D model for numerical simulations. Another approach is experimental and is based on a replica of bat's noseleaf made of elastic materials and connected through metal rods to actuators mounted behind the prototype baffle [8,9]. The sound from an air coupled transducer is conveyed through a conical structure and directed into the artificial noseleaf mounted at the end of the conical structure. Baffles are moved by servomotors in accordance with the emitted signal and the change of shape of the baffles is then comparable with the duration of the emitted signal (i.e., 2 ms). The experiments evaluate the beam pattern of the emitter at several frequencies and, in the case of the dynamic baffle, as a function of time [9]: the acoustic emission from the assembled transducer-actuator-baffle is measured for several orientations. This is an attempt to understand the role of noseleaves in bat echolocation and how the diffraction of sound from the final emitter can be implemented and used in the development of a sonar system.

4.3.2 Biomimetic receivers

The work on bat-inspired receivers is more developed than that on biomimetic emitters. In particular, receivers' directional properties can be altered by designing and mounting external baffles around an ultrasonic microphone used as a receiving transducer. In such a way, the beam pattern of the receiver can be closer to that associated with a bat ear. In addition, the spatial filtering through receivers affects the signal that is recorded as a function of the orientation of the acoustic signal and of its frequency: therefore, the design of external structures for the receivers is linked to localisation strategies and therefore to the implementation of ultrasonic sonar systems.

For simplicity, the term *receiver* will, from now on, refer to the assembly of microphone and external baffle.

Receivers associated with a directional beam pattern have a twofold advantage: they can focus on a narrow area in front of them corresponding to the main axis of the beam pattern, as well as making it possible to discriminate between orientations which are associated with different attenuations of the incoming sound.

The possibility of using external baffles to modify the omnidirectional (at least within half a sphere) beam pattern of conventional commercial microphones into a more directional one has inspired the usage of a genetic algorithm to attempt the systematic design of such baffles [10,11]. This procedure places flat discs around a receiver and led to the placement of up to 10 of them in order to provide the receiver with a beam pattern similar to that of the bat specie *Rhinolophus ferrumequinum*. This improved the directional sensitivity of the receiver by maximising the angular resolution of the receiving transducer. The initial placement of the discs should then evolve into a continuous surface to which each of the placed disks belongs using a region coverage method. This study is theoretical and is supported by acoustic simulations where sound losses due to absorption in the reflectors (i.e., the flat discs) are considered. In the simulations, the transducer is static and reflectors in increasing number are fixed for all possible target positions.

The attempt to use a genetic algorithm in designing receiver structures has for the moment been isolated. Small disc reflectors are a far too simple approximation of a pinna model: therefore, a more realistic model of bat external receivers should instead be investigated [10]. The authors of [10] themselves conclude that the outcome of the research on applying such an algorithm to evolve surfaces for biomimetic receiver structures is not feasible [11]. In conclusion, the design of external receivers providing microphones with a directional beam pattern is not supported, at the moment, by a systematic procedure. There is no mathematical theory defining a list of rules in the design of a receiver that will be associated with a specific beam pattern. Because of this, the design of biomimetic receivers has explored a more empirical approach consisting of the reproduction of the shape of bats' external ears, typically with simplified and more regular surfaces. Some approaches, though, have managed to associate physiological features in bat ears with the directionality of their beam patterns, which make bats accurate in determining target location.

External ear shapes of several bat species are therefore replicated and mimicked to implement artificial ones. Typically, the investigation is conducted using techniques such as micro-CT scanning to recover an accurate digital 3D model of bat ears. These models can be imported into finite-element packages where acoustic simulations can calculate the beam pattern associated with them. As simulating sound sources all around the receiver would be computationally too expensive, the receiver is typically considered as a sound source: because of the reciprocity principle, its beam pattern would be the same as if it was a receiver. A sound source is therefore placed at the base of the ear shape, or in a cavity underneath it modelling the ear canal, and a pressure load is applied at the source location. The pressure load can also be applied on a planar surface thus resembling the tympanic membrane. In finite element models, the pressure load is then calculated in each point of the model and time histories are stored at its boundaries. Extrapolation in the frequency domain of the time histories allows the recovery of the beam pattern in the far field.

The ear models obtained from scanning bat ears can be imported into 3D printers to machine an accurate artificial version of them. These prints are then mounted on microphones and integrated as parts in a sonar system, ready for measurements and experiments in a laboratory environment. Artificial bat pinnae can therefore reproduce relatively accurately the complexity of real bat ear shapes.

On the other hand, knowledge of bat external ears from micro-CT scanning or X-ray tomography inspires the design of pinnae which are simplified versions of the real ones. In the case of dynamic receiver shapes, elastic materials or rubber film are cut and folded appropriately to diffract the incoming sounds similarly to a real bat ear. In addition, the material makes it possible to modify the shape of the receiver in synchrony with the sound emission and reception. Static receivers can be obtained from regular shapes like cones and parabolic surfaces with oblique truncation: features such as external perimeter, presence of tragus and ridges on the inner surface of the receiver strongly contribute to the final beam pattern.

A deeper description of simulating the beam pattern associated with a bat ear and bat head is in [12]. Micro-CT scanning is performed of the *P. discolor* bat head, see Figure 4.3, to recover its digital model [12]. A Boundary Element Method (BEM) is then used to simulate the head-related transfer function (HRTF) of the head. With respect to Finite Element Methods where discretisation of the whole computational domain is necessary, BEM requires only the boundaries of the model to be discretised. The HRTF takes into account the filtering effect of ears, noseleaves and the shape of the bat head in defining the spatial filtering of the incoming sounds received at one of the ear canals (monaural HRTF). Simulation results of the sound attenuation as a function of orientation and frequency are in good agreement with measurements. Importantly, the FEM model makes it possible to take into account alternatively the single contribution of ears, noseleaves and head shape on the overall HRTF or a combination of those. In this way, it is seen that the ear has the

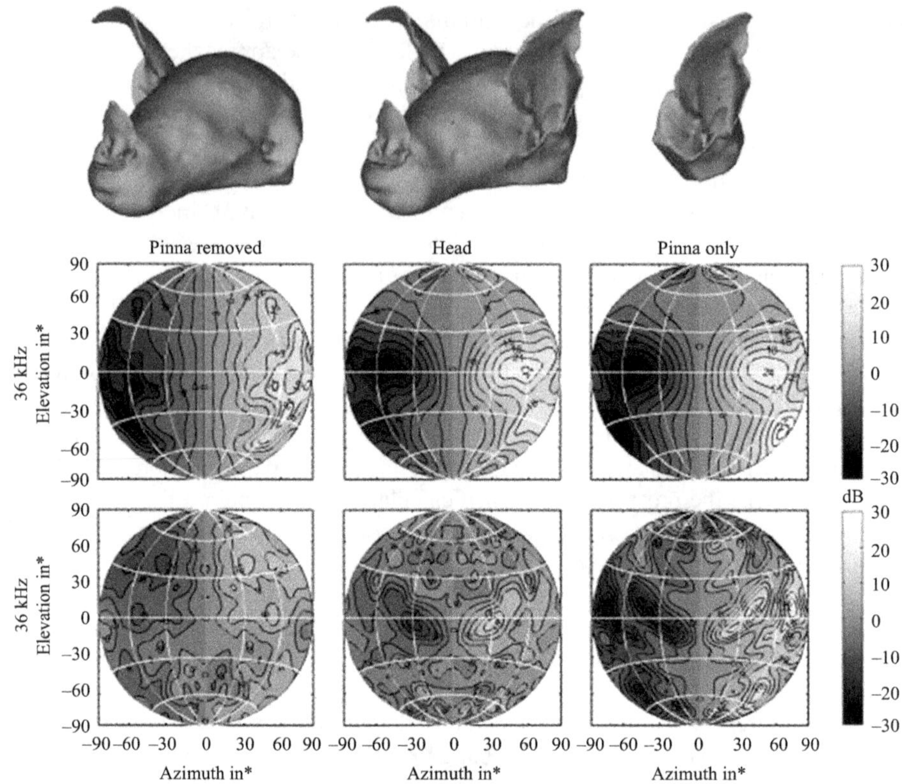

Figure 4.3 Top: artificial bat head from micro-CT scanned P. discolor*'s head. Bottom: polar maps of distribution of Interaural Intensity Differences (IIDs). Reprinted with permission from [12]*

predominant filtering effect in the overall HRTF. Interaural intensity differences (IID) patterns are also calculated from the simulated beam pattern by reversing in the azimuth domain the HRTF with the only contribution of the external ear.

Similarly, biological samples of bat ears are taken using X-ray micro-tomography [13]. The raw data is then processed to generate a computational model of the bat ear: the model is composed of a mesh of cubes which are classified as air or hard material (ear). This model is processed in a Finite Element package where the ear structure is treated as an emitter to recover the beam pattern associated with it. Finally, the bat ear model is used as an inspiration to build artificial simplified versions of real bat pinnae. These shapes are recovered from obliquely truncated horns which are then augmented with a ripple pattern on the inner surface of the pinnae and a tragus. Several external receiver shapes are recovered by including or not the tragus as well as the ripple pattern. These features are reckoned to have a significant effect on the beam pattern associated with real bat ears. Their effect on the final beam pattern was therefore tested on these artificial receiver shapes. This work [13] in particular describes a set of methods that are necessary to study acoustical properties of bat external ears.

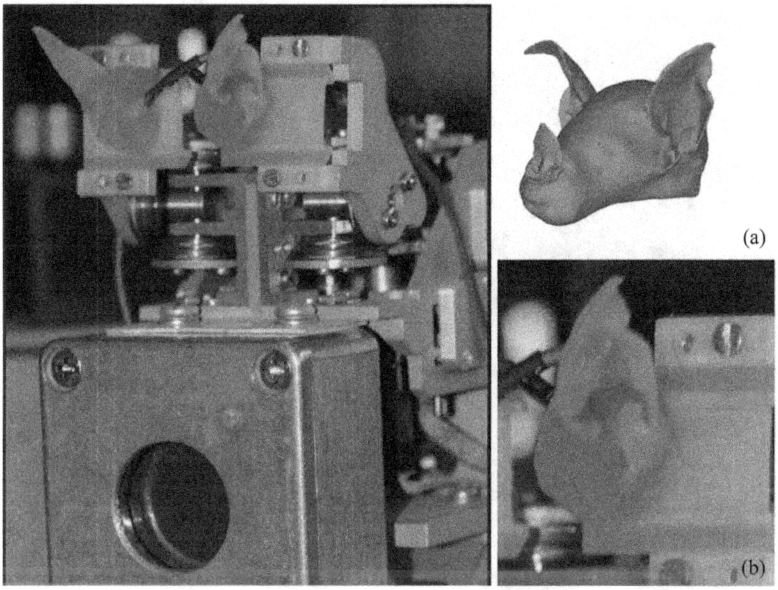

Figure 4.4 Biomimetic robotic head. The transmitter is a Polaroid transducer while receivers are microphones equipped with artificial bat ears (b) from a 3D model of a bat head (a). Reprinted with permission from [15]

Indeed, these strategies [12,13] represent a procedure to get insight into the complex shapes of bat external receivers and how to recover computational models to use in acoustic simulations and 3D printing.

Biomimetic receivers have been built using external structures resembling scans of real bat ears [14–17] or truncated cones as simplified versions of the latter [14,16]. Artificial pinna structures modelled like truncated cones [14] introduce acoustic features in the head-related transfer function (HRTF) not present in the beam pattern of common transducers. The HRTF is measured for different assembly microphone-pinna shape, where the pinna shape is altered with and without one or both features: ridges in the internal surface of the pinna and a tragus in front of the opening of the pinna. The presence of the pinna enhances the spatial resolution of the HRTF in target localisation performance and increases the elevation discrimination power with respect to the beam pattern of the bare transducer. Both the truncated cones and the artificial bat ears were used on a sonar system equipped with one emitter (bare transducer) and two receivers. The sonar system is a robotic bat head whose receivers can tilt and pan, see Figure 4.4, where artificial bat ears are mounted on electret omnidirectional microphones to mimic the beam pattern associated with real bat ears.

Bat-inspired truncated cones augmented with ripples in the inner side of the cone and with a tragus were used on the same biomimetic robot, see Figure 4.5.

In addition to external receivers inspired by bat ears, also arrays of microphones were used to replicate the beam pattern of one microphone equipped with a biomimetic external structure. To do so, a beamforming algorithm [18] altered filter parameters in a filter-and-sum method post processing the recordings of an

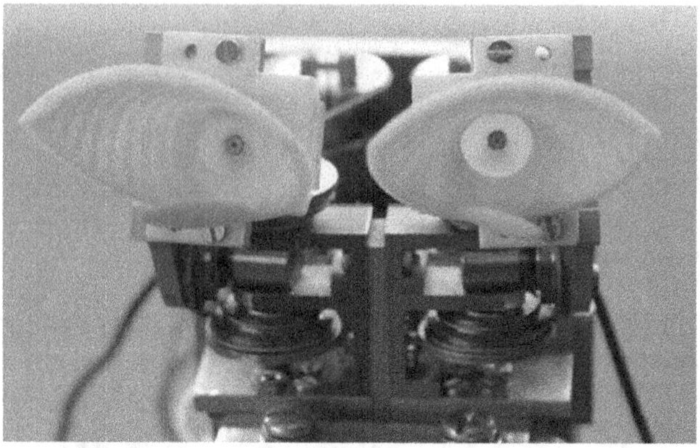

Figure 4.5 *Truncated cones modelling bat ears and mounted on electret microphones. The ridges and the tragus are visible on each cone. Reprinted with permission from [14]*

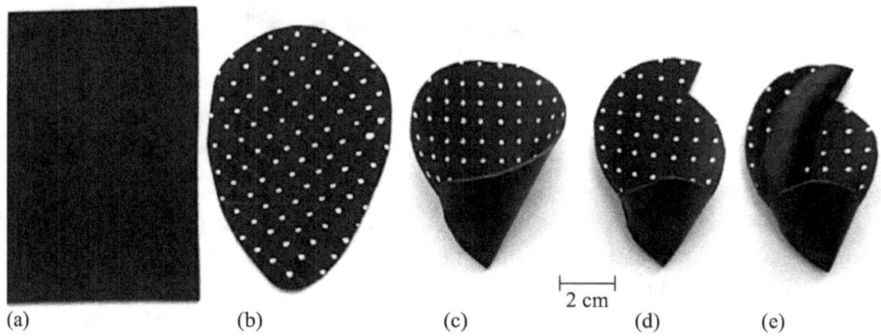

Figure 4.6 *Steps to machine a dynamic baffle mimicking a bat ear. Reprinted with permission from [20]*

array of (16 + 16=) 32 microphones [16,19]. The effect of the beamforming algorithm on the microphone array recordings was compared to spatial analogue filtering provided by bat-inspired pinnae [16] and directional information as the mutual information between received echo spectra and target direction was found to be very similar.

The dynamic behaviour of biomimetic receivers has also been investigated. From the observation that horseshoe bats change the shape of their outer ears in timescales comparable to the duration of their calls, a time-variant diffraction process is induced on a deforming baffle for an ultrasonic receiver [6]. The hypothesis that this dynamic process can enhance the information carried by the received signals is investigated in a biomimetic microphone baffle inspired by the shape and the movements of the external ear of the horseshoe bat [6]. The baffle is made of rubber and shaped like an obliquely truncated cone with features (a ridge, a flap and an incision along its rim) added to the basic cone structure [20,21], see Figure 4.6.

The acoustic behaviour is characterised with a Brüel & Kjær microphone coupled to the baffle and sound being played at different orientations around it while an actuation mechanism deforms the baffle from an upright position to a half-bent one. The beam pattern at the upright position has a main lobe which splits into secondary lobes as bending increases. The results show time-variant features that can improve target localisation in engineered sonar systems and can give insight on the behaviour of bats as well as explaining the function of physiological details of their external ears.

Finally, static receiver shapes have been developed using simplified versions of real bat ears [22]. The external ear of *Rousettus leschenaultii* is the inspiration for the size of truncated cones augmented with an elliptical external perimeter and an outward bending of the upper edge of the cone. Omnidirectional Brüel & Kjær microphones are placed at the lower opening of these structures: the beam pattern associated with the microphone equipped with the final receiver shape is comparable to that of the bat ear, see Figure 4.7. In this way, parameters such as the bending of the upper edge and the elliptical external perimeter are shown to be fundamental to associating the final external receiver with a beam pattern comparable to *R. leschenaultii*'s ear as a narrow main lobe is created. Another approach to develop static biomimetic receivers is based on eigen ears [23] that are defined from analysis of real bat ears from 100 species. In particular, the average in cylindrical coordinates of the 100 bat ears returns a basic shape, shown in Figure 4.8, from which all the other bat ears can be recovered by linear combination of the eigen ears weighted with a scalar value. This work represents a way to characterise the diversity in pinna shapes. Both studies on static receiver shapes [22,23] aim to give two quantitative descriptions of complex bat ears or biomimetic inspiration from them. Such methods target physiological features [23] and the acoustical behaviour [22].

4.4 Bat-inspired sonar systems and localisation methods

Bat-inspired sonar systems and localisation methods attempt to estimate the position (in terms of azimuth, elevation and distance) of a reflecting target by combining deterministic data and formulations as well as statistical formulations. The location of a target is described in spherical coordinates where target distance is calculated from the measurement of time of flight while target orientation is estimated from knowledge of the spatial filtering given by the receivers. Most of the localisation techniques in this chapter deal with an estimation of target orientation with respect to a sonar system. Indeed, while target distance can be determined straight from a measurement of time of flight, estimation of target orientation is more difficult and several solutions have been reported. In addition, the calculation of target distance is enhanced when using broadband signals such as chirps (typically down-swept frequency modulated as bats do) as compensation for sweep rate in the received signal allows us to consider the contribution of all frequencies in determining the time of flight. This makes it possible to increase significantly the

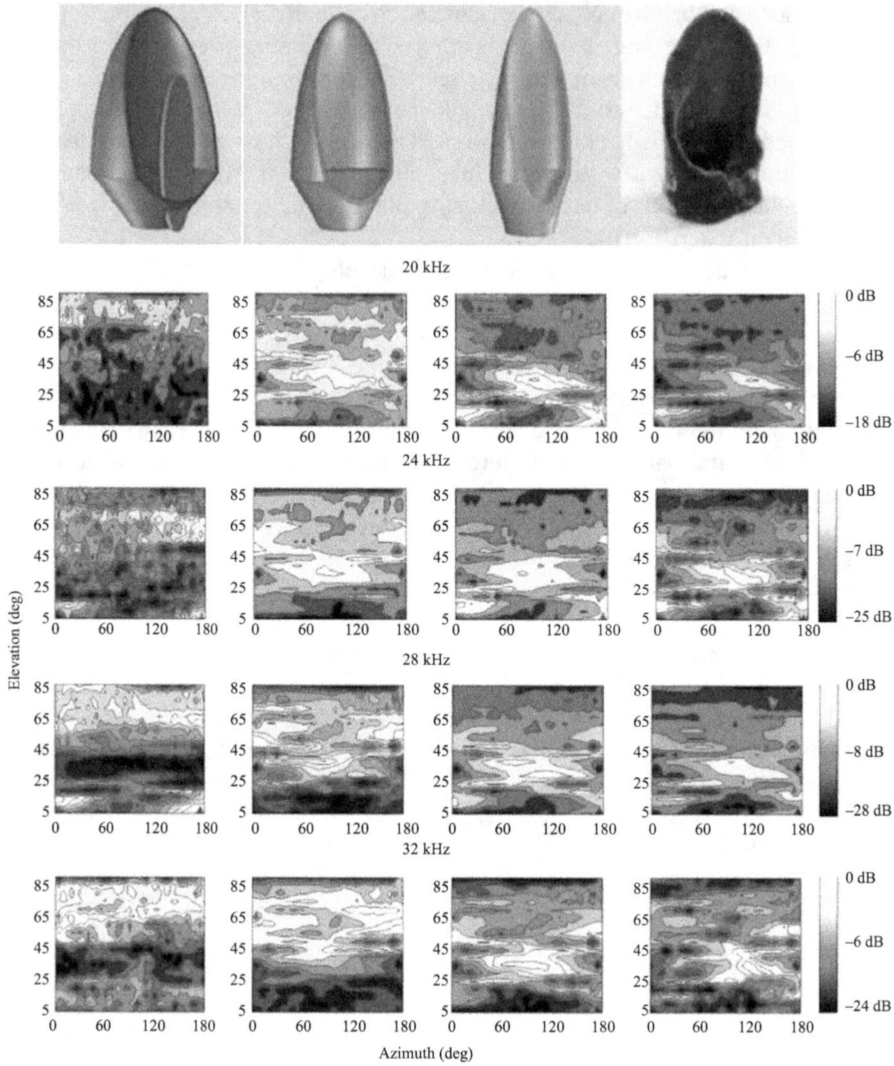

Figure 4.7 Bat-inspired regular shapes with the printed artificial bat ear and the associated beam patterns measured at different frequencies. Reprinted with permission from [22]

signal-to-noise ratio. Spectrograms are used to calculate the energy distribution in a broadband signal [24,25].

External baffles like those described in the previous paragraphs for emitters and receivers are strictly linked to spatial filtering to improve estimation of target orientation, or even to make it possible. Indeed, these static and dynamic baffles improve the spatial resolution of the sonar system by making the beam pattern of the receivers (or the emitter) either more directional or less ambiguous in determining

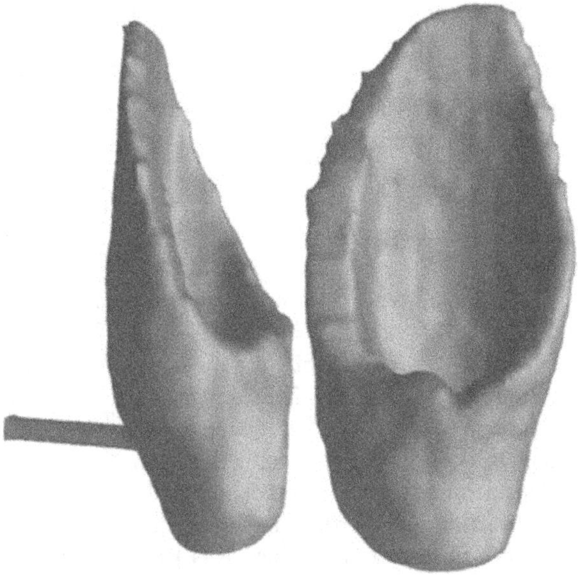

Figure 4.8 Truncated cone obtained from averaging the external ear shape of 100 bat species. Reprinted with permission from [23]

the direction of the reflected sound. This since, across frequencies, there are only few orientations that can be associated with the same sound attenuation.

A low cost implementation of a sonar system with broad bandwidth transducers estimates distance and bearing angle of reflectors by means of a bat-inspired processing of the reflected echoes [26]. One transducer is used as emitter and two as receivers in a configuration resembling the displacement of the mouth and ears on a bat's head. The echoes are processed through a filter bank inspired by how signals are processed in the bats' cochlea: the case of *Eptesicus fuscus* is analysed [27,28]. In this procedure, frequency components in the received echoes are separated as each filter is centred at one frequency so that all filters span the whole frequency range of the echoes: then compensation for the signal sweep rate is performed and all frequency components in the echoes are aligned and then summed over all frequencies returning the instantaneous energy associated with the reflection from one target. This procedure is similar to that by Matsuo [24] and allows discrimination between closely spaced objects reflecting sound back to the receivers. Target distance is then calculated from the instantaneous energy [26] while 2D bearing angle is estimated by comparing the spectrogram of the received echoes to a set of templates of the echo spectrogram from different angles. This template-matching procedure returns the bearing angle of a target as that minimising the Euclidean distance between the measured spectrogram and each template. The estimation of the bearing angle is robust.

The robotic bat head CIRCE [29] is used in target localisation experiments. This is a sonar head including a transmitter (Polaroid transducer) and two receivers (microphones on which a replica of a bat ear is mounted) to replicate the echolocation system of many bat species as reported in [29]. The research purpose is two-fold as experiments conducted with the CIRCE head make it possible to gain insight into bats' acoustic behaviour and, at the same time, to implement a robot model of bat echolocation for ultrasonic localisation problems. The CIRCE head approximates technical features of bat echolocation. The transducer used as the emitter is made with an EMFi foil, has a fairly flat response in the range 20 kHz to 200 kHz and is capable of 80–100 dB SPL at 1 m [7]. The same transducer is used as a receiver and coupled with pinna shapes. The external pinna modifies the directional properties of the receiver and is 3D printed from CAD models. In particular, pinnae are augmented from basic conical shapes with internal ridges and tragus [14]. The CIRCE sonarhead is employed in target localisation tests which take into account spectral information to estimate target location [15,16,19,30,31]. Indeed, the augmented pinnae or the bat ear replica are used to improve the receivers' directional properties.

The head of a *P. discolor* bat is scanned with a microCT machine [14] and this model is 3D printed: a microphone is fitted into the print in order to artificially improve the spatial directional properties of the receivers by the bat ear replica [15]. The active head related transfer function (AHRTF) is introduced as the direction dependent spectral filtering performed by a binaural active echolocation system. The received signals in the frequency domain for left and right receivers, S_L and S_R, in the frequency domain are the result of a series of frequency-dependent spatial filter operations

$$\begin{aligned} S_L &= H_L(\theta_L, \varphi_L, f) \cdot H_a(r, f) \cdot H_{rfl}(\theta_{rfl}, \varphi_{rfl}, f) \cdot X \\ S_R &= H_L(\theta_R, \varphi_R, f) \cdot H_a(r, f) \cdot H_{rfl}(\theta_{rfl}, \varphi_{rfl}, f) \cdot X \end{aligned} \tag{4.3}$$

with

$$X = H_a(r, f) \cdot H_{tr}(\theta, \varphi, f) \cdot X_c(f) \tag{4.4}$$

where H_{tr}, H_a, H_{rfl}, H_L and H_R model the filtering due to transmitter, air, reflector and left and right receivers, and X_c the spectrum of the emitted call. The AHRTF retains information about spatial filtering of the transmitter and the two receivers

$$H_{AHRTF}(\overline{\theta}, \overline{\varphi}, f) = \begin{bmatrix} H_L(\overline{\theta}_L, \overline{\varphi}_L, f) \cdot H_{tr}(\overline{\theta}, \overline{\varphi}, f) \\ H_R(\overline{\theta}_R, \overline{\varphi}_R, f) \cdot H_{tr}(\overline{\theta}, \overline{\varphi}, f) \end{bmatrix} \tag{4.5}$$

According to this, the AHRTF changes its magnitude depending on the reflector location. The AHRTF is measured for both left and right receivers [15] and used in three-dimensional localisation tasks. The change in the reflected echoes from reflectors depending on the direction of the reflectors themselves is at the base of the binaural localisation method. As the filtering due to the reflector and air absorption do not dominate the received spectrum, templates are collected and stored for a set of angles. The spectrum of the echo reflected from an unknown placed reflector is then compared to the templates. The reflector angle is estimated as the maximum a posteriori probability

$$\left(\widehat{\overline{\theta,\varphi}}\right) = \max_{\left(\overline{\theta,\varphi}\right)} \left[P\left(\left(\overline{\theta,\varphi}\right)|S_x\right)\right] = \max_{\left(\overline{\theta,\varphi}\right)} \left[\frac{P\left(S_x|\left(\overline{\theta,\varphi}\right)\right)P\left(\overline{\theta,\varphi}\right)}{P(S_x)}\right] \tag{4.6}$$

The localisation method is tested using three reflectors: a wooden ball, crumpled paper and a block reflector from toy building blocks. The ball reflector is also the one used to recover the spectral templates.

This work makes use of spectrograms to determine the distance of a target and of knowledge of the spatial filters of the two receivers to estimate target orientation. In particular, it is shown that the beam patterns associated with both the transmitter and the receivers should be taken into account in target localisation with sonar systems as they are both dominant in the spatial filtering of the echoes [15]. Spatial filtering of only the receivers is possible if the transmitter is purported to be omnidirectional [3].

Using complex external artificial pinnae for the receivers enhances the spatial filtering as a function of the orientation of the reflector and significantly improves the localisation performance of the sonar system [17]. The localisation performance of a target using conventional transducers (Polaroid ones) as receivers is compared [17] to that with artificial bat pinnae when processing the filtered echoes through the same localisation method [15]. It is noted that the narrow beamwidth of the individual transducers limits the area where the resulting spectral values can add spatial information, that is, conventional transducers in this case lack peripheral sensitivity. On the other hand, using artificial bat pinnae provides the AHRTF with higher peripheral sensitivity thus contributing to the sonar localisation information transfer.

The CIRCE head along with two artificial *P. discolor* pinnae is used to test another localisation method that implements a maximum likelihood estimator to select the orientation of multiple reflectors [32]. In particular, three-dimensional localisation of multiple reflectors is achieved of three reflectors simultaneously. In this work, the binaural echolocation is investigated taking into account the complete sound path, that is, the transmitter radiation pattern, air and reflector filtering, and the HRTF of both pinnae. The spectral code for each angle is extracted using a wooden ball as reflector thus defining the templates as the direction-dependent filtering characteristics of the sonar system [32] which are dominated by the transmitter and receiver beam pattern [15].

The received spectral code of an unknown reflector is compared to the spectral templates previously collected. The reflector orientation (azimuth and elevation angles) is estimated by selecting the maximum a posteriori probability (4.6).

This is a probabilistic model based on the similarity between the measurement spectrum and the spectra of a set of templates, and it is implemented on a binaural sonar system. Results show that all three targets are correctly located by the method within an area corresponding to a high acoustic energy from the transmitter: as the transmitter is a Polaroid transducer, this performance is directly linked to its radiation properties, i.e., the angular extension of the main lobe of the transmitter beam pattern.

Plastic replicas of bat ears are mounted on two Knowles microphones while the emitter is a Polaroid transducer [19,30]. This sonar system implements a localisation method based on a filtering reproducing the signal processing performed in the cochlea and on a Simultaneous Localisation and Mapping (SLAM) algorithm [33,34]. The sonar system collects data during the navigation of a robot on which the sonar system is mounted. The collected data are passed as inputs to the SLAM algorithm and then used to map the path the robot travelled during the experiment and build a map of the environment. In particular, the output of the cochlea filter model is the input to the SLAM. The whole process is called BatSLAM [19,30]. The initial filtering of the received echoes through a model of the bat's cochlea makes it possible to separate each frequency component in the frequency modulated signal and then align all the components to recover the spectrogram and the energy distribution of the received signal [26–28]. The sonar system, equipped with one emitter and two receivers, is mounted on a mobile robot; the sonar localisation works along with a map reconstruction algorithm, SLAM, and it supports robot navigation.

Effects of medium disturbances on a biomimetic ultrasonic sonar system are taken into account [35]. The medium distortion provoked by air currents, presence of fire and presence of water spray are considered while the sonar system is emitting and receiving. The localisation method [15] is based on the AHRTF: as the spatial filtering of the echoes is dominated by the transmitter and the receiver beam patterns, more weight is given to the filtering due to air absorption and distance by transmitting and receiving ultrasonic broadband signals through air currents, flames and spray water [35]. A reflecting plate is placed at 1 m from the sonar system and the spectrum of the echo is extracted from the echoes which are collected in case of no disturbance and in each of the three distortion situations (air currents, flames, spray water). The covariance matrix is calculated from the spectrum of the echoes in the frequency range 30 kHz to 80 kHz using the logarithmic magnitude of the echo spectra. Values in the covariance matrix are very high in the presence of flames and air currents thus indicating that these disturbances have a large impact on target localisation performance. Also, the lower frequencies are the most affected by the disturbance. On the other hand, a reference medium (that is, no distortion) and water spray are associated with low values in the covariance matrix. Target localisation tasks are performed using the binaural AHRTF of the bat *P. discolor* [15], that is, using replicas of bat pinnae for the receivers, and then using the Polaroid transducer for both receivers. The localisation performance significantly improves when *P. discolor*'s ears are the receivers' beam patterns. This suggests that a biomimetic emitter would also enhance 3D localisation of targets in the presence of medium disturbances.

The radiation properties of the transmitter have to be taken into account when analysing the localisation performance of a binaural sonar system with a highly directional transmitter. This is particularly true when a Polaroid transducer is used as transmitter [15,17,32] so that the active head related transfer function is defined to include the spatial filtering effects.

Localisation methods inspired by bat echolocation using only the spatial filtering of the receivers are possible provided that the transmitter is almost omnidirectional over a set of orientations. In this case, the emitter ensonifies an area with the same acoustic energy and the spatial filtering of the echoes is dominated by the receiver beam pattern only. Technical solutions to build an omnidirectional emitter or at least to enlarge the beam pattern of probes are: cylindrical PVDF film transducers [36], twisted acoustic lenses [37] and spherical piezoelectric transducers [38]. PVDF film transducers [36] are cylindrical therefore their pressure field is constant for orientations 0°–180° along a plane normal to the height of the cylinder. Nevertheless, because of the geometry of the cylinder itself, the spreading of acoustical pressure in the vertical direction is very directional: the beam pattern decreases 40 dB within ±30° off the main axis. Twisted lenses [37] provide an enlarged beam pattern, though the difference between the maximum and the minimum values of the beam pattern is 20–30 dB. Finally, radial modes of spherical piezoelectric transducers [38] are reported; the directional properties of these probes are though not detailed.

The knowledge *a priori* of the receivers' beam pattern is used to estimate the orientation of a target with respect to a sonar system in a localisation method called Beam Based Method (BBM) [3,39]. A biomimetic sonar system is composed of an omnidirectional emitter and two receivers whose beam pattern is directional and characterised by a prominent main lobe. In this case, the attenuation of the echoes recorded by a sonar system is dominated by the spatial filtering provided by the two receivers' external structures which return a set of values, one for each frequency in the signal, corresponding to the signal amplitude in the frequency domain. The comparison between received and emitted signals returns the attenuation provided by the receivers' directivity which, for each frequency, provides a set of echo directions. The contribution of spectra at all frequencies is considered for the reflected signals at the left $S_L(f)$ and right $S_R(f)$ receivers as well as for the emitted signal $S(f)$

$$\begin{cases} T_L = \displaystyle\int_{f_{st}}^{f_{en}} |S_L(f)|df \\ T_R = \displaystyle\int_{f_{st}}^{f_{en}} |S_R(f)|df \\ T = \displaystyle\int_{f_{st}}^{f_{en}} |S(f)|df \end{cases} \tag{4.7}$$

The filtering due to the beam pattern of the left and right receivers is dominant in the ratios

$$\begin{cases} R_L = \dfrac{T_L}{T} \\ R_R = \dfrac{T_R}{T} \end{cases} \tag{4.8}$$

The orientations corresponding to the beam pattern values in the left and right receivers closest to the values in (4.8) are chosen as potential true orientations of the target. Geometrical relationships between all of the chosen orientations of the target with respect to the left and right receivers are taken into account to reduce ambiguities and choose one orientation only as the estimated one

$$\begin{cases} \theta_R = \tan^{-1}\left[\dfrac{d_R \cos\varphi_L \sin\theta_L - d}{d_L \cos\varphi_L \cos\theta_L}\right] \\ \varphi_L = \varphi_R \end{cases} \tag{4.9}$$

In (4.9), d_L and d_R are the distance of the target from the left and right receiver, respectively, and d is the distance between the two receivers. The orientation, defined for the left and right receivers, fulfilling (4.9) will be the estimated orientation of the target with respect to the sonar system.

The choice of the receivers' beam pattern affects the efficiency of this method in discriminating between the orientations that correspond to similar attenuation values of the received signals. In particular, beam patterns with monotonic slopes both in azimuth and elevation are ideal for this purpose [40]. The beam pattern associated with regular structures made of truncated cones and provided with a main lobe for most frequencies is depicted in Figure 4.9.

The beam pattern associated with the *R. leschenaultii* bat is considered and the BBM is tested on parts of the beam pattern itself. As suggested, the BBM estimates accurately target direction if the search is restricted to angular extensions where the beam pattern is approximately monotonic over orientations. This leads to the conclusion that appropriate beam patterns, and therefore receiver structures, can be designed to be optimal for the BBM and a bat-inspired sonar system.

The work by Matsuo [41] aims to reproduce the accuracy bats show in locating objects even when they are flying. An experimental setup with a sonar system including one transmitting transducer and two microphones is used to estimate the position in 2D of a moving target by processing the echoes from the target to the microphones of the broadband ultrasonic signal emitted by the transducer [41]. The Doppler shift in this case provoked by the moving object affects the accuracy in determining the 2D position of the target. Waveforms were transformed into spectrograms by simulating the process of the mammalian cochlea [24,25]: the waveforms are convoluted with Gaussian chirplets whose centre frequencies are consistent with the sweep rate of the emitted signal.

$$F(f_j, t) = exp\left(-\frac{t^2}{\alpha}\right)exp\left(2\pi i\left(f_j t + \frac{1}{2}st^2\right)\right) \tag{4.10}$$

In (4.10), f_j is the centre frequency of the jth bandpass filter, t is time, s the sweep rate of the emitted signal and α is a parameter related to the width of the window function.

The output of the Gaussian filter chirplets is the spectrogram which is then converted into the range-frequency pattern by compensating for the emission sweep rate. Hence, the average pattern is calculated from the range-frequency pattern by integration over all the frequencies, see Figure 4.10.

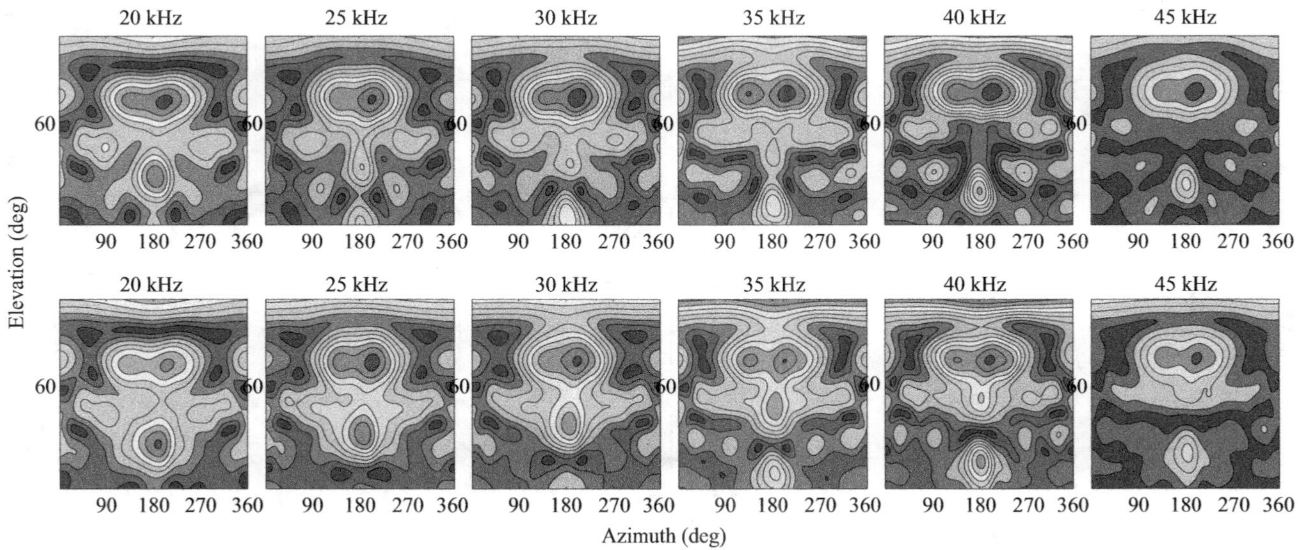

Figure 4.9 Beam pattern at several frequencies of a bat-inspired structure composed of regular structures. Reprinted with permission from [40]

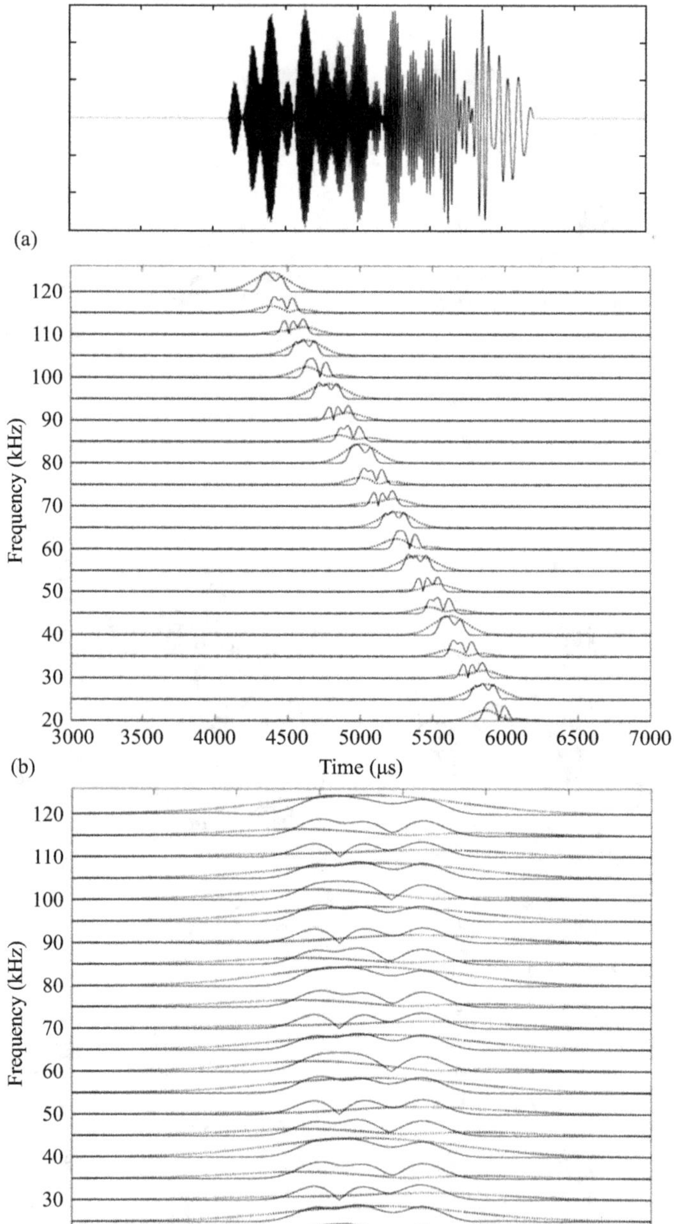

Figure 4.10 Received signal (a) is converted into its spectrogram (b) and the range pattern distribution (c). Reprinted with permission from [24]

The time delay is calculated around the onset of the averaged pattern. The distance between the two microphones d is used along with the time delay to estimate the 2D angle of the target according to

$$\theta = \sin^{-1}\left(\frac{\Delta r}{d}\right) \tag{4.11}$$

where Δr is the difference between the distance of the target from the left and from the right receiver while d is the distance between the two receivers. The Doppler shift affects the accuracy in measuring the location of a target and this is compared to the situation when the object is static.

A strategy to estimate the elevation angle of a target is accomplished with a signal with a tri-modal spectrum [42]. This spectrum is associated with a multi-beam directivity over elevation orientations and a multi-peak spectrum. The ratio between beam values in the elevation plane provides a set of templates which are calculated using the values of the second and first harmonic of the spectrum. These templates are used then with data collected in experiments to estimate the elevation angle; the method shows sub-degree accuracy around the elevation angle where the two beams, corresponding to the frequencies at which the signal spectrum has peaks, intersect. This strategy is proposed as a possibility through which mouth emitting bats (like *E. fuscus*) can estimate the vertical angle of a reflecting target.

4.4.1 Bat-inspired signal processing (cochlea models, discrimination of closely spaced objects and estimation of time delays using chirps, bat-inspired waveforms, target identification)

Inspiration from bat echolocation has led to the formulation of models for signal processing of broadband signals collected by two receivers [11,24,25,43]. In these models the activity of the hair cells in the cochlea is modelled by means of filters (band pass filters or Gaussian chirplets) centred at different frequencies according to the incoming waveforms' frequency range and distribution.

The RoBat is composed of a 6-DOF biomimetic sonarhead and a 3-DOF mobile platform [11,43]. The biomimetic sonarhead uses Polaroid ultrasonic transducers as transmitter and receivers. RoBat is used to estimate Doppler shift with bat cochlea models for processing the received echoes.

The bat cochlea is modelled with two different approaches. In the cochlea model a bandpass filter bank splits by means of a set of parallel filters with different central frequencies the incoming signal into a number of sub-signals, while the neural activity of the hair cells in the bat cochlea is rendered through a half-way rectifier and low-pass filter. In the first model the distribution of central frequencies of the passband filter bank is uniform while in the second model the distribution is non-linear. Also, the quality factor is defined in the second model to take into account bats' capability to estimate Doppler shifts around the frequency centred in the acoustic fovea of CF-FM bats [43].

An artificial neural network and a weighted average are then used to interpolate the output of the filter channels and to calculate the Doppler shift of the frequency.

Another implementation of the mammalian cochlea, that is, like that of some bat species as described in [24,25], has already been described previously and makes use of Gaussian chirplets as pass band filters to decompose the received waveforms into signals centred around different frequencies. Depending on the centre frequencies of the chirplets [24,25] the waveforms are convoluted with Gaussian chirplets whose centre frequencies are consistent with the sweep rate of the emitted signal, see (4.10).

The output of Gaussian filter chirplets is the spectrogram which is then converted into the range-frequency pattern by compensating for the emission sweep rate. Hence, the average pattern is calculated from the range-frequency pattern by integration over all the frequencies. This model is used to estimate Doppler shift [41] and to distinguish between closely spaced objects whose broadband echoes overlap [24,25].

Other works make use of bioinspired techniques to identify targets from analysis of acoustic echoes [44,45]. Inspiration from the bat *Eptesicus nilssonii* led to insight gained into the possibility of using multi-harmonic waveforms to increase spatial resolution with sonar and radar systems [44]. Indeed, typically sonar and radar systems do not use multi-harmonic waveforms; as some bat species (*E. nilssonii*) intentionally emit such waveforms the same approach can be followed.

4.5 Conclusions

The typical bat-inspired sonar system consists of one emitter and two receivers. Such an assembly serves the purpose of producing and receiving ultrasonic broadband acoustic signals to locate reflecting targets. Methods have been implemented and tested to measure the distance of a reflecting target and to estimate the orientation of a target with respect to the assembled sonar system. The methods are also inspired by the bat echolocation process and make use of bat-inspired signal processing and filtering: the bat cochlea is emulated to separate individual frequencies in the received signals and to compensate for the signal sweep rate, while the design of artificial bat pinnae as external receivers provide the receiving system with spatial filters. The latter introduces amplitude modulations of the incoming signals and thus allows us to discrimination between the orientations of the echoes.

The development of bat-inspired air-coupled sonar systems in the last 15 years is characterised by both the machining of appropriate structures for spatial filtering and the implementation of localisation methods. Sonar systems comprise a transmitter and two receivers, mimicking the displacement of mouth and ears on a bat head. A broadband signal is emitted and echoes from one or more reflectors are recorded through the receivers. The processing of binaural spectral information of echo signals in combination with time delays allows 3D localisation of reflectors. Spatial filtering is thus accomplished by means of external receiver structures inspired by bats (bat ear replica or simplified pinnae) which provide the sonar system with additional and fundamental filtering for estimating target orientation (works by Schillebeeckx, De Mey, Guarato, Ma, Mueller). Time cues are also considered in conjunction with signal processing resembling the frequency separation in the cochlea of the received signals. Such an operation allows the contribution of all the frequencies of the spectrogram to enhance the accuracy in determining time of flight and hence target distance (works by Carmena, Peremans, Matsuo). Finally, the signal processing techniques for target localisation reviewed in this chapter are developed according to the information that a sonar system that makes use of spatial filtering and time domain data can provide (works by Reijniers, Schillebeeckx, Steckel, Guarato, Kuc).

The work in the last 10–15 years on sonar systems inspired by bat echolocation is very much directed toward the development of analogue filtering to enhance localisation capabilities and spatial resolution. A lot of work has been done on the design of baffles for receivers, both static and dynamic. Further work is now being developed on baffles for the emitters as the complementary part to the receivers' filtering in a biomimetic sonar system.

References

[1] Müller, R. and Kuc, R., "Biosonar-inspired technology: goals, challenges and insights", *Bioinspiration & Biomimetics*, vol. 2, no. 4, Oct. 2007.

[2] Biber, C., Ellin, S., Shenk, E., and Stempeck, J., "The polaroid ultrasonic ranging system", *Audio Engineering Society Convention*, New York, 1 Oct. 1980.

[3] Guarato, F., Windmill, J., and Gachagan, A., "A beam based method for target localization: inspiration from bats' directivity and binaural reception for ultrasonic sonar", *Journal of the Acoustical Society of America*, vol. 133, no. 6, pp. 4077–4086, Jun. 2013.

[4] Firzlaff, U. and Schuller, G., "Spectral directionality of the external ear of the lesser spear-nosed bat, *Phyllostomus discolor*", *Hearing Research*, vol. 181, no. 1, pp. 27–39, Jul. 2003.

[5] Wotton, J.M. and Simmons, J.A., "Spectral cues and perception of the vertical position of targets by the big brown bat *Eptesicus fuscus*", *Journal of the Acoustical Society of America*, vol. 107, no. 2, pp. 1034–1041, Feb. 2000.

[6] Meymand, S.Z., Pannala, M., and Müller, R., "Characterization of the time-variant behavior of a biomimetic beamforming baffle", *Journal of the Acoustical Society of America*, vol. 133, no. 2, pp. 1141–1150, Feb. 2013.

[7] Streicher, A., Muller, R., Peremans, H., and Lerch, R., "Broadband ultrasonic transducer for a artificial bat head", *2003 IEEE Symposium on Ultrasonics*, vol. 2, pp. 1364–1367, Honolulu, HI, 5 Oct. 2003.

[8] Gupta, A.K., Fu, Y., Webster, D., and Müller, R., "Bat noseleaves as an inspiration for smart emission baffle structures", *ASME 2013 Conference on Smart Materials, Adaptive Structures and Intelligent Systems*, American Society of Mechanical Engineers, Snowbird, UT, 16 Sep. 2013.

[9] Caspers, P., Fu, Y., and Mueller, R., "A design for a biomimetic dynamic sonar head", *Journal of the Acoustical Society of America*, vol. 137, no. 4, p. 2334, Apr. 2015.

[10] Kim, D., Carmena, J.M., and Hallam, J.C., "Towards an artificial pinna for a narrow-band biomimetic sonarhead", *International Conference on Evolvable Systems*, Edinburgh, pp. 113–122, Springer, 17 Apr. 2000.

[11] Carmena, J.M., Kämpchen, N., Kim, D., and Hallam, J.C., "Artificial ears for a biomimetic sonarhead: from multiple reflectors to surfaces", *Artificial Life*, vol. 7, no. 2, pp. 147–169, Apr. 2001.

[12] De Mey, F., Reijniers, J., Peremans, H., Otani, M., and Firzlaff, U., "Simulated head related transfer function of the phyllostomid bat *Phyllostomus discolor*", *Journal of the Acoustical Society of America*, vol. 124, no. 4, pp. 2123–2132, Oct. 2008.

[13] Müller, R. and Hallam, J.C., "Knowledge mining for biomimetic smart antenna shapes", *Robotics and Autonomous Systems*, vol. 50, no. 4, pp. 131–145, Mar. 2005.

[14] Schillebeeckx, F., De Mey, F., and Peremans, H., "Bio-inspired sonar antennae: enhancing directivity patterns for localization", *Second IEEE RAS & EMBS International Conference on Biomedical Robotics and Biomechatronics*, pp. 366–371, Scottsdale, AZ, 19 Oct. 2008.

[15] Schillebeeckx, F., De Mey, F., Vanderelst, D., and Peremans, H., "Biomimetic sonar: binaural 3D localization using artificial bat pinnae", *The International Journal of Robotics Research*, vol. 30, no. 8, pp. 975–987, Jul. 2011.

[16] Steckel, J., Schillebeeckx, F., and Peremans, H., "Biomimetic sonar, outer ears versus arrays", *2011 IEEE Sensors*, Limerick, pp. 821–824, 28 Oct. 2011.

[17] Schillebeeckx, F., Vanderelst, D., Reijniers, J., and Peremans, H., "Evaluating three-dimensional localisation information generated by bio-inspired in-air sonar", *IET Radar, Sonar & Navigation*, vol. 6, no. 6, pp. 516–525, Jul. 2012.

[18] Bai, M.R. and Ou, K.Y., "Head-related transfer function (HRTF) synthesis based on a three-dimensional array model and singular value decomposition", *Journal of Sound and Vibration*, vol. 281, no. 3, pp. 1093–1115, Mar. 2005.

[19] Steckel, J. and Peremans, H., "A novel biomimetic sonarhead using beamforming technology to mimic bat echolocation", *IEEE Transactions on Ultrasonics, Ferroelectrics, and Frequency Control*, vol. 57, no. 7, pp. 1369–1377, Jul. 2012.

[20] Müller, R., Pannala, M., Reddy, O.P., and Meymand, S.Z., "Design of a dynamic sensor inspired by bat ears", *Smart Materials and Structures*, vol. 21, no. 9, Aug. 2012.

[21] Pannala, M., Meymand, S.Z., and Müller, R., "Interplay of static and dynamic features in biomimetic smart ears", *Bioinspiration & Biomimetics*, vol. 8, no. 2, May 2013.

[22] Guarato, F., Andrews, H., Windmill, J.F., Jackson, J., Pierce, G., and Gachagan, A., "Features in geometric receiver shapes modelling bat-like directivity patterns", *Bioinspiration & Biomimetics*, vol. 10, no. 5, Sep. 2015.

[23] Ma, J. and Müller, R., "A method for characterizing the biodiversity in bat pinnae as a basis for engineering analysis", *Bioinspiration & Biomimetics*, vol. 6, no. 2, May 2009.

[24] Matsuo, I., Kunugiyama, K., and Yano, M., "An echolocation model for range discrimination of multiple closely spaced objects: transformation of spectrogram into the reflected intensity distribution", *Journal of the Acoustical Society of America*, vol. 115, no. 2, pp. 920–928, Feb. 2004.

[25] Matsuo, I. and Yano, M., "An echolocation model for the restoration of an acoustic image from a single-emission echo", *Journal of the Acoustical Society of America*, vol. 116, no. 6, pp. 3782–3788, Dec. 2004.

[26] Reijniers, J. and Peremans, H., "Biomimetic sonar system performing spectrum-based localization", *IEEE Transactions on Robotics*, vol. 23, no. 6, pp. 1151–1159, Dec. 2007.

[27] Saillant, P.A., Simmons, J.A., Dear, S.P., and McMullen, T.A., "A computational model of echo processing and acoustic imaging in frequency-modulated echolocating bats: the spectrogram correlation and transformation receiver", *Journal of the Acoustical Society of America*, vol. 94, no. 5, pp. 2691–2712, Nov. 1993.

[28] Peremans, H. and Hallam, J., "The spectrogram correlation and transformation receiver, revisited", *Journal of the Acoustical Society of America*, vol. 104, no. 2, pp. 1101–1110, Aug. 1998.

[29] Peremans, H. and Reijniers, J., "The CIRCE head: a biomimetic sonar system", *International Conference on Artificial Neural Networks*, Warsaw, pp. 283–288, Springer, 11 Sep. 2005.

[30] Steckel, J. and Peremans, H., "BatSLAM: simultaneous localization and mapping using biomimetic sonar", *PLoS One*, vol. 8, no. 1, 24 Jan. 2013.

[31] Steckel, J. and Peremans, H., "Biomimetic sonar for biomimetic SLAM", *IEEE Sensors*, Taipei, pp. 1–4, 28 Oct. 2012.

[32] Schillebeeckx, F. and Peremans, H., "Biomimetic sonar: 3D-localization of multiple reflectors", *2010 IEEE/RSJ International Conference on Intelligent Robots and Systems (IROS)*, Taipei, pp. 3079–3084, 18 Oct. 2010.

[33] Dissanayake, M.G., Newman, P., Clark, S., Durrant-Whyte, H.F., and Csorba, M., "A solution to the simultaneous localization and map building (SLAM) problem", *IEEE Transactions on Robotics and Automation*, vol. 17, no. 3, pp. 229–241, Jun. 2001.

[34] Choset, H. and Nagatani, K., "Topological simultaneous localization and mapping (SLAM): toward exact localization without explicit localization", *IEEE Transactions on Robotics and Automation*, vol. 17, no. 2, pp. 125–137, Apr. 2001.

[35] Steckel, J., Vanduren, W., and Peremans, H., "3D localization by a biomimetic sonar system in a fire-fighting application", *Fourth International*

Congress on Image and Signal Processing (CISP), Shanghai, vol. 5, pp. 2549–2553, IEEE, 15 Oct. 2011.

[36] Toda, M., "Cylindrical PVDF film transmitters and receivers for air ultrasound", *IEEE Transactions on Ultrasonics, Ferroelectrics, and Frequency Control*, vol. 49, no. 5, pp. 626–634, May 2002.

[37] Dahl, T., Ealo, J.L., Papakonstantinou, K., and Pazos, J.F., "Design of acoustic lenses for ultrasonic human-computer interaction", *2011 IEEE International Ultrasonics Symposium*, Orlando, FL, pp. 64–67, 18 Oct. 2011.

[38] Kim, J.O., Lee, J.G., and Chun, H.Y., "Radial vibration characteristics of spherical piezoelectric transducers", *Ultrasonics*, vol. 43, no. 7, pp. 531–537, Jun. 2005.

[39] Guarato, F., Windmill, J., and Gachagan, A., "Sonar localization using receiver beam profile features", *2012 IEEE International Ultrasonics Symposium*, Dresden, 7 Oct. 2012.

[40] Guarato, F., Windmill, J., Gachagan, A., and Harvey, G., "An investigation of acoustic beam patterns for the sonar localization problem using a beam based method", *Journal of the Acoustical Society of America*, vol. 133, no. 6, pp. 4044–4053, Jun. 2013.

[41] Matsuo, I., "Localization and tracking of moving objects in two-dimensional space by echolocation", *Journal of the Acoustical Society of America*, vol. 133, no. 2, pp. 1151–1157, Feb. 2013.

[42] Kuc, R., "Echolocation with bat buzz emissions: model and biomimetic sonar for elevation estimation", *Journal of the Acoustical Society of America*, vol. 131, no. 1, pp. 561–568, Jan. 2012.

[43] Carmena, J.M. and Hallam, J.C., "Estimating Doppler shift using bat-inspired cochlear filter bank models: a comparison of methods for echoes from single and multiple reflectors", *Adaptive Behavior*, vol. 9, no. 3–4, pp. 241–261, Sep. 2001.

[44] Balleri, A., Griffiths, H.D., Baker, C.J., Woodbridge, K., and Holderied, M.W., "Analysis of acoustic echoes from a bat-pollinated plant species: insight into strategies for radar and sonar target classification", *IET Radar Sonar and Navigation*, vol. 6, no. 6, pp. 536–544, Jul. 2012.

[45] Kuc, R., "Improving object classification with biomimetic sonar", *Journal of the Acoustical Society of America*, vol. 110, no. 3, pp. 1263–1266, Sep. 2001.

Chapter 5

Analysis of acoustic echoes from bat-pollinated plants

Alessio Balleri[1], Hugh Griffiths[2], Chris Baker[3] and Marc Holderied[4]

5.1 Introduction

In recent years, with the development of high range resolution radar and sonar systems, the desire to be able to identify targets under all conditions has become of great importance. This is an activity carried out with considerable success by echolocating bats that are able to detect, select and attack prey even in a dense clutter environment [1–3]. Bats have evolved echolocation as a means of detecting, identifying and attacking prey for over 50 million years and hence there is potentially a great deal that can be learnt from understanding how they use these capabilities and how this might usefully be applied to radar and sonar systems [4–6]. Although identification and classification of targets is a very important task, and much has been published on these topics, there has been less research effort, within the radar and sonar communities, to understand and utilise the successful techniques used by echolocating bats.

Classification of insects by bats is performed through the combined information available in both the time and the frequency domains [7,8]. The amount of power reflected from insects gives the bats information on their size and the bigger the insect, the greater the fraction of power that is reflected. Information can also be obtained by analysing modulations in the amplitude of the echo. It has been shown that there is a positive correlation between body size and wing beat period in insects. Also, the periodic amplitude modulation by moving the wings is directional, giving the bats information about the look angle. Doppler shift and micro-Doppler modulations provide a further source of information aiding identification. Indeed it has been shown that each species of insect has a unique micro Doppler signature even when the wing beat frequency is the same [7,8].

[1]Centre for Electronic Warfare, Information and Cyber, Cranfield University, Defence Academy of the UK, Shrivenham SN6 8LA, UK
[2]Department of Electronic and Electrical Engineering, University College London, London WC1E 6BT, UK
[3]Aveillant Ltd, 300 Science Park, Milton Road, Cambridge CB4 0XL, UK
[4]School of Biological Sciences, University of Bristol, 24 Tyndall Avenue, Bristol BS8 1TQ, UK

Although several studies have examined how bats recognise moving targets, the literature shows a lack of knowledge on how echolocating mammals perform classification of static targets. Echolocating nectar feeding bats are able to detect and select flowers of bat-pollinated plants even in highly cluttered environments. They play an important role in the process of pollination of plants because by visiting the flowers for nectar they ensure the pollen transfer between different individual plants. Because it is in the interest of both the plant and the bat that pollination takes place it has been hypothesised that co-evolution has shaped bat-pollinated flowers in order to ease classification [9–11]. As a result of this, it is thought that these flowers have evolved to ease classification by bats, and that their echo-acoustic signatures contain critical information that aids the bat in choosing the most suitable flowers.

Nectar feeding bats and their flowers represents a combination highly suited to the study of target classification. Firstly, flowers are motionless and silent so that bats cannot rely on Doppler information or passive echolocation, and secondly their habitat is often a densely cluttered environment which requires a high level of skill in distinguishing between flowers worth feeding from and other scenario content. Finding and approaching a flower is a gradual process that involves all the bats' senses. While bats can only use vision at dusk, they can rely on their sense of smell and on echolocation to search for flowers in darkness. Nectar feeding bats have a highly developed sense of smell and they are highly attracted by the scent of sulphur-compounds which are produced by many bat-pollinated plants [12]. Although the sense of smell certainly plays an important role, in particular for long range attraction, latest research results support the fact that nectar feeding bats largely rely on echolocation to plan their approach flight and to detect and select the flowers in the proximity of the plant.

Previous studies have shown that bats are able to detect and select objects by echolocation alone [13–16]. An interesting example is the flowers of the bat-pollinated vine *Mucuna holtonii* plant which releases the greatest amount of nectar reward only on the first visit by a bat. When the bat lands on the flower to reach the nectar, it ignites an explosion of the flower's vexillum which functions as a trigger for the pollen and the nectar release. Behavioural experiments have shown that bats can successfully select unvisited flowers within an inflorescence in darkness and this is a task that can only be achieved by echolocation. The scent of the flower, in fact, remains unchanged before and after the explosion. It has also been shown that modifying the echo-acoustic properties of these flowers, by inserting a small pad of cotton in the cavity of the vexillum (i.e., without altering the scent and visual aspect of the flower) results in a reduction of the success rate of the bat [13,14]. There have been further studies to demonstrate that bats can recognise targets by echolocation alone. A few individuals of *Glossophaga soricina* bats were successfully trained to discriminate between an unscented hollow hemisphere and an unscented paraboloid, irrespective of size and in darkness, to show that bats have the ability to generalise object shapes by echolocation [15]. In another experiment, bats were able to discriminate between unscented hollow hemispheres of different sizes in darkness [16].

Choosing the most suitable flowers to be visited within an inflorescence is a task in which echolocation certainly plays a fundamental role and, although classification of flowers in such an environment is demanding, nectar-feeding bats succeed in their pollination task. In investigating the features of these flowers that aid the bats search for nectar, the strategy underpinning the task of classification of static targets by bats may be understood and this can offer useful insights to radar and sonar engineers working on target classification.

In this chapter, we analyse a real set of data containing high range resolution profiles (HRRP) of unpollinated corollas of *Cobaea scandens* and of an inflorescence of *Rhytidophyllum auriculatum*, which are plants typically pollinated by bats. These were collected by transmitting a synthetic wideband linear chirp with an acoustic radar capable of a very high range resolution. Two *C. scandens* flowers taken at different stage of maturity were measured to investigate if and how the information contained in the flowers' echo acoustic signatures changes as the flower wilts and hence stops producing nectar and loses attractiveness. These were then modified, by manually removing the petals, in order to study feature-specific contributions to the scattering. The acoustic signatures of the *C. scandens* and the *R. auriculatum* do not change as a result of pollination by a bat and indeed the same individual flower can be pollinated multiple times as long as it contains nectar. The analysis presented in this chapter is subtly different from investigating pre and post bat fed plants but it demonstrates the acoustic changes as the flower degrades, and hence it is a valid experiment for this bat-pollinated plant species. A photo of a *C. scandens* and a photo of the *R. auriculatum* are shown in Figure 5.1(a) and (b). The signatures are analysed and results are discussed with respect to the task of classification of manmade targets.

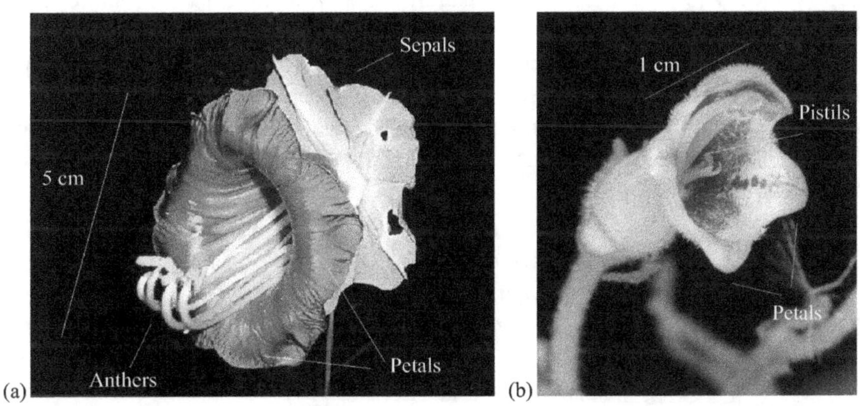

Figure 5.1 (a) Photo of a desiccated C. scandens *flower with protruding anthers, a bell-shaped corolla of largely merged petals whose unmerged petal ends fold back, and a ring of partly merged sepals at its base. (b) Photo of a* Rhytidophyllum auriculatum *flower. Reprinted with permissions from [17,18]*

5.2 Analysis of the signature of *C. scandens* corollas

The data consists of HRRP of a few individual flowers of *C. scandens*. These were collected at the School of Biological Sciences of the University of Bristol in 2008. The flowers to be ensonified were impaled on a thin pin that was placed at the centre of a turntable rotating with an angular step resolution of 1°. At each step of the turntable, a Linear Frequency Modulated waveform (LFM, or down-chirp) spanning the frequencies between 50 and 250 kHz was transmitted towards the flower with a piezoelectric custom-built loudspeaker. Table 5.1 reports the values of the loudspeaker beamwidths at 50, 100, 150 and 200 kHz. For each frequency the level of the main lobe measured with respect to the maximum level achievable in the range of frequencies between 10 and 220 kHz is also given [19]. A detailed description of the loudspeaker (about 1.5 cm × 2 cm in size) can be found in [20]. The echoes from the flowers were recorded with a 1/4-inch precision condenser microphone (G.R.A.S. type 40 BF) with a nominal flat frequency response between 4 Hz and 100 kHz (±2 dB) and a sensitivity of 4 mV/Pa. At 100 kHz the nominal −3 dB beamwidth of the microphone was greater than 30°. A detailed description of the microphone can be found in [21]. The echoes were sampled at a rate of $f_s = 500$ kHz using a National Instruments PCIe-6251 card. During the experiments, both the loudspeaker and the microphone were placed into a custom-built artificial bat head (about 2 cm × 1 cm long) in order to reproduce the spatial arrangement of the two ears (receivers) and the mouth (transmitter) of the bat. A photograph of the artificial bat head is given in Figure 5.2. Although the artificial head was capable of hosting one loudspeaker and two microphones, only one microphone was used to record the data analysed in this chapter. The setup allowed measurements with a range resolution of approximately 1.5 mm. A picture of the setup is shown in Figure 5.3. A measurement was taken, all being stationary, at each angular step of the turntable so to form a multi-perspective image of the flower. Three different horizontal images were recorded, each one representing HRRPs over 180° of unpollinated *C. scandens* in three different cases: (i) a flower under ideal conditions for pollination, (ii) a desiccated flower and (iii) a flower whose petals were removed by hand. The raw data was recorded so that it could be processed offline.

Table 5.1 Beamwidth of the loudspeaker (calculated at −3 dB) at 50, 100, 150 and 200 kHz and the corresponding level of attenuation with respect to the maximum level obtained between 10 and 220 kHz [19]

Frequency	Main lobe width (°)	Main lobe level (dB)
50 kHz	15	−7
100 kHz	10	−7
150 kHz	11	−2
200 kHz	8	−20

It is worth noting that nectar feeding bats transmit waveforms that cover a bandwidth between 20 and 140 kHz and that commonly include harmonics. The waveforms are additionally diversified during the pollination task and their duration, bandwidth and design can significantly change within an echolocation buzz [22].

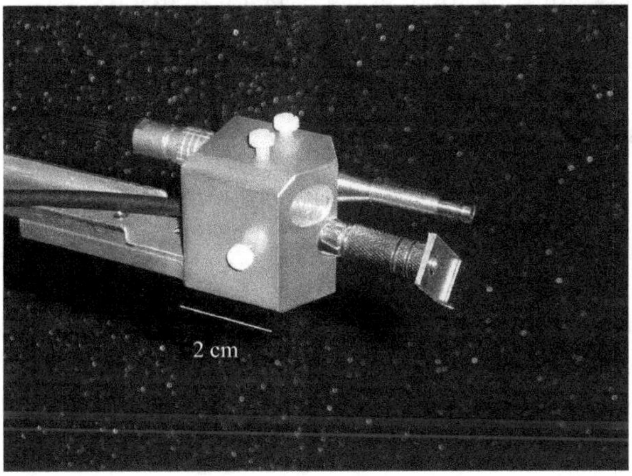

Figure 5.2 Photograph of the artificial bat head used for the experiments. Reprinted with permissions from [17]

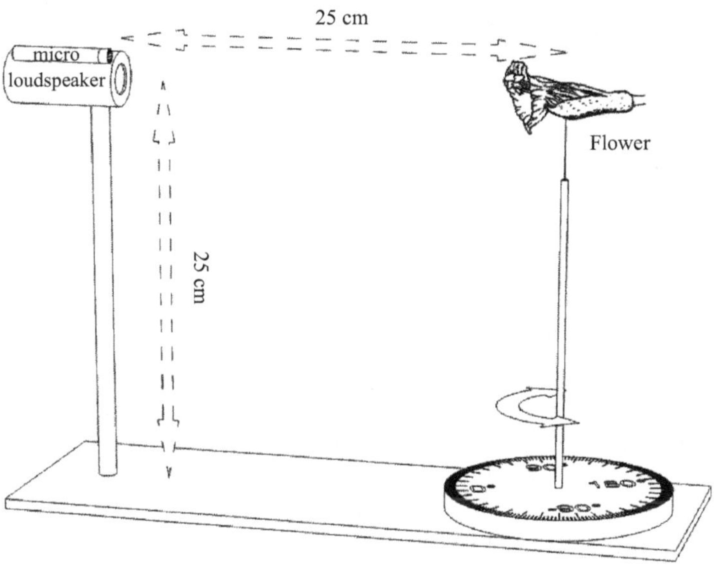

Figure 5.3 Schematic drawing of the experimental setup. Note that the depicted flower is not a C. scandens. *Taken from [9]. The geometry in the sketch corresponds to an aspect angle of 0°, e.g., the corolla facing the artificial bat-head. Reprinted with permissions from [17]*

Although the use of the linear chirp transmitted in these experiments may not be fully representative of a real scenario, it allowed a range resolution similar to that achieved by bats [23], and provided a flat frequency response to allow an estimate of the impulse response of the flowers at the frequencies where the distortions introduced by the electronics were not significant (from 50 kHz up to at least 150 kHz). These were quantified by measuring the width of the response from a metallic flat plate and by looking at the difference between the theoretical range resolution ($c/2B = 0.84$ mm) and the measured range resolution (1.5 mm). The data was gathered on the assumption that the echoes measured in the experiment provide an acceptable representation of the information displayed to a bat before any neural processing.

5.3 Signature of *C. scandens* flowers

The measurements consist of recorded data values $s_j(n)$, where $j = 1, \ldots, N_j$ indicates the jth angular perspective, $N_j = 180$ is the number of measured angular perspectives and n represents the discrete time index. The analytic signal $\hat{s}_j(n)$ of $s_j(n)$ defined as

$$s_j(n) = Re\{\hat{s}_j(n)\} \tag{5.1}$$

was computed using the Hilbert transform method and matched filtered to the analytic signal of a copy of the transmitted chirp. The output of the matched filter $z_j(n)$ represents the HRRP corresponding to the jth angular perspective. The background static clutter was removed by removing the mean value from any range bin $\overline{n}$, i.e., by removing the mean clutter component of the function $z_j(\overline{n})$, estimated as

$$m(\overline{n}) = \frac{1}{N_j}\sum_{j=1}^{N_j} z_j(\overline{n}) \tag{5.2}$$

For each angular perspective the target HRRP $\mathbf{x_j}$ was formed with the element of $\mathbf{z} = (z(1)\ z(2)\ \ldots\ z(N))^T$ corresponding to range bins from 18 to 32 cm, i.e., those range bins that contained the echo from the flowers. Similarly a noise profile $\mathbf{n_j}$ was formed with the elements of $\mathbf{z}$ corresponding to the range bins that contained noise only.

The images showing the magnitude of the HRRPs and those showing the magnitude of their corresponding Fourier Transform for each angular perspective are given in Figures 5.4–5.6. These are related to the angular perspectives ranging between $-90°$ and $+90°$, where the $0°$ line corresponds to the case when the flower was directly facing the microphone and the loudspeaker (see Figure 5.3). Normalisation was performed across all images with respect to the highest value. The magnitude of the HRRPs of the individual of *C. scandens* that was at the peak of attractiveness to a bat for pollination is shown in Figure 5.4. This exhibits strong reflections between ranges of 21 cm and 23 cm from the microphone which originates from the petals of the corolla, and weaker reflections between 20 cm and

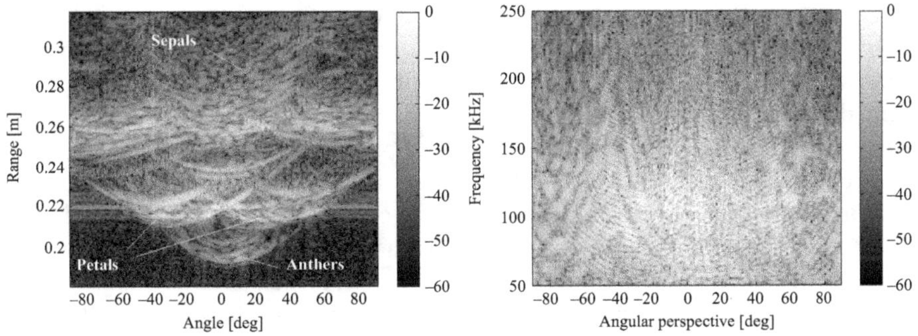

Figure 5.4 Magnitude of HRRPs and magnitude of their corresponding Fourier transform over 180° for a C. scandens *ready for pollination. 0° is to the front of the flower. HRRPs direction is horizontal, i.e., from left to right with respect to the flower's natural orientation. Scaled in relative [dB]. Reprinted with permission from [17]*

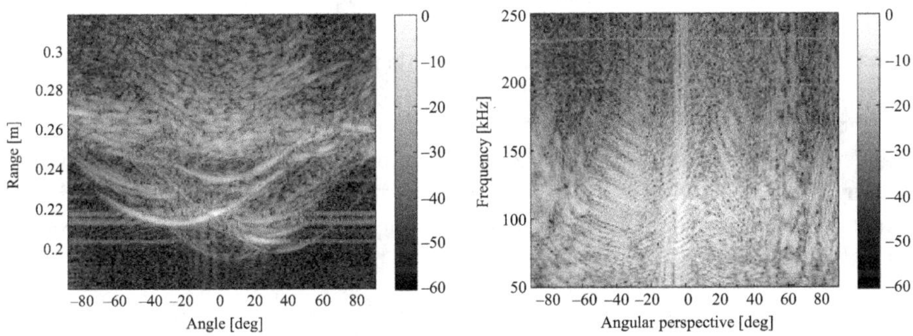

Figure 5.5 Magnitude of HRRPs and magnitude of their corresponding Fourier transform over 180° for a desiccated flower of C. scandens. Reprinted with permission from [17]

21 cm originated from the anthers protruding from the corolla (see Figure 5.1). A weaker level of scattering originates from the bell-shaped inner part of the flower's corolla between 25 and 27 cm. This is the part of the corolla that contains nectar and from which the pistil grows. Weaker scattering due to the echo generated by the ring of sepals that cover the external back side of the flower is also visible from about 28.5 cm to 29.5 cm. It can be observed that the petals can scatter a considerable amount of power over a wide range of angles between −60° and +60°. The plot of the magnitude of the HRRPs Fourier transform shows a complex structure of peaks and troughs which is somewhat difficult to interpret. As the flower wilts, i.e., stops producing nectar, it loses attractiveness to bats. By examining the

change in the structure of the HRRPs together with changes in classification performance from case to case we can better understand which features are responsible for classification. The time and frequency plots related to the HRRPs of the desiccated *C. scandens* that had started to wilt due to an overnight frost are displayed in Figure 5.5. The HRRPs plot shows that the scattering due to the petals between 21 and 23 cm has lost detail compared to that shown in Figure 5.4. Wide angular scattering that spans between −60° and 0° is still present at about 22 cm and is probably due to one petal that was still in good shape at the time of recording. However, most of the scattering from the petals is no longer visible leading to a marked loss of information. This is because the petals tend to fold back during the wilting process and the retro-reflective effect that is generated when they are in the upright position diminishes. The loss in turgor might also affect reflective properties of the plant tissue. The scattering from inside the corolla does not show any loss of complexity but it is weaker than before with respect to the maximum value within the image. Differences in the frequency domain remain complex and difficult to interpret. To investigate the image of the flower without petals at the corolla, the front ends of the petals from the same individual were removed manually using a pair of scissors. Figure 5.6 shows the HRRPs image obtained from this modified flower. The scattering from the petals disappears over all angles. A fraction of power is still scattered at 0°, probably originated by some residuals of petals that might have not been removed completely or simply by the line delimiting the aperture of the corolla. The fact that the image is very similar to the one obtained from the wilting flower strongly suggests that the scattering from the petals tends to disappear when the flower is no longer fresh. Scattering from the inside of the flower is now more visible as compared to Figure 5.5 but no relevant change in the complexity of the echo could be discerned.

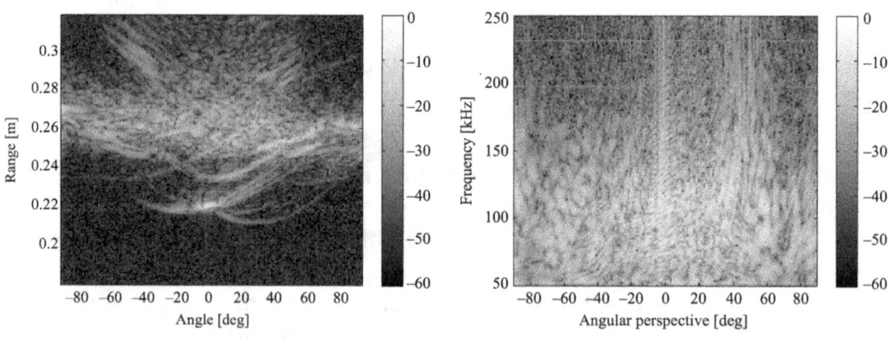

Figure 5.6 Magnitude of HRRPs and magnitude of their corresponding Fourier transform over 180° for a modified desiccated C. scandens*; the petals and the pollen sacks were removed by hand from the flower. Reprinted with permission from [17]*

Thus we might hypothesise that the bat can recognise the differences between the three cases and categorise them as three different objects.

5.3.1 Power reflection as a potential cue

One cue available to bats is the overall power of the echo. The mean power of the noise was estimated from the noise profiles $\mathbf{n_j}$ as

$$P_n = \frac{1}{N_i N_j} \sum_{i=1}^{N_i} \sum_{j=1}^{N_j} |n_j(i)|^2 \tag{5.3}$$

and similarly the mean power of the range bin containing target plus noise was estimated from all perspectives $\mathbf{x_j}$ as

$$P_{x+n} = \frac{1}{N_i N_j} \sum_{i=1}^{N_i} \sum_{j=1}^{N_j} |x_j(i)|^2 \tag{5.4}$$

In the equation N_i and N_j are the number of rows and columns of an image, respectively. The signal-to-noise ratio (SNR) for the three targets was then estimated as

$$SNR = \frac{P_{x+n}}{P_n} - 1 \tag{5.5}$$

The results are reported in Table 5.2 and show a difference of more than 3 dB between the fresh flower and the other two. The SNR was estimated independently for each perspective as

$$P_{x+n}(j) = \frac{1}{N_i} \sum_{i=1}^{N_i} |x_j(i)|^2 \quad j = 1, 2, \ldots, N_j \tag{5.6}$$

As expected, in the case of the full flower the mean power of the echo presents higher values over a larger angular view due to the presence of the petals. There are differences in the power depending on the perspective that might be used by bats as a clue for classification [16] (see Figure 5.7).

Table 5.2 Values of the estimated signal-to-noise ratios (SNRs) corresponding to the three C. scandens *flowers*

	Full flower	**Desiccated**	**Modified**
SNR	24.5 dB	21.1 dB	20 dB

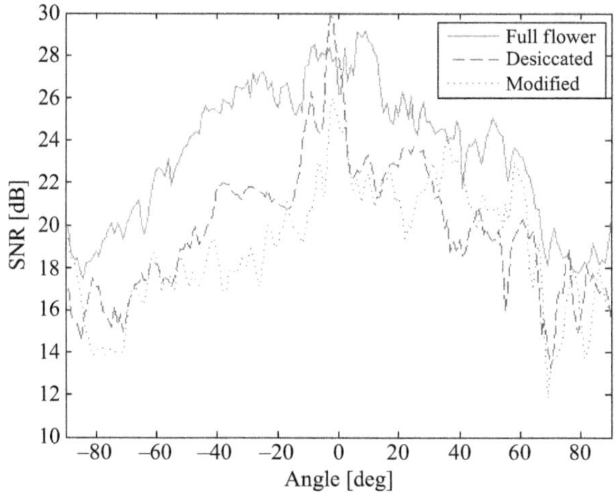

Figure 5.7 Signal-to-noise ratio as a function of the angle perspectives. Estimated as in (5.6). Reprinted with permission from [17]

5.4 Analysis of a *R. auriculatum* inflorescence

By visiting flowers for nectar, the bat is responsible for pollen transfer between different individual plants and hence plays a key role in plant pollination. In other words, although the short-term interest of the bat is solely efficient feeding, it is in the long-term interest of both the bat and plant species for pollination to take place successfully.

As a result of co-evolution, in addition to the individual flowers, the entire plants will have developed to provide clues in the echo responses obtained by the bat. In particular, bats have to distinguish between good flowers, wilting flowers and buds as well as timing their feed for maximum effect (efficient eating and energy consumption). It would seem likely, therefore, that characteristics indicating the difference between these flower states are embedded in echoes of the bats cry to facilitate discrimination. An examination of the echoes from a plant inflorescence is useful to evaluate their dominant features and better understand how the bat might utilise them to discriminate between flowers of different stage of maturity and to plan the approach into selected flower corollas. Ultimately, this knowledge may inform on how to interpret the target information available in man-made synthetic aperture radar and sonar images. To investigate this, in this chapter, we study echoes from an inflorescence extracted from an *R. auriculatum* plant. The *R. auriculatum* hook is a bat-pollinated plant that grows in the Caribbean region and produces small flowers whose nectar is extremely attractive to bats (Figure 5.1(b)). Echoes from this plant are ideal for investigating the information sensed by the bats, as well as the relationship between the echo and the maturity status of different individual flowers.

Firstly, two datasets containing HRRPs of an open flower and a bud of *R. auriculatum* are compared. Of course we cannot be certain that the bat processes

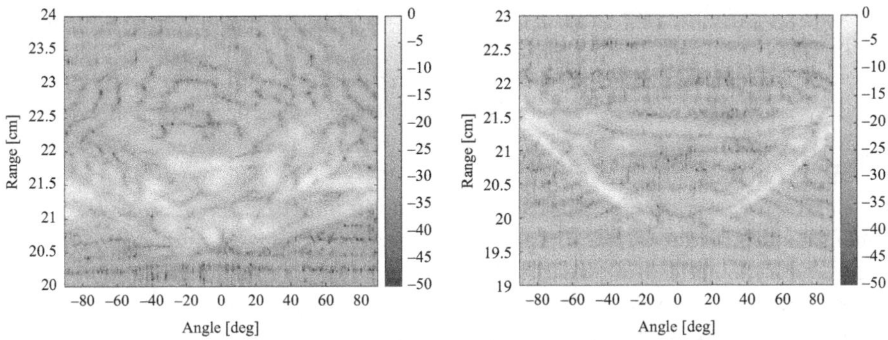

Figure 5.8 Magnitude of the HRRPs of (a) a R. auriculatum *open flower and (b) a* R. auriculatum *bud. The structure of the bud is very different from that of the open flower. This may be thought as the flowers way to attract the bat attention. The grey colour scale indicates the echo strength in [dB] normalised to the maximum echo value. Reprinted with permission from [18]*

received echoes as HRRPs but previous biological analysis has proposed bats to be sensitive to target range and range differences between parts of a target [24]. The bat has to be able to discriminate between nectar-bearing open flowers and closed buds. Although the form of the bats neural signal processing is largely unknown, we expect the HRRPs of the bud to therefore be significantly different from those of open flowers. Closed buds are physically smaller than open flowers and consequently the amount of energy they reflect is lower than for open flowers. Thus overall echo strength can be one factor in helping the bat distinguish between open flowers and closed buds. HRRP data was collected at the School of Biological Sciences at the University of Bristol in 2009 using an equivalent setup as that described in the previous sections. The flowers were impaled on a 1.5 mm diameter metallic pin placed at the centre of the turntable and were ensonified with LFM waveforms from 200 to 50 kHz to achieve a range resolution of approximately 1.5 mm. This was experimentally measured as the width of the main lobe of the cross-correlation function between the transmitted chirp and the echo generated by a metallic flat plate. A measurement of the background was removed from the echoes before subtracting the data mean value. The Hilbert transform of each echo was then cross-correlated with that of the transmitted waveform to generate the HRRPs.

Figure 5.8(a) shows the HRRPs plotted as a function of orientation angle for an open *R. auriculatum* flower. The complex structure of the scattering from the corolla is visible over the entire angular range. There are discernible regions of both high and low reflectivity that persist over large angles. Whilst complex, the HRRPs are far from random, but a route to correct recognition is not obvious. The flower is broadly bell-shaped. This results in a relatively large echo being reflected over the wide range of angles, making them easier to detect against the background.

The petals contribute significantly to the amount of reflected energy. Indeed, the petals are themselves visible over a wide angular window that spans at least 90°. Figure 5.8(b) shows HRRPs when the *R. auriculatum* bud was ensonified. The structure of the HRRPs of the bud is very different from that of the open flower. Again, scattering is present over all angles but the level of complexity in the HRRPs is reduced giving them a more sparse appearance. The image suggests only one or two discrete scatterers are now contributing to the HRRPs. At an angle of 0° (when the sensor directly faces the bud), the scattering is weaker than at any other angles. This is due to the ensonified surface now being smaller than at any other angles and sloping away from the direction of the sensor. There is also scattering caused by the sepals that cover the back of the corolla. This is visible at around 21.5 cm and exists over all angles between 90° and −90°. These results illustrate that specific parts of the flower contribute to the overall information sensed by the bat. The HRRP responses for the open and closed flower show how information is modified as the flower changes appearance both in terms of the total reflected signal as well as the detailed structures shown in the HRRPs.

However, the bat has a much more complex task in approaching and selecting flowers suitable for feeding. It has to do this in an area of highly dense scattering caused by many *R. auriculatum* flowers of different ages as well as other plants not suitable for feeding. The bat has to be able to process echoes to gather the right information for successful feeding [14]. It seems likely that the bats flight trajectory in approaching the flower is also a contributory element in extracting information in the recognition of suitable nectar-bearing open flowers. Because the plant species perpetuates through the resulting pollination, the spatial and shape arrangements between individual flowers, buds and calyxes (i.e., flowers without the corolla) have all evolved to give the bat the necessary information to succeed in the task of flower recognition. Differences between echo responses from buds,

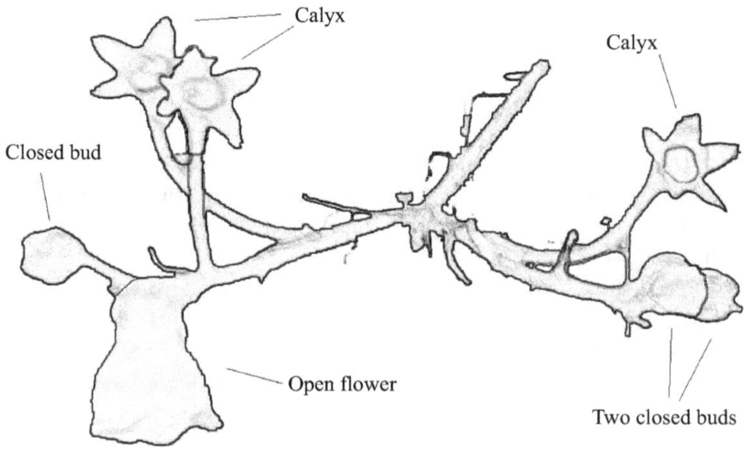

Figure 5.9 Sketch of the inflorescence of the R. auriculatum *plant composed of an open flower and three buds. Reprinted with permission from [26]*

calyxes and open flowers allow the bat to detect and recognise the correct target (the open flower). The bell-shaped flowers act as efficient retro-reflectors over a broad range of angles correspondingly causing a large echo at all angles. This alone helps make them easier to detect. Closed buds and calyxes scatter less energy back towards the bat. These differences between echo strengths are exploited by the bat and help to plan the approach trajectory to the nectarium (the part of the corolla that contains the nectar). Figure 5.10 shows HRRPs as a function of azimuth and elevation angle relative to the inflorescence of a *R. auriculatum* plant (see Figure 5.9). The measurements were made with the same apparatus described earlier. The vertical-axis is the distance in metres between the artificial bat-head and the centre

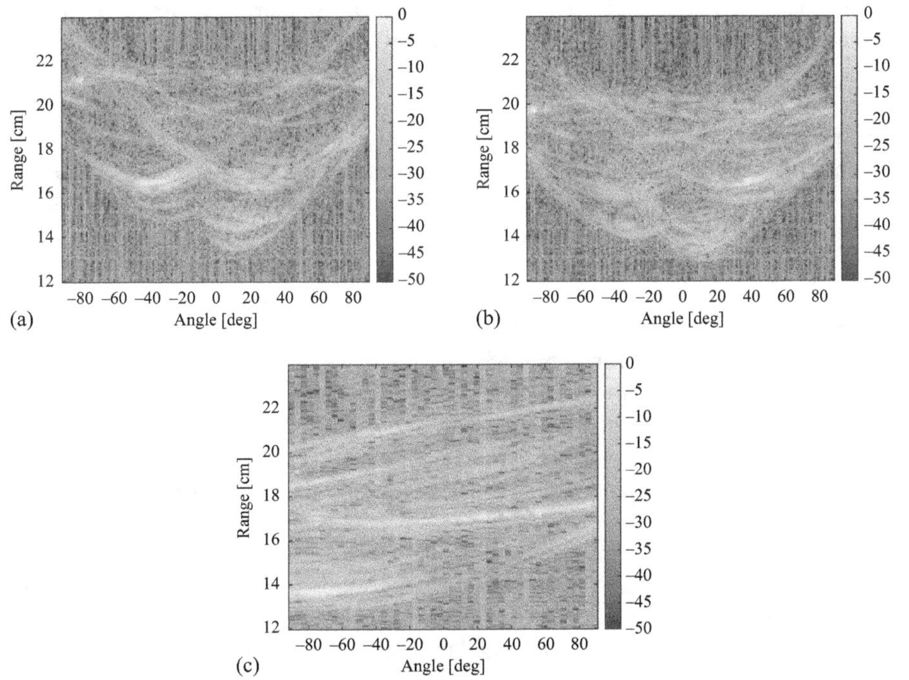

Figure 5.10 HRRPs of the inflorescence of a R. auriculatum *plant composed of an open flower and three buds measured: (a) horizontally from a vertical angle of 0°, (b) horizontally from a vertical angle of −25° and (c) vertically from a horizontal angle of 0°. The 0° aspect angle corresponds to the artificial bat-head being placed at the same height as the centre of rotation and facing the 0° direction. The image shows the more complex structure and higher directionality of flowers in the horizontal and vertical planes. The flower also protrudes and this facilitates separation from buds, calyxes and the background. The grey colour scale indicates the echo strength in [dB] and it has been normalised to the maximum echo value. Reprinted with permission from [26]*

of rotation of a turntable located at a distance of approximately 20 cm. The horizontal axis is the angle between the horizontal turntable and the bat-head. At 0° the bat-head directly faces the plant. The grey colour coding indicates echo strength and has been normalised to the maximum echo value.

Figure 5.10(a) shows, at a 0° depression angle, that the scattering from the two buds located on the left of the open flower (as viewed from the front) is clearly visible at a distance between 16 and 18 cm. The bud scattering is present over a large angular window between −80° and 0°. The scattering from the bud located on the same side as the open flower superimposed with echoes from the dead branches is visible between −30° and +80° at a distance from 16 to about 19 cm. The open flower protrudes maximally and hence is visible at the nearer range of about 14 cm. Its reflections are more directional than those of the buds.

In Figure 5.10(b) (−25° depression angle) the scattering from the two buds located on the right of the open flower is clearly visible between 16 and 18 cm and again is present over a large angular ambit from −80° and 0°. However, the scattering from the two buds on the left-hand side is much weakened. This illustrates the strong dependence of the scattering on relative orientation between the bat and the ensonified object. The open flower still protrudes maximally and is visible at a range of about 14 cm. As before, its reflections are more directional than those associated with the buds.

The vertical plane HRRPs are shown in Figure 5.10(c). The figure shows that scattering from the buds and the dead branches with calyx is visible at a distance of 16 cm and persists over almost all perspectives. The open flower is visible at a distance of 14 cm and over a smaller angular window between −25° and −10°. This is typical of the comparatively higher directionality of floral echoes in the vertical plane. The bell shape of the flower provides the bat with a strong echo response over a broad range of azimuth and elevation angles. The figures also show the open flower protruding which helps to facilitate separation from buds, calyxes, and background clutter.

Nectar-feeding bats have a remarkable ability to extract information from echoes by exploring a number of vertical and horizontal perspectives. They use a lot of hovering flight which can certainly help them build up a profile of acoustic flow field images across different angular profiles [25]. This aids selection of an appropriate approaching angle into the corolla. Information is exploited in a wide variety of ways. The analysis of HRRPs carried out above suggests that the magnitude of echoes and the detailed and angular-dependent echo structures might all combine to enable successful recognition.

To explore this further in the context of synthetic sensing, HRRPs of a car and a tank are measured in the same way as for the flowers. Figures 5.11 and 5.12 show photos of the two (scaled) targets used together with their corresponding HRRPs. The HRRPs were collected over an angular ambit of 180° at 1° intervals. Across this broad range of angles, it is clear that the echo responses from the two objects are very different but also exhibit a complexity that makes recognition challenging. To examine how recognition might be affected by the relative orientation between the sensor and the target, classification performance discriminating the two targets

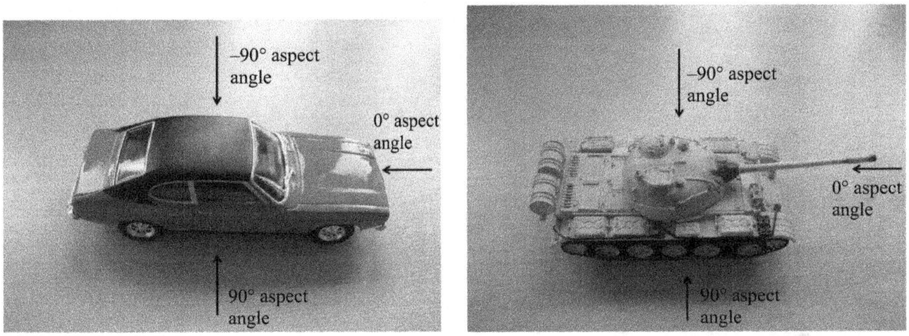

Figure 5.11 Photo of the scaled Ford car and the scaled T55 tank used for the experiments. Reprinted with permission from [18]

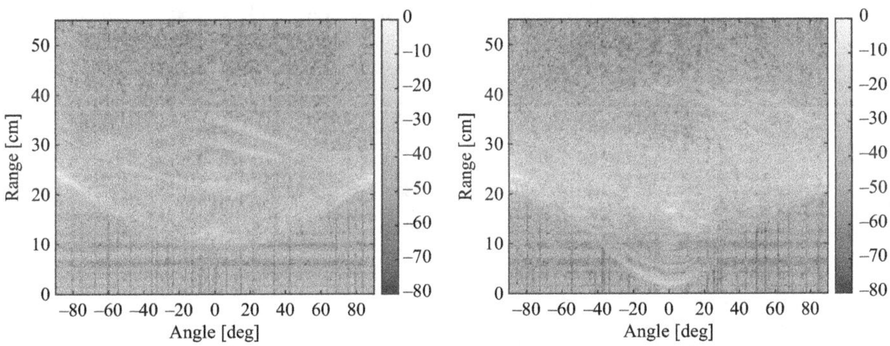

Figure 5.12 Amplitude of the HRRPs of (a) the scaled Ford car and (b) the scaled T-55 tank over an angular window between $-90°$ and $+90°$. The 0° aspect angle corresponds to the artificial bat-head being placed at the same height as the centre of rotation and facing the front side of the target. The grey colour scale indicates the echo strength in [dB], normalised to the maximum echo value. Reprinted with permission from [18]

is computed as a function of look-angle. Classification performance is assessed using a k-NN classifier. In this specific case the N_{train} profiles separated by a constant angular step $s = 30°$ were extracted from each object to train the classifier. These were used to form a training set U. The remaining N_{test} HRRPs in each class were used to form the test set W. Features were extracted from the profiles and used for training. The $N_f = 20$ features were extracted from each HRRP using the Principal Components Algorithm (PCA) similar to [17] to generate the feature-based training set D and test set E. This was achieved through a linear transformation on $D = A^T U$ and $E = A^T W$, where A is the matrix whose columns represent the N_f strongest eigenvectors of the cross-correlation matrix S calculated as $S = UU^H$.

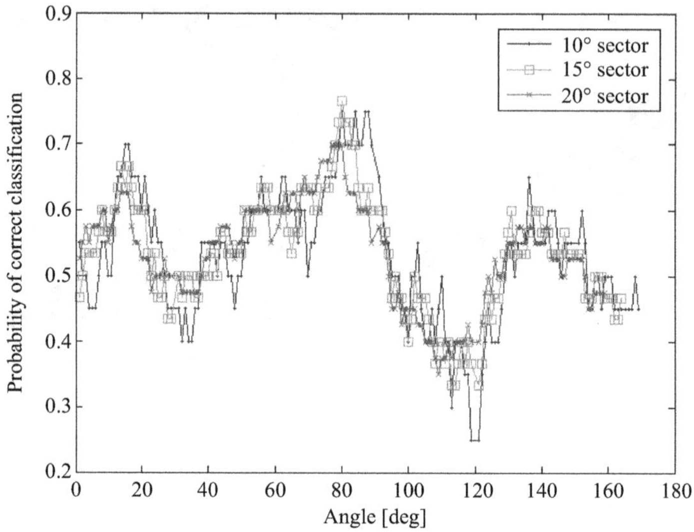

Figure 5.13 Classification performance as a function of angle for three different sector widths: 10°, 15° and 20°. The image shows that there are angles that provide best separation between the two targets. If selectable, these would represent a more optimum choice for target classification. Reprinted with permissions from [18]

Classification performance was assessed as a function of the width of the target sector. For each target sector, the Euclidean distance between a test feature vector and the elements of the training set was calculated. The $k = 3$ lowest distances were then selected and the test vector was assigned to the class that owned the highest number of training vectors between those which generated the k lowest distances. Figure 5.13 shows a plot of the results as a function of angle for 10°, 15° and 20° sector sizes. Each angle on the x-axis corresponds to the first HRRP belonging to the target sector. For example the performance at 0° for the 10° large sector corresponds to the case when the test set was formed with all HRRPs from 0° to 9°. The training set remained constant.

Results show that classification performance can vary significantly as a function of illumination angle with the sector width only having a small influence. Thus there are angles that provide best separation of the two targets that, if selectable, would represent a more optimum choice.

5.5 Conclusion

Results show that the echo acoustic signatures of bat-pollinated flowers contain features, in both the time and frequency domains, that change with the stage of maturity of the flower. These features can be used to achieve high level classification performance and to guide the flight approach of the bat into the corolla of the flower. The analysis of the flower signatures as a function of the aspect angle has shown that the contribution of flower petals is likely to be a key feature for classification. This supports the hypothesis that the scattering from the petals and its impact on the properties of the echo acoustic signatures as a function of the state of the flower could indicate which flower is most suitable for pollination. This would indicate a nectar reward, and could therefore potentially be the basis for the classification strategy being used by echolocating bats. The results of the power reflectivity analysis show that differences in the backscattered power can also be used to aid classification as the power differences are in the bat's detectable range. The task of flower classification by bats shows close similarities with classification of static targets by radar and sonar systems on a moving platform. For these systems, multi-perspective classification has been investigated and results have shown that there can be an increase in classification performance when a decision is made jointly using a number of HRRPs [27,28]. The investigations in this chapter obviously involve a higher level of intelligence and suggest the use of prior knowledge that is used to adjust the trajectory so that target classification is based only on those perspectives that contain the main target features. This is potentially an important technique with practical application in radar and sonar target classification. This will form the basis for future theoretical and experimental work in these areas.

References

[1] G. Jones and E. C. Teling, "The evolution of echolocation in bats," *Trends in Ecology and Evolution*, vol. 21, no. 3, pp. 149–156, 2006.

[2] G. Jones and M. W. Holderied, "Bat echolocation calls: adaptation and convergent evolution," *Proceedings of the Royal Society of London Series B-Biological Sciences*, vol. 274, no. 1612, pp. 905–912, 2007.

[3] J. D. Altringham, *Bats Biology and Behaviour*. Oxford: Oxford University Press, 1996.

[4] C. Baker, M. Vespe, and G. Jones, "50 million years of waveform design," *Waveform Diversity and Design in Communications, Radar and Sonar, 2006. The Institution of Engineering and Technology Forum on*, pp. 7–21, Nov. 2006.

[5] A. Balleri, K. Woodbridge, C. Baker, and M. Holderied, "Classification of flowers by bats: comparison with the radar case," in *2009 International Waveform Diversity and Design Conference*, Kissimmee, FL, Feb. 2009, pp. 1–3.

[6] A. Balleri, K. Woodbridge, C. Baker, and M. Holderied , "Flower classification by bats: radar comparisons," *Aerospace and Electronic Systems Magazine, IEEE*, vol. 24, no. 5, pp. 4–7, May 2009.

[7] H.-U. Schnitzler, D. Menne, R. Kober, and K. Heblich, "The acoustical image of fluttering insects in echolocating bats," in *Neuroethology and Behavioral Physiology: Roots and Growing Points*. Heidelberg: Springer, pp. 235–250, 1983.

[8] G. von der Emde and H.-U. Schnitzler, "Classification of insects by echolocating greater horseshoe bats," *Journal of Comparative Physiology A: Neuroethology, Sensory, Neural, and Behavioral Physiology*, vol. 167, pp. 423–430, 1990.

[9] D. von Helversen, M. W. Holderied, and O. von Helversen, "Echoes of bat-pollinated bell-shaped flowers: conspicuous for nectar-feeding bats?" *Journal of Experimental Biology*, vol. 206, no. 6, pp. 1025–1034, 2003.

[10] M. W. Holderied and O. von Helversen, "Binaural echo disparity' as a potential indicator of object orientation and cue for object recognition in echolocating nectar-feeding bats," *Journal of Experimental Biology*, vol. 209, no. 17, pp. 3457–3468, 2006.

[11] R. Simon, M. W. Holderied, C. U. Koch, and O. von Helversen, "Floral acoustics: conspicuous echoes of a dish-shaped leaf attract bat pollinators," *Science*, vol. 333, no. 6042, pp. 631–633, 2011.

[12] O. von Helversen, L. Winker, and H. Bestmann, "Sulphur-containing 'perfumes' attract flower-visiting bats," *Journal of Comparative Physiology A*, vol. 186, pp. 143–153, 2000.

[13] D. von Helversen and O. von Helversen, "Acoustic guide in bat-pollinated flower," *Nature*, vol. 398, pp. 759–760, 1999.

[14] D. von Helversen and O. von Helversen, "Object recognition by echolocation: a nectar-feeding bat exploiting the flowers of a rain forest vine," *Journal of Comparative Physiology A: Neuroethology, Sensory, Neural, and Behavioral Physiology*, vol. 189, pp. 327–336, 2003.

[15] D. von Helversen, "Object classification by echolocation in nectar feeding bats: size-independent generalization of shape," *Journal of Comparative Physiology A: Neuroethology, Sensory, Neural, and Behavioral Physiology*, vol. 190, pp. 515–521, 2004.

[16] R. Simon, M. W. Holderied, and O. von Helversen, "Size discrimination of hollow hemispheres by echolocation in a nectar feeding bat," *Journal of Experimental Biology*, vol. 209, pp. 3599–3609, 2006.

[17] A. Balleri, H. D. Griffiths, C. J. Baker, K. Woodbridge, and M. W. Holderied, "Special section on biologically-inspired radar and sonar systems – analysis of acoustic echoes from a bat-pollinated plant species: insight into strategies for radar and sonar target classification," *IET Radar, Sonar and Navigation*, vol. 6, no. 6, pp. 536–544, Jul. 2012.

[18] C. J. Baker, G. E. Smith, A. Balleri, M. Holderied, and H. D. Griffiths, "Biomimetic echolocation with application to radar and sonar sensing," *Proceedings of the IEEE*, vol. 102, no. 4, pp. 447–458, Apr. 2014.

[19] *Results of private conversation with Dr. Marc Holderied and Ms. Violene Dapreaux*. School of Biological Sciences, University of Bristol, Bristol, UK, 2010.

[20] A. Streicher, R. Muller, H. Peremans, and R. Lerch, "Broadband ultrasonic transducer for a artificial bat head," *2003 IEEE Symposium on Ultrasonics*, vol. 2, pp. 1364–1367, 2003.

[21] "Datasheet of 1/4-inch free-field microphone type 40bf – high level," accessed on 24 February 2017. [Online]. Available: http://www.gras.dk/40bf.html

[22] C. Baker, H. Griffiths, and A. Balleri, "Biologically inspired waveform diversity," in *Waveform Design and Diversity for Advanced Radar Systems*. Stevenage: The Institution of Engineering and Technology, pp. 149–172, 2012.

[23] W. M. Masters and K. A. S. Raver, "Range discrimination by big brown bats (*Eptesicus fuscus*) using altered model echoes: implications for signal processing," *Journal of Acoustical Society of America*, vol. 107, pp. 625–637, Jan. 2000.

[24] J. A. Simmons and J. E. Gaudette, "Special section on biologically-inspired radar and sonar systems – biosonar echo processing by frequency-modulated bats," *IET Radar, Sonar and Navigation*, vol. 6, no. 6, pp. 556–565, Jul. 2012.

[25] A. Volz, *Echolocation and flight behaviour of neo-tropical nectar-feeding bats (Chiroptera, Glossophaginae) during flower approach.* Master Thesis, University of Erlangen, Erlangen, Germany, 2006.

[26] A. Balleri, H. Griffiths, K. Woodbridge, C. Baker, and M. Holderied, "Impact of flight trajectory on the detection and selection of flowers by nectar-feeding bats," in *Second International Workshop on Cognitive Information Processing (CIP)*, Elba, Italy, Jun. 2010, pp. 350–353.

[27] M. Vespe, C. Baker, and H. Griffiths, "Multi-perspective target classification," *IEEE International Radar Conference 2005*, Arlington, VA, May 2005, pp. 877–882.

[28] M. Vespe, C. Baker, and H. Griffiths , "Radar target classification using multiple perspectives," *IET Radar, Sonar Navigation*, vol. 1, no. 4, pp. 300–307, Aug. 2007.

Chapter 6

The biosonar arms race between bats and insects

Thomas R. Neil[1] *and Marc W. Holderied*[1]

6.1 Introduction

The electronic warfare between an interceptor and a target is exquisitely mirrored within the natural world, where every night across the globe, a 65 million-year-old acoustic battle between bats and flying insects resumes. Bats, armed with ultrasonic biosonar, target and pursue a myriad of night flying insects. The insects meanwhile have evolved numerous counter adaptations to avoid becoming a bats latest meal. The relationship between bats and insects is one of the most sophisticated predator-prey interactions and has been a focus of scientific research for decades, with every subsequent study revealing more fascinating aspects to this age old arms-race.

Perhaps the best known insects that have evolved to try and stifle the attacks of insectivorous bats are moths. Moths and butterflies belong to the order Lepidoptera, originating around 150 million years ago [1,2]. Bats instead first originated between 60 and 95 mya [3,4]. The evolution of biosonar occurred later in bats, around 65 to 50 mya [5–7]. Equipped with powered flight and the ability to locate prey items in the night sky by their echo reflections, bats quickly occupied the niche of a vertebrate predator of night flying insects and became the dominant nocturnal predators that we know today. It is this predation by bats that drove moths and other insects to develop a host of anti-bat defences, including the ability to hear bat biosonar calls. Subsequently, bats have adapted their own hunting techniques to counter the counter adaptations of insects, thus creating a fascinating evolutionary arms race between predator and prey.

6.2 Bat biosonar

Bats emit high intensity sound pulses in the ultrasonic frequency range and then listen for the echo reflections of objects in the vicinity, such as habitat elements and prey. They interpret these echoes to create a representation of their environment allowing them to orient in swift flight, as well as to detect and capture prey.

[1]School of Biological Sciences, University of Bristol, 24 Tyndall Avenue, Bristol BS8 1TQ, UK

6.2.1 Implementing the sonar equation

The key question in the bat-insect arms race is whether or not a bat is able to detect a certain insect at a certain relative direction and distance by biosonar. This can be answered using the active sonar equation. This equation is used to estimate the signal-to-noise ratio (SNR) of returning sonar sound pulses (echoes) [8]. It includes the emitted sound pulse source level (SL), two-way transmission loss (2TL), a measure of the backscatter properties of the target, known as target strength (TS), the noise level (NL) and the directivity index (DI). The SNR in a dB scale is therefore given as:

$$\mathrm{SNR} = \mathrm{SL} - 2\mathrm{TL} + \mathrm{TS} - (\mathrm{NL} - \mathrm{DI}) \tag{6.1}$$

If the SNR of the returning echo is higher than the detection threshold of the auditory system of the listening animal then the echo will be detected. The SL is the sound pressure level (SPL) at a specific reference distance. In biosonar research, this distance is generally accepted as 10 cm in front of the sound source. In order to successfully forage using biosonar bats must emit calls with a very high SL due to the effects of high 2TL and the low TS of their insect targets. Some bats have been found to echolocate with calls reaching SLs of up to 140 dB peak-equivalent SPL [9], and most bat species seem to use similar maximum SL [10]. Sound pulses in biosonar are directional and as they propagate away from the echolocating animal they experience attenuation due to geometric spreading and frequency-dependent absorption. Bat biosonar calls suffer from high TL due to the high absorption of acoustic energy at ultrasonic frequencies in air, subsequently affecting their maximal detection range. Two-way transmission losses (2TL) of 78 dB SPL have been recorded in bats calling at 30 kHz at a range of 5.1 m. At 60 kHz a transmission loss of 91 dB has been measured [11].

The amount of sound energy that is reflected from the target is determined by the target strength (TS). TS is the ratio of the intensity of the reflected wave at one reference distance to the incident sound wave. TS increases with increased target size and impedance difference between the target and the medium of sound propagation (air). TS for a moth with a wingspan of 2 cm is typically around −30 dB standardised to 10 cm [12].

Many sources of noise in the environment that bats forage in contribute to NL: noise can come in the form of the echolocation calls of other bats or from singing insects, ultrasound signals resulting from wind and fast-flowing water, and wind noise as the bats fly forwards. NL also includes internal noise in the auditory system. The directivity index DI describes the directionality of the sonar beam. Bats can adjust the directionality of their calls to moderate the level of noise. The directivity of the calls enables bats to improve the overall SNR by reducing the noise levels via the narrowing of the acoustic opening angle of the ear, which limits the direction over which the ear is able to receive noise [13]. Bats are also able to change their DI's by changing the frequency of their calls or the size of the transmitting aperture [14]. A doubling of either of these parameters results in a DI that is 6 dB higher if all other variables remain the same. Some bats have been shown to alter their DI's by adjusting the frequency of their echolocation calls [15,16]. Daubenton's bats calling at a frequency of 55 kHz have been shown to change their DI's from 11 dB in the lab to 16 dB in the field, presumably by simply increasing the gape of their mouth in the field [17].

To calculate the ability of prey to hear bat calls the echo generation and return is irrelevant. So the sonar equation can be simplified by applying transmission loss only once and to discount target strength. If the target prey item is on the acoustic axis of the bat (e.g., during approach), the directivity index can also be ignored. For off-axis target positions, the directivity index of sound emission can be used to calculate the respective angle-dependent reduction in source level.

6.2.2 Adaptive bat biosonar

The structure of bat biosonar calls varies between species, with emitted calls containing frequencies between 8 and 215 kHz, whilst pulse repetition rates are between 3 and 200 Hz [18,19]. The majority of bat species possess the ability to use biosonar to navigate in the dark, yet mainly in the taxa formerly classified as 'microbats' echolocation has evolved to become a highly effective method of locating and tracking their prey.

In addition to call structures that are species-specific, each bat species has the ability to modify these typical calls to create a repertoire of different call types for different uses. For example a bat species' ultrasonic pulses during insect capture can be divided into three sequential localisation phases: search, approach, and terminal (buzz) (Figure 6.1) [20]. During the search phase a bat that hunts for flying prey will typically call at a relatively low repetition rate, 3–15 Hz, use calls of a comparatively long duration around 5–30 ms, with narrow bandwidth at low frequencies, whilst flying relatively straight. Once it detects a target, a foraging bat enters the approach phase. It will turn towards the prey and gradually increase the repetition rate of its pulses as it gets closer, factoring in the reduction in echo delay as it approaches the target. Often the frequency bandwidth of calls is gradually increased with higher frequencies being incorporated as the bat gets closer. This is considered to improve the echo resolution, as the wavelengths of these higher frequencies are shorter with respect to target size. The final part of the attack phase is called the terminal phase. During this time the bat will often increase its call rate to some 100–200 Hz, decrease its pulse duration to around 0.5–1.0 ms, and use the highest frequency and bandwidth [21–23]. The sound impression this rapid succession of calls during terminal phase creates in a human has given it its name of terminal or feeding buzz. Because the informational requirements during these three tasks are similar for all bat species hunting similar prey, such species show similarities in how they adjust their calls. In summary, the calls of bats are highly plastic. Bats are able to vary the call source level, repetition rate, frequency, bandwidth, frequency modulation and directionality of their calls, enabling them to adapt their calls to varying situations [24,25].

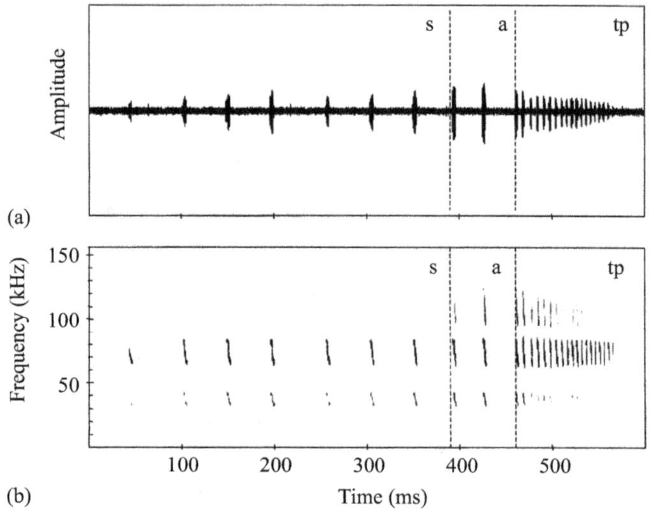

Figure 6.1 Oscillogram (a) and spectrogram (b) showing the echolocation calls emitted by a free flying bat, Pteronotus quadridens, *during a hunting sequence. Vertical dashed lines show the transitions between the search(s), approach (a) and terminal (tp) phases of the attack. © 2003 Oxford University Press. Reprinted with permission from Macías S, Mora EC. "Variation of echolocation calls of* Pteronotus quadridens *(Chiroptera: Mormoopidae) in Cuba". Journal of Mammalogy 2003;84(4):1428–1436*

6.3 Prey defences

Arthropod prey can respond to the predation pressure mediated by biosonar in a number of ways that can be classified into two groups. Passive defences work even if the prey is unaware that it is currently under attack by a bat. Active defences, on the other hand, require the prey to hear the attacking bat's calls hence allowing it to actively respond with adequate defensive actions.

6.3.1 Passive defences

Not all nocturnal insect species possess ears that can pick up ultrasound. These insects cannot actively take avoiding actions in response to bat ultrasound, so must use alternative passive strategies for increasing their survival.

6.3.1.1 Behavioural adaptations

Moths that are unable to hear have been shown to consistently fly quickly or erratically to reduce their chances of capture [26]. Certain moth species have been found to fly within 0.5 m of vegetation, attempting to hide from bats amongst the clutter generated by leaves and branches. This sort of behaviour can be considered a type of 'acoustic crypsis', by which moths try to avoid detection by hiding

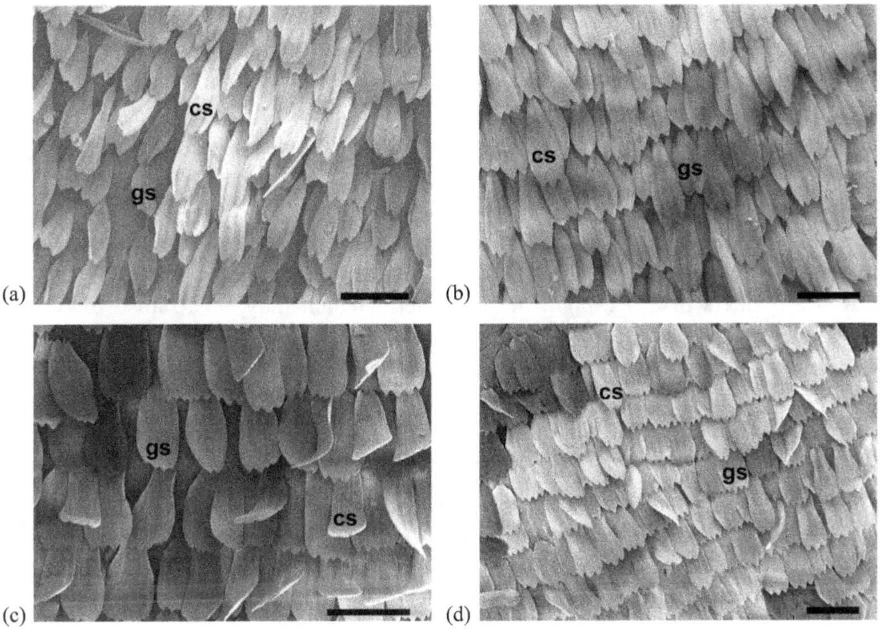

Figure 6.2 Scanning electron microscope image of the hind wing of two moth species showing the arrangement of ground scales (gs) and cover scales (cs). A tiger moth, Spilosoma niveus *(a). A clouded buff,* Rhypariodes amurensis *(b). Scale bar = 100 μm. © 2011 Zeng et al. Reprinted with permission from Zeng JY, Xiang N, Jiang L, et al. "Moth wing scales slightly increase the absorbance of bat echolocation calls". PLoS ONE 2011;6(11)*

amongst echoes generated by other objects [27]. When flying close to vegetation, the attack rate on the moths from bats was found to be low, suggesting the bats struggled to identify them as targets.

6.3.1.2 Ultrasound absorption by wing scales

The wings of moths are made up of a layer of chitin and are covered in tiny overlapping scales, usually <0.25 mm in length (Figure 6.2) [28]. The scales of moth wings are thought to bring a number of potential benefits to moths, including: (a) warning potential predators of their toxicity by bright colours [29], (b) providing camouflage or mimicry as a defence against visual predation [30], (c) providing a means of protection from adhesive webs through shedding [31], and (d) reducing target strength by increasing absorption of ultrasonic bat calls [32].

The scales on the wings of moths can be divided into two distinct layers, ground and cover scales (Figure 6.2). Experiments in micro-reverberation chambers showed that there was a significant decrease in the absorption capabilities of moth wings with the scales removed, which could potentially reduce the distance by which bats can detect moths by around 5%–6% [32]. Interestingly, there was no difference in the absorption capabilities of two diurnal butterflies with their scales removed, suggesting that the ability of moths to reflect weaker echoes might be a direct result of predation pressure from bats. Work using acoustic tomography to visualise the reflectivity of moth wings in response to echolocation has shown that moths may exploit a type of 'acoustic camouflage', whereby they selectively rest on rough surfaces in order to make them less conspicuous to predators [33].

The mechanisms by which the scales help in acoustic absorption from bat calls have not been quantified. The structure and positioning of the scales means that there are pockets of air trapped beneath them, which may act to absorb some of the ultrasound. It has been theorised that these large hollows of interconnected airspace may act to dissipate the incoming acoustic energy as heat [34]. The layout of the scales on moth wings is not unlike those of perforated panels, which are commonly used for sound insulation. It is possible that these scales are acting as a micro perforated panel absorber, with the airspaces behind acting to absorb acoustic energy through the oscillation of air molecules.

6.3.2 Acoustic decoys

Another type of passive defence that is employed by moths is the use of acoustic decoys. Much like the eye spots on many butterflies and fishes that draw away the attacks of visual predators from vital body parts, some moths have decoys that divert a bat's attack away from its body.

As bats are not visual hunters, these decoys are assumed to be reflecting some sort of diversionary acoustic signature, which the bats focus on and attack. The elongated hindwings of some saturniid moths have long been suggested as such a decoy [35], with the long wing elements acting as either a contrasting component of the returning echoes or an alternative target altogether (Figure 6.3). Work on predation by big brown bats on luna moths with their tails intact and with their tails removed showed that bats would indeed target the tails over the body of the moth, improving the chances of survival of intact moths by around 47%, suggesting that the elongated moth wings were in fact acting as some sort of acoustic decoy [36]. The echoes produced by the long tails of luna moths have been characterised using high-speed videography and pulse compression sonar processing. The echoes produced by the tails were actually found to be comparatively weak compared to the echoes coming from the large wings and thorax, particularly when the wings were positioned at right angles to the sound source. However, the twisted morphology of the hind wing tails created echoes that persisted throughout the entire wing beat cycle. These persistent echoes may work to induce target localisation errors in hunting bats and disrupt the accuracy of their tracking [37].

6.3.3 Active defences and ear evolution in insects

The prevalence of biosonar in bats hunting nocturnal arthropods leads to the independent evolution of ultrasound sensitive ears in many groups of insects, giving

*Figure 6.3 Photo of a Madagascan luna moth (*Argema mittrei*), showing the elongated hindwing lobes which act as acoustic decoys during a bat attack. © 2017 Neil TR. 'Madagascan luna moth'. Reprinted with permission*

the prey lifesaving information about its attacker. Acoustically sensitive tympanal ears have evolved independently six times in moths and butterflies alone [38]. Ultrasound hearing is also present amongst many other insect orders such as: Neuroptera (green lacewings), Coleoptera (tiger and scarab beetles), Mantodea (mantids), Diptera (tachinid flies), and Orthoptera (crickets, katydids and locusts). However, hearing in these insects probably pre-dated bat echolocation, indicating that the original purpose of hearing in these insects was likely within species communication. Nevertheless, these insects were also under sustained selective pressure from bat predation, resulting in frequency tuning of ears to bat calls and the development of adequate defensive behaviours.

A key aspect of insect hearing is sound localisation. Being able to determine the direction of a sound source is vital if you are trying to avoid being that sound source's dinner. However, the small size of many insects and therefore their auditory organs relative to the wavelengths of bat biosonar calls presents a challenge in determining sound localisation, resulting in numerous elegant hearing mechanisms that allow sound localisation with ears that are sometimes smaller than the wavelengths of the sounds they aim to localise.

6.3.3.1 Pressure receivers

The tympanal ears of insects can be classed by the mechanism by which they detect sound. Some ears act as simple pressure receivers. In these systems, the tympanal membrane forms on one side of an air enclosed chamber and sound pressure is only received at the external surface. These types of ear are generally found in insects that are relatively large compared to the wavelengths of relevant sound frequencies. For example, bat biosonar calls at 30 kHz have a 10 mm wavelength, and a 100 kHz call is at 3 mm. In this instance diffractive effects are generated by the insects' body, resulting in pressure differences that can be detected with pressure receiving ears on either side of the insect's body. In sound pressure receivers usually the two ears are acoustically independent. Many insects are able to determine sound localisation with this ear morphology if the pressure difference between the two ears (interaural difference) is sufficient. Sound pressure between the ears has been shown to vary some 20–40 dB in moths resulting from the diffraction of sound at frequencies of bat calls (30–60 kHz) [39]. These large differences in sound amplitude are a result of the relatively large size of the moths used in the study (60–85 mm wing span) in comparison to the sound wavelengths. However, such large interaural sound pressure differences are not essential for correct sound localisation. Interaural differences in sound pressure of around 6 dB are sufficient for sound localisation in moths of a more intermediate size [40], whilst sound pressure differences of just 1–2 dB have been shown to be sufficient for orientation towards calling songs in grasshoppers [41].

The wavelengths of the calling songs in katydids are similar in size to their body dimensions, yet their ears are located away from their body, with a pair of tympanal membranes in each of their front legs. Measurements made using a thin microphone probe have demonstrated that there is insufficient diffraction around the front legs of katydids to provide directional cues to aid in sound localisation at biologically relevant frequencies [42]. The katydids instead utilise the diffractive effects of the body, the inner surface of the ear drum is connected to the lateral side of the thorax via a tracheal tube, guiding sound from the ear to the thorax [43]. The tracheal tube in many katydids is horn-shaped, with the eardrum acting as the mouth of the horn and the bell at the thoracal spiracle, decreasing the cross-sectional area from the point of sound entry (surface of the thorax). This acts to increase the sound pressure at the narrow end of the trachea. Experiments have shown that at frequencies used in the animals' calling songs (>10 kHz), the geometry of the trachea increases the amplitude acting on the inner surface of the eardrum more than 10-fold. The trachea therefore facilitates the same diffraction based directionality to the katydid ear that it would have if the ear were located on the thorax of the animal [43].

6.3.3.2 Pressure difference receivers

Pressure difference receivers are typically found in insects with body sizes much smaller than the wavelength of sound at behaviourally relevant frequencies. There is a greater diversity in their anatomy, and their modes of operation are more complicated than simple pressure receivers. In contrast to pressure receivers, pressure difference receivers respond to the action of sound pressure on both sides of the tympanal membrane.

An ear functioning by this mechanism has the potential to give the organism directional hearing so long as the ear is subject to two or more acoustic inputs, with some inputs conducting sound pressure waves to the internal side of the tympanum [44]. Often, the two sound inputs are the two tympana on opposite sides of the animal's body. Some of the sound will pass through the thin tympanal membranes and, via air sacs, reach the inner surface of the other tympanum. A sound will hence reach the outer surface of a tympanum directly and the inner surface via a somewhat longer internal route through a tracheal tube or across air sacs. This inner route can result in various degrees of attenuation or amplification, but, more importantly, the longer internal route results in a time delay of the sound. Sound is a periodic oscillation of local pressure, and a tympanum will move if the pressures acting on its two surfaces differ. If both sounds have similar amplitude and oscillate simultaneously (in phase), there will be very little to no movement. But if one copy gets delayed, the high- and low-pressure phases of the opposing sound waves no longer match in time, resulting in an oscillating pressure difference between the two surfaces of the tympanum, which will set it in motion. Depending on the direction of sound incidence, these so-called phase shifts and amplitude variations act differently on the ipsilateral (directed towards the sounds source on one side of the insect) and contralateral ears (opposite ear directed away from sound). If these phase shifts and amplitude changes are sufficiently adjusted between the two ears, this creates interaural tympanal vibration differences that suffice for directional hearing even when the hearing structures themselves are smaller than a full wavelength of the relevant sound.

The hearing organ of locusts consists of tympanal cavities that are acoustically coupled by air-filled tracheal sacs, permitting interaural sound transmissions [45]. Two sclerotised rings sit within a recess on the abdomen, encircling the tympanal membrane on either side of the body. In such a system, pressure waves reach the outside but also travel from the contralateral side of the animal to act on the internal side of each tympanal membrane. Owing to the fact that the conduction of internal sound decreases with increasing frequency, the ear of the locust is predicted to act as a pressure receiver when subject to high frequency waves, but as a pressure difference receiver in response to lower frequencies [45,46]. Work on the biophysics of directional hearing in both the locust *Schistocerca gregaria* and the much smaller grasshopper *Chorthippus biguttulus* confirmed that the air sacs between the tympanal cavities in locusts was indeed acting as an acoustic low-pass filter [47]. Meanwhile it was concluded that directionality cues were poor at

5 kHz in *C. biguttulus*, due to an insufficient interaural transmission delay. Later phonotactic behavioural work however, found that *C. biguttulus* could reliably respond to directional cues at 5 kHz [48]. Experiments using more sophisticated methods revealed transmission delay times in *C. biguttulus* that were substantially different from those calculated previously, but were similar to those of the larger locust [49]. The results demonstrate the importance of interaural sound transmission delays, and show that, like the locust, the two-input model is an appropriate description of the mechanical function of the acoustics of both the locust and the grasshopper ear.

A second example of a pressure difference receiver in insects occurs in the field cricket, *Gryllus bimaculatus* [50]. This distance between tympana in this species is around 0.8 cm, whilst the wavelength of its calling song is approximately 7 cm. The morphology of the auditory organs in the field cricket is relatively complex, with four acoustic inputs and one internal interaural transmission channel. The anterior tympanal membrane of the field cricket is subject to inputs from both internal and external sounds originating from a contralateral sound input. A septum at the midline of the tracheal connection between the tympana may play a role in delaying internal interaural transmission time [51].

6.3.3.3 Particle velocity receivers

The pressure waves constituting sound are created by oscillations of the many particles of the propagation medium (e.g., air). Non-tympanal hearing organs exist that pick up such oscillations of air particles, however there is little evidence that suggests the prevalence of such hearing acts as a means of defence against predation by bats in those insects that possess them. Changes in air particle velocity are detected by long, lightly articulated filiform hairs or antennae and are found in mosquitoes, bees and fruit flies. For example, the antennae of honeybees are sensitive to air pressure changes, and are used to communicate with each other via sound pressure changes produced by the vibration of other honeybees' wings [52].

6.3.4 Active responses to ultrasound in moths

Despite the widespread presence of hearing in insects, it is the moths that have garnered the majority of attention in the bat insect predator-prey relationship. Perhaps the greatest asset in the moths' defensive arsenal against bats is the ability to hear the echolocation calls of bats.

The theory that hearing in moths evolved as a direct result of bat predation is supported by the fact that the primary or sole function of ears in moths is for the detection of echolocating bats [38]. Moths that take evasive action in response to the ultrasound produced by attacking bats have a significant survival advantage [53–55]. The ears of moths are particularly adept at detecting the high intensity ultrasound calls of bats [56]; moreover, they are typically tuned to the specific frequencies used by their sympatric bat species [57,58]. In the presence of an innocuous but constant sound (e.g., cicada calling), the ears of moths will quickly exhibit sensory adaptation whilst remaining sensitive to the pulsed echolocation calls of bats [59].

Ears in moths evolved from vibration-sensitive stretch receptors located in various positions across the body. The primary function of these stretch receptors is to give the moth information pertaining to the position of various parts of its body, enabling for the coordination of movement. Originally a structurally simple area of thin cuticle innervated by a single nerve, some of these areas experienced thinning, evolving to become sensitive to the vibrations of air. Moths have stretch receptors all over their body, resulting in ears evolving in many strange places across difference species. The most common location of the ear is the thorax, however they can also be found on the abdomen, legs and in some species, the mouthparts. The ears are typically made up of a tympanum and one to four auditory receptor neurons (afferents), making for a neuro-physiologically simple system that nevertheless provides the moth with essential acoustic feedback about its environment. Moths cannot discriminate between frequencies because all auditory afferents have the same frequency tuning, regardless of the number of receptors present [57]. However, each afferent does have a different sound pressure level threshold [60]. Differences in afferent thresholds increase the range of amplitudes that the ear is able to respond to. This is because the least sensitive afferent is able to continue processing increasing amplitudes once the most sensitive afferent reaches its maximum firing capacity. In general, moth ears are tuned to frequencies ranging from 20 to 50 kHz, although there are some species which are sensitive to frequencies over 200 kHz [61,62].

The evolution of ears allowed moths to react to bat calls and take action in order to improve their chances of survival. The distance at which a moth can hear an approaching bat is determined by the bat call's source level SL, the hearing threshold of the moth and the one way transmission loss TL [63]. The maximum distance at which a bat can hear a moth is determined by the call's SL, the two way TL and the target strength TS. The moth is at an immediate advantage as it only suffers from one-way as opposed to the bat's two-way TL. However, bats have a lower hearing threshold than moths allowing them to partly compensate for their disadvantage at TL.

6.3.4.1 Evasive flight manoeuvres

The responses of moths to bat-like calls have been well documented [64,65]. Evasive responses are graded as they change both qualitatively and quantitatively with increasing sound amplitude. This adaptive response allows moths to adjust their reaction to the severity of the threat. Most moths are able to detect the calls of a bat before the bat detects the echoes of the moth [60,66], in these circumstances moths can avoid detection by making small changes to their flight trajectory. Moths that have two ears may be able to use differences in sound arrival time or intensity between the ears to get information about the direction of the sound source. Moths have been shown to make slight phonotactic course adjustments in response to low-amplitude ultrasonic calls [53,64,67,68].

However, as sound amplitude increases, as would be the case in a close encounter with a bat, the responses of the moths are more extreme and erratic, taking desperate measures to try and avoid being eaten [60]. A moth that finds itself under pursuit by a bat will take advantage of its superior manoeuvrability, performing behaviours such as dives, loops, zig-zags, and spirals in an attempt to shake off the attacking bat [53].

6.3.4.2 Answering the call: ultrasound production

Some moth species have the ability to produce ultrasound of their own, meaning that not only can they hear a bat, they can answer them too. Most such species are capable of producing their own sound in the form of short, repetitive clicks, which are produced by the tymbal organ. The tymbal organ is controlled by modified steering muscles, contraction and relaxation of these muscles causes the tymbal organ to buckle, producing clicks [69]. The tymbal organ in some species is further adorned with tiny ridges known as microtymbals. The purpose of these microtymbals is to produce microclicks, enhancing the complexity of the sounds that the moth is able to produce [70]. Some moth species produce sound as a means of intraspecific communication [71], a number of Arctiid and Geometrid moths however, produce sound as part of their defence against bats (Figure 6.4). The moths produce sound in response to bat calls and will match the frequency of their clicks to those of bats that are in the approach phase of their attack [72]. Sound production in these moths is generally accepted as a means of defence against bat biosonar; however, there is still some debate as to how the returning clicks of the moth exert their effects on the bat. Three hypotheses exist, each of which have their

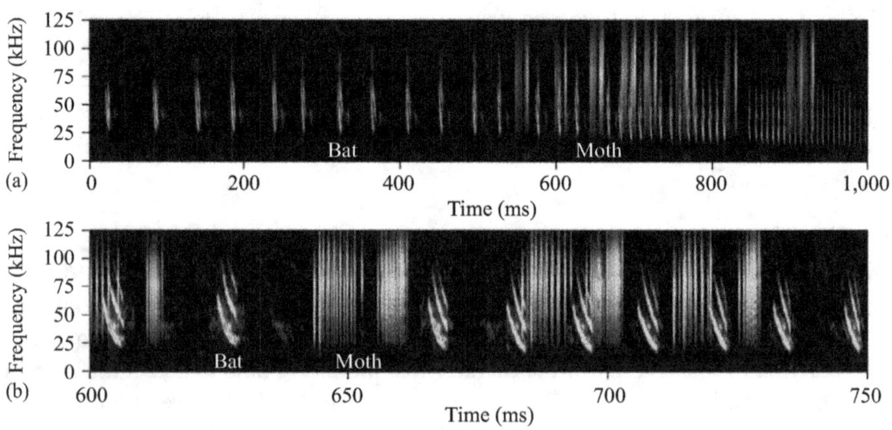

Figure 6.4 Spectrogram of the clicks produced by the moth Eubaphe unicolor in response to a playback of a bat echolocation attack sequence (a). Close up of the same sequence, showing the details of the acoustic structure of moth clicks, which begin in the approach phase of the attack. © 2014 Springer. Reprinted with permission from Corcoran AJ, Hristov NI. "Convergent evolution of anti-bat sounds," Journal of Comparative Physiology A 2014;200(9):811–821

supporters, the startle hypothesis, the jamming hypothesis, and the acoustic aposematism hypothesis [73].

6.3.4.3 Startle the attacker

The startle hypothesis predicts that the ultrasound produced by the moth triggers the mammalian startle reflex in attacking bats [74]. This theory is supported by the fact that moths will tend to emit their returning calls during the terminal phase of the bat's attack, increasing the likelihood of disrupting the bats final approach and not allowing enough time for the bat to respond to the startling calls of the moth. However, empirical evidence suggests that bats quickly become habituated to the clicks of moths, with any startle response quickly diminishing. Free-flying bats were found to quickly learn to ignore the low duty cycle clicks of *Cycnia tenera* after initially being startled by them after only one to three interactions [75]. The long-term effectiveness of startling a bat seems to be negligible to the moth in this instance. However, recent work on the high duty cycle clicking moth *Bertholdia trigona* has resurrected the credentials of this theory, with startle reactions from bats in response to moth clicks persisting for much longer periods. Bats continued to display startle-like effects after up to 40 interactions with *B. trigona*; the persistence of the startle response in bats suggests that it could play a key role in moth survival, particularly in response to naïve bats [76].

6.3.4.4 Sonar jamming

The jamming hypothesis stipulates that the ultrasound produced by the moths interferes with the returning echoes of bat calls, causing confusion and hindering the bat's ranging ability. Arctiid moths that are able to produce a greater number of clicks per unit time are more capable sonar jammers than those that produce fewer clicks. The moth species *B. trigona* produces a high number of high duty cycle clicks, increasing the likelihood of these return calls falling within the 2 ms window in which the bat is expecting target echoes. This has been found to significantly increase the chance of the bat being deterred from the attack, with bats backing out >90% of attacks on *B. trigona* compared to ~20% in *Euchaetes egle*, which produces far fewer clicks per unit time [77,78].

The exact mechanism by which moths are able to jam bat sonar are yet to be explained. The phantom echo hypothesis predicts that bats might mistake the clicks of a moth as returning echoes from non-existent objects [79,80]. Some moth clicks have similar temporal and spectral properties to bat echolocation calls, potentially rendering the bat thoroughly confused as it attempts to home in on its prey, although there is not much empirical evidence in support of this theory.

Another theory is that the clicks interfere with the bat's ability to calculate the distance between itself and its prey. The ranging hypothesis suggests that the clicks disrupt the neural mechanism which encodes information regarding target distance [81]. When moth clicks occurred within the 2 ms window preceding the arrival of echoes, the bat's ranging ability has been found to degrade by as much as 400% [81]. Simulated moth clicks were shown to interfere with the neural response of single units in the nuclei of the lateral lemniscus in the big brown bat. This brain region is implicated in target ranging. The interference was again found only when the clicks fell within the 2 ms window preceding the simulated echo [82]. Moths cannot actively time their clicks to fall within this window, so species producing many clicks in rapid succession enhance their chances of being able to jam the echolocation call of bats.

A third possibility is that moth clicks act in the same way as white noise to reduce or mask the presence of echoes [83], essentially causing the moth to become invisible to the bat.

A key phase of the bat approach to capturing prey is the final part of the attack. During these final moments the bats produce a terminal buzz, increasing the rate of echolocation to gain increased information about its target before it strikes. If the phantom echo hypothesis were accurate, bats would be expected to veer away from these phantom echoes much as they would to avoid clutter and other objects during flight. The ranging hypothesis predicts that bats would miss their target due to interference from the clicks to their ranging ability and possibly make numerous capture attempts of the target. If the clicks were acting like white noise to mask the echoes of the moth, it would be expected that the bat would completely lose the ability to track the moth at all. Experiments using *B. trigona* and naïve big brown bats found that they will approach control noctuid moths in the typical way expected of bats, approaching rapidly and directly before making their attack. However, when attacking *B. trigona* their approach differed markedly. Initially, the bats were startled by the clicks of the moths, but they quickly became habituated to the sound. Following habituation the bats continued to pursue the moths, but consistently missed their targets by a distance predicted by the ranging interference hypothesis. There was no evidence that the bats were responding to phantom targets or echoes, or that they lost the ability to track the moth all together. The results of the study support the theory that the clicking produced by moths interferes with the ranging ability in bats, suggesting that *B. trigona* are able to confuse pursuing bats as to their exact location [76].

6.3.4.5 Acoustic aposematism

Many moths possess toxic chemicals as a deterrent to would be predators. The chemicals are generally acquired from their host plants, consumed by the caterpillars of the species during the larval stage of development. Moths possess many chemicals in their defence against predators, but two of the most common are pyrrolizidine alkaloids and cardiac glycosides. In low concentrations, the pyrrolizidine alkaloids cause the host animal to have a bitter taste, whilst in high concentrations a vertebrate eating an animal containing them will experience tissue damage and possible death [84]. It is the effect of bad taste though which is most important when considering how predators might get conditioned to avoid certain prey species. Some will learn to reject prey simply by smell or a bite, whilst others require ingestion to develop a negative response to a particular prey item [29]. Arctiid moths acquire alkaloids both from their host plants as caterpillars [85], and as adults by absorbing them from alkaloid containing plants that have withered [86]. Cardiac glycosides inhibit enzymes associated with ion transport in cells and can cause emesis (vomiting).

Arctiid moths obtain cardenolides as larvae through the consumption of cardenolide containing food plants, which are then sequestered into the body tissue and remain present into its adult form [87]. Both pyrrolizidine alkaloids and cardiac glycosides work as effective deterrents to predation by bats [77].

Acoustic aposematism suggests that there is a link between the chemical defences of some moth species and their ability to produce ultrasonic clicks, with bats learning to associate particular sounds with unpalatable prey and avoiding them [88–90]. Naïve bats have been shown to quickly associate toxic sound producing moths with bad taste, and will quickly start to avoid all sound producing moths, regardless of whether they are chemically protected or not [91]. This leads to the reasonable assumption that some species should exploit acoustic mimicry, with palatable moths producing similar sounds to unpalatable moths in order to benefit from the decreased chance of predation by bats. There are certain species of moth that do produce sound but have no chemical defences that would make them unpalatable to bats. *Euchaetes egle* is one such species, their seasonal phenology overlaps with the unpalatable species *C. tenera*. *Euchaetes egle* produces a similar acoustic signature to *C. tenera*, suggesting that it could be an acoustic mimic [92]. Research on bats that have had interactions with *C. tenera* showed they will immediately stop eating them, and will not target *E. egle* either. However, upon disabling the ability of *E. egle* to produce sound via the removal of their tymbals, the bats started to target and eat the mimics. Clearly, it is the production of sound in moths that is the cue that the bats are using to discern whether prey is palatable or not [91]. It has been shown that bats can in fact distinguish between toxic moths and mimics [93]. However, the bats seem to generalise the warning signals and will generally avoid all sound producing moths once they have learnt to associate them with bad taste, regardless of whether they are a mimic or not.

6.3.5 Defences in other insects

Although moths are by far the most extensively studied, many other insects have evolved ears either as a means of bat detection or have pre-existing ears that have since gone on to act as bat detectors.

6.3.5.1 Mantids

Ears are found in 65% of mantids (praying mantis). They are composed of two tympanic membranes but function as a single unit. Mantids are generally diurnal, but will sometimes make dispersal flights at night [94]. The ears of mantids are generally sensitive to ultrasound between 25 and 50 kHz [95]. Mantids have been found to react to bat calls at distances of as much as 10 m. Work on the praying mantis *Parasphendale agrionina* showed how the ears mediated evasive manoeuvres in response to bat echolocation calls, with the intensity of manoeuvres graded with distance to the sound source. Weak calls resulted in simple turns, whilst strong calls eliciting a fleet of elaborate evasive responses including steep diving turns and spirals. The ability of this mantis to be able to hear and react to bats was found to significantly increase its chances of survival [96,97].

6.3.5.2 Crickets

Members of the family Gryllidae have ears that are located just below the "knee" on their forelegs. Females of the species *Teleogryllus oceanicus* use their ears as both a means of predator avoidance, and to listen out for potential mates. The males of the species use a low-frequency calling song to attract females, which the females are sensitive to. The ears of the female also have a second high-frequency sensitivity peak, allowing them to detect the calls of echolocating bats and take evasive action if necessary [98].

6.3.5.3 Katydids

The ears of katydids (bush crickets) have been shown to be sensitive to frequencies ranging from 10 to 50 kHz, a much broader range than the mating calls they use to communicate with each other (18–23 kHz) [99], suggesting that they additionally use their ears to detect the calls of echolocating bats. Research has shown that they are able to detect the calls of gleaning bats [100,101], responding by completely shutting down their own acoustic courtship displays in order to avoid detection. Additionally, evasive flying manoeuvres have been recorded in the katydid *Neoconocephalus ensiger*, closing all four wings and beginning to freefall within 50 ms of hearing an ultrasonic pulse [102].

6.3.5.4 Beetles

The tiger beetle, *Cicindela marutha*, possesses ears on its first abdominal segment. During stationary flight, the detection of ultrasound has been shown to alter their behaviour, with head rolls and changes in wing beat frequency occurring that should alter their flight trajectories in such a way as to make them more difficult for bats to catch [103]. Several species of tiger beetle produce ultrasonic clicks of their own in response to ultrasound. The clicks are produced from contact between the flapping hind wings and the hardened forewings during flight. The functionality of these clicks has yet to be determined in tiger beetles and it is currently unknown whether they are warning, jamming or startling bats.

In response to ultrasound, some night-flying scarab beetles will perform head turns characteristic of flight path changes and drop to the ground [104]. Their ears are located on the cervical membranes and evolved independently of those ears in tiger beetles [105].

6.3.5.5 Lacewings

The ears of the green lacewing are unusual because their organs for detecting sound are on their wing, made up of a swelling that is located on the radial vein of the forewings [106]. The green lacewing adopts a similar evasive strategy to *N. ensiger*, folding in its wings and dropping rapidly in response to the sonar signals of bats. The green lacewing does however attempt to increase its chances of survival further. It will suddenly open its wings out in response to the terminal phase of a bats' call (the final phase of attack; Figure 6.5) attempting to make its trajectory less predictable [107,108].

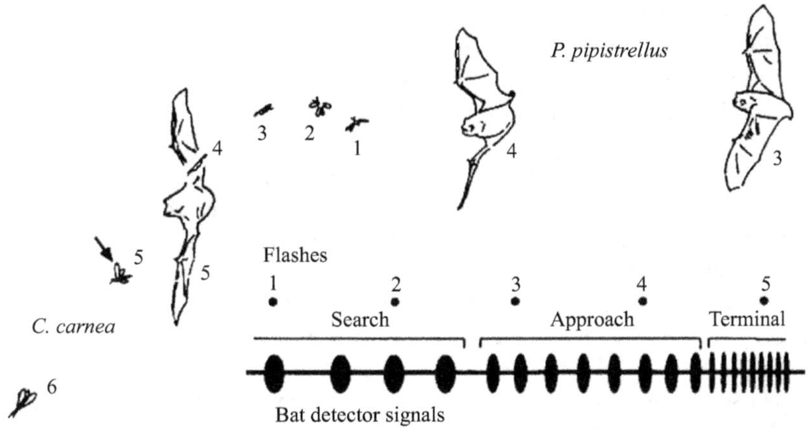

Figure 6.5 Evasive manoeuvres of the green lacewing in response to a bat attack, photographed in a flight cage using stroboscopic flashes (70 ms intervals). A schematic of the bat calls during the attack is included showing the change in call repetition rate during the attack. The lacewing closes its wings and dives during the approach phase (4), but opens them again during the terminal phase of the attack (5). © 2001 Oxford University Press. Reprinted with permission from Miller LA, Surlykke A. "How some insects detect and avoid being eaten by bats: tactics and countertactics of prey and predator," Bioscience 2001;51(7):570–581

6.4 Bat countermeasure to insect defences

A predator-prey interaction such as the one between bats and insects can only be called an arms race if the two are continually evolving new adaptations and counter-adaptations [109]. So, have bats evolved any strategies to counter insect defences?

6.4.1 Allotonic frequency hypothesis

The allotonic frequency hypothesis predicts that bats might lower the frequency at which they emit echolocation calls so that they shift outside of the sensitive range of insect hearing [56]. Equally, bats might be expected to switch to ultra high calling frequencies, making them harder for eared insects to detect. Both strategies have been observed in various bat species and each has its advantages. Ultra-high frequencies yield greater resolution but low frequencies having the potential for greater echolocation range [109].

Several bat species have been found to produce echolocation calls at much lower frequencies than would be expected based on their size. The spotted bat, *Euderma maculatum*, is a small-to-medium sized vespertilionid bat. Based on the established scaling relationship between size and call frequency in this family of bats, it would be expected to call at frequencies of around 30 kHz [110]. However, these bats call at frequencies of around 10 kHz, producing echolocation calls outside of the sensitivity range of their sympatric moth species [111]. Analysis of faecal matter suggests that these bats consume more moths than would be expected from the moth abundance alone [112]. Although seemingly aiding bats in their pursuit of prey, calling at low frequencies does not unambiguously qualify as counter adaptation as part of an arms race. All bat species calling at such low frequencies are open air aerial hawkers, where calling at lower frequencies increases their biosonar detection range, as lower frequencies attenuate less rapidly in air [113].

Most bats produce low duty cycle calls, with emitted calls and returning echoes being separated in time to avoid pulse-echo overlap. Other bats produce high duty cycle, constant frequency calls, separating call from echo using Doppler shift initiated frequency differences [89,114,115]. Such high duty cycle bats call at a higher frequency than equivalently sized low duty cycle bats. This has been proposed as a means of avoiding detection by moths, because bat species producing the highest frequency high duty cycle calls consume the most moths [116]. For example, the African hipposiderid, *Cleotis percivalis*, produces calls at around 212 kHz [117], likely making them inaudible to their prey. Calling at such high frequencies results in extreme atmospheric attenuation, meaning the bats detection range is greatly reduced. Yet this still cannot count as unambiguous proof of a counter adaptation, because such high frequency calls also give higher resolution echo information regarding the bat's immediate surroundings, an advantage to this particular species of bat as it tends to forage in highly cluttered habitats.

6.4.2 Stealth aerial hawking

The vespertilionid bat *Barbastella barbastellus* has been shown to use stealth echolocation as a method of countering the hearing of its moth prey [66]. This bat produces echolocation calls that are 10–100 times lower (94 dB SPL) in source level (SL) than the similar sympatric vespertilionid *Nyctalus leisleri* (127 dB SPL). Both species call at similar frequencies (*B. barbastellus* 28 kHz; *N. leisleri* 33 kHz). The result of this low amplitude, stealth echolocation is a reduction in the distance at which the moth can detect the approaching bat by one order of magnitude (*B. barbastellus* ~4 m; *N. leisleri* >30 m), giving the bat a much greater chance of successfully capturing its prey. That reduction in SL does come at a cost to the bat however, with the distance over which the bat can detect a moth decreasing 3- to 4-fold, from ~20 m in *N. leisleri* to around 5 m in *B. barbastellus*.

So what is the benefit of the low amplitude calls of *B. barbastellus*? Molecular diet analysis of the two species show that the diet of *B. barbastellus* is around 85% eared moths, whilst the range in the diet of *N. leisleri* is just 0–56%. A respective analysis of who detects whom first (Figure 6.6) shows that its low amplitude calls make *B. barbastellus* the only bat that can detect its prey before its prey can hear them. The use of stealth echolocation in open air space brings no obvious benefit to the bat other than increasing its ability to predate on eared moths,

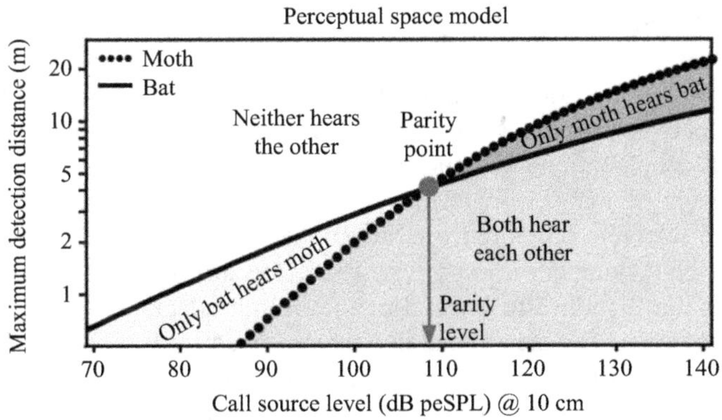

Figure 6.6 A perceptual space model showing that at low source levels bats can hear moth echoes at greater distances than moths can hear bat calls. Whilst at high source levels the opposite is true. The parity level describes source levels which have an equal detection distance for both bat and moth. © 2010 Elsevier. Reprinted with permission from Goerlitz HR, ter Hofstede HM, Zeale MR, Jones G, Holderied MW. "An aerial-hawking bat uses stealth echolocation to counter moth hearing," Current Biology 2010;20(17):1568–1572

suggesting that this has evolved solely to circumvent the ability of insects to detect bat echolocation calls. The Namibian long-eared bat (*Laephotis namibensis*) is a small insectivorous bat that might use a similar strategy. Its calls are unusual in that they combine elements that are normally associated with two different hunting strategies: they are relatively short (2.6 ms), a strategy used by bats that hunt within a cluttered habitat, but are also narrow bandwidth (13.5 $\pm$ 2.9 kHz) and low in frequency (22.1 $\pm$ 0.6 kHz), resembling calls usually used by bats hunting in open habitats [118]. It could be that this unique call strategy is an adaptation to be less audible to eared prey, which is supported by the fact that it feeds almost exclusively on eared moths. Its call sources are unknown to date though.

The moths that are preyed upon by *B. barbastellus* appear to be defenceless against stealth echolocation, with no counter adaptation having evolved. Moths are under selective pressure from bats to increase their auditory sensitivity, but are also under opposing pressure from the associated costs that increased sensitivity would bring. If moths began responding to innocuous sounds it would likely reduce the time that they can spend foraging and decrease the likelihood of mating opportunities. The rare-enemy effect predicts that the rarity of a particular predator strategy will affect the selective pressure acting on the species being targeted. If the interaction is not frequent enough, then there will be insufficient pressure for the moth to adapt [119]. It seems in this particular aspect of the bat-insect arms race, the bat is truly on top, having not only an advantage over its prey species, but also to loud calling bats by being able to target prey not obtainable by others.

6.5 Conclusion

The evolutionary arms-race between bats and insects is a classic example of coevolution between predator and prey. The counter adaptations of night flying insects to bat echolocation calls and subsequent adaptations of bats to combat the hearing abilities of many insects makes it one of the few examples of species responding to selective pressure from one another. A true arms-race of the sensory systems and behaviour across species. There are already examples of some of the strategies that bats use to try and counter moth adaptations found within the technological world. The stealth echolocation utilised by *B. barbastellus* is analogous to low-probability-of-intercept (LPI) radar, whilst the use of bat-inspired harmonics in waveform design has been shown to potentially bring advantages in enhancing radar performance in unfavourable conditions [120].

Although a great deal has been learnt about the ongoing acoustic battle between bats and insects, there is of course still much to uncover about this fascinating interaction. Much of the work done thus far has focused on moths, yet numerous night flying insects are preyed upon by bats, most of which have yet to be studied in similar detail. More advanced techniques have facilitated the study of bat-insect interactions in the field, with 3D microphone arrays and 3D infrared videography revealing even more about how the real-world encounters between species are played out [121]. There are likely to be many more secrets to this particular war that could advance the development of radar and sonar, and we are only just beginning to unearth them.

References

[1] Misof B, Liu S, Meusemann K, *et al.* "Phylogenomics resolves the timing and pattern of insect evolution," *Science* 2014;346(6210):763–767.
[2] Corcoran AJ and Hristov NI. "Convergent evolution of anti-bat sounds," *Journal of Comparative Physiology A* 2014;200(9):811–821.
[3] Simmons NB, Seymour KL, Habersetzer J, and Gunnell GF. "Primitive Early Eocene bat from Wyoming and the evolution of flight and echolocation," *Nature* 2008;451(7180):818–821.
[4] Bininda-Emonds OR, Cardillo M, Jones KE, *et al.* "The delayed rise of present-day mammals," *Nature* 2007;446(7135):507–512.
[5] Macías S and Mora EC. "Variation of echolocation calls of *Pteronotus quadridens* (Chiroptera: Mormoopidae) in Cuba", *Journal of Mammalogy* 2003;84(4):1428–1436.
[6] Teeling EC. "Hear, hear: the convergent evolution of echolocation in bats? *Trends in Ecology & Evolution* 2009;24(7):351–354.
[7] Conner WE and Corcoran AJ. "Sound strategies: the 65-million-year-old battle between bats and insects," *Annual Review of Entomology* 2012;57:21–39.
[8] Møhl B. "Target detection by echolocating bats," In: *Animal sonar.* New York, NY: Plenum Press; 1988. pp. 435–450.

[9] Surlykke A and Kalko EKV. "Echolocating bats cry out loud to detect their prey," *PLoS ONE* 2008;3(4):e2036.

[10] Holderied MW and von Helversen O. "Echolocation range and wingbeat period match in aerial-hawking bats," *Proceedings of the Royal Society B: Biological Sciences* 2003;270(1530):2293–2299.

[11] Lawrence BD and Simmons JA. "Measurements of atmospheric attenuation at ultrasonic frequencies and the significance for echolocation by bats," *Journal of the Acoustical Society of America* 1982;71(3):585–590.

[12] Waters DA and Jones G. "Echolocation call structure and intensity in five species of insectivorous bats," *Journal of Experimental Biology* 1995;198(2): 475–489.

[13] Surlykke A, Nachtigall PE, Popper AN, and Fay RR. *Biosonar*. Vol. 51. New York, NY: Springer; 2014.

[14] Kounitsky P, Rydell J, Amichai E, *et al.* "Bats adjust their mouth gape to zoom their biosonar field of view," *Proceedings of the National Academy of Sciences USA* 2015;112(21):6724–6729.

[15] Jakobsen L and Surlykke A. "Vespertilionid bats control the width of their biosonar sound beam dynamically during prey pursuit," *Proceedings of the National Academy of Sciences USA* 2010;107(31):13930–13935.

[16] Jakobsen L, Brinkløv S, and Surlykke A. "Intensity and directionality of bat echolocation signals," In: Melcón ML and Moss CF (eds). *How nature shaped echolocation in animals*. Lausanne: Frontiers Media SA. 2014; pp. 72–80.

[17] Surlykke A, Pedersen SB, and Jakobsen L. "Echolocating bats emit a highly directional sonar sound beam in the field," *Proceedings of the Royal Society of London B: Biological Sciences* 2009;276(1658):853–860.

[18] Schnitzler H-U and Kalko EK. "Echolocation by insect-eating bats: we define four distinct functional groups of bats and find differences in signal structure that correlate with the typical echolocation tasks faced by each group," *Bioscience* 2001;51(7):557–569.

[19] Simmons JA, Fenton MB, and O'Farrell MJ. "Echolocation and pursuit of prey by bats," *Science* 1979;203(4375):16–21.

[20] Griffin DR, Webster FA, and Michael CR. "The echolocation of flying insects by bats," *Animal Behaviour* 1960;8(3):141–154.

[21] Kalko E. "Foraging behavior, capture techniques, and echolocation in European pipistrelle bats," *Animal Behavior* 1995;50:861–880.

[22] Kalko EK and Schnitzler H-U. "The echolocation and hunting behavior of Daubenton's bat, *Myotis daubentonii*," *Behavioral Ecology and Sociobiology* 1989;24(4):225–238.

[23] Surlykke A and Moss CF. "Echolocation behavior of big brown bats, *Eptesicus fuscus*, in the field and the laboratory," *Journal of the Acoustical Society of America* 2000;108(5):2419–2429.

[24] Ghose K and Moss CF. "The sonar beam pattern of a flying bat as it tracks tethered insects," *Journal of the Acoustical Society of America* 2003;114(2): 1120–1131.

[25] Ghose K and Moss CF. "Steering by hearing: a bat's acoustic gaze is linked to its flight motor output by a delayed, adaptive linear law," *The Journal of Neuroscience* 2006;26(6):1704–1710.

[26] Lewis FP, Fullard JH, and Morrill SB. "Auditory influences on the flight behaviour of moths in a Nearctic site. II. Flight times, heights, and erraticism," *Canadian Journal of Zoology* 1993;71(8):1562–1568.

[27] Rydell J. "Bat defence in lekking ghost swifts (*Hepialus humuli*), a moth without ultrasonic hearing," *Proceedings of the Royal Society of London B: Biological Sciences* 1998;265(1404):1373–1376.

[28] Ghiradella H. "Light and color on the wing: structural colors in butterflies and moths," *Applied Optics* 1991;30(24):3492–3500.

[29] Rothschild M, Reichstein T, Von Euw J, Aplin R, and Harman R. "Toxic lepidoptera," *Toxicon* 1970;8(4):293–296.

[30] Schaefer HM and Stobbe N. "Disruptive coloration provides camouflage independent of background matching," *Proceedings of the Royal Society of London B: Biological Sciences* 2006;273(1600):2427–2432.

[31] Eisner T, Alsop R, and Ettershank G. "Adhesiveness of spider silk," *Science* 1964;146(3647):1058–1061.

[32] Zeng J, Xiang N, Jiang L, *et al.* "Moth wing scales slightly increase the absorbance of bat echolocation calls," *PLoS ONE* 2011;6(11):e27190.

[33] Clare EL and Holderied MW. "Acoustic shadows help gleaning bats find prey, but may be defeated by prey acoustic camouflage on rough surfaces", *eLife* 2015;4:e07404.

[34] Cox TJ and D'Antonio P. *Acoustic absorbers and diffusers: theory, design and application*. London: CRC Press; 2009.

[35] Weeks AC. "Theory as to evolution of secondaries of moths of the genus catocala," *Journal of the New York Entomological Society* 1903;11(4): 221–226.

[36] Barber JR, Leavell BC, Keener AL, *et al.* "Moth tails divert bat attack: evolution of acoustic deflection," *Proceedings of the National Academy of Sciences USA* 2015;112(9):2812–2816.

[37] Lee W-J and Moss CF. "Can the elongated hindwing tails of fluttering moths serve as false sonar targets to divert bat attacks?" *Journal of the Acoustical Society of America* 2016;139(5):2579–2588.

[38] Yack J and Dawson J. "Insect ears," *The Senses: A Comprehensive Reference* 2008;3:35–54.

[39] Payne R, Roeder K, and Wallman J. "Directional sensitivity of the ears of noctuid moths," *Journal of Experimental Biology* 1966;44(1):17–31.

[40] Hoy RR and Fay RR. *Comparative hearing: insects*. Heidelberg: Springer Science & Business Media; 2012.

[41] von Helversen D and Rheinlaender J. "Interaural intensity and time discrimination in an unrestraint grasshopper: a tentative behavioural approach," *Journal of Comparative Physiology A* 1988;162(3):333–340.

[42] Michelsen A. "Directional hearing in crickets and other small animals," *Fortschritte der Zoologie* 1994:195.

[43] Lewis B. "Directional cues for auditory localization," In: Lewis B (ed). *Bioacoustics: a comparative approach*. London: Academic Press; 1983. pp. 233–257.

[44] Autrum H. "Über lautäusserungen und schallwahrnehmung bei arthropoden II," *Journal of Comparative Physiology A: Neuroethology, Sensory, Neural, and Behavioral Physiology* 1940;28(3):326–352.

[45] Miller LA. "Directional hearing in the locust *Schistocerca gregaria* Forskål (Acrididae, Orthophera)," *Journal of Comparative Physiology* 1977;119(1): 85–98.

[46] Michelsen A. "The physiology of the locust ear," *Zeitschrift für vergleichende Physiologie* 1971;71(1):63–101.

[47] Michelsen A and Rohrseitz K. "Directional sound processing and interaural sound transmission in a small and a large grasshopper," *Journal of Experimental Biology* 1995;198(9):1817–1827.

[48] Pollack GS. "Neural processing of acoustic signals. In: RR Hoy, AN Popper, RR Fay (eds). *Comparative hearing: insects*. Berlin, Heidelberg, New York: Springer; 1998. pp. 139–196.

[49] Schul J, Holderied M, Helversen D, and Helversen O. "Directional hearing in grasshoppers: neurophysiological testing of a bioacoustic model," *Journal of Experimental Biology* 1999;202(2):121–133.

[50] Michelsen A, Popov A, and Lewis B. "Physics of directional hearing in the cricket *Gryllus bimaculatus*," *Journal of Comparative Physiology A* 1994; 175(2):153–164.

[51] Löhe G and Kleindienst H-U. "The role of the medial septum in the acoustic trachea of the cricket *Gryllus bimaculatus*," *Journal of Comparative Physiology A* 1994;174(5):601–606.

[52] Michelsen A, Towne WF, Kirchner WH, and Kryger P. "The acoustic near field of a dancing honeybee," *Journal of Comparative Physiology A* 1987; 161(5):633–643.

[53] Roeder KD. "The behaviour of free flying moths in the presence of artificial ultrasonic pulses," *Animal Behaviour* 1962;10(3–4):300–304.

[54] Acharya L and Fenton MB. "Bat attacks and moth defensive behaviour around street lights," *Canadian Journal of Zoology* 1999;77(1):27–33.

[55] Fullard JH and Napoleone N. "Diel flight periodicity and the evolution of auditory defences in the Macrolepidoptera," *Animal Behaviour* 2001;62(2): 349–368.

[56] Fullard JH. "The sensory coevolution of moths and bats," *Comparative hearing: insects*. New York, NY: Springer; 1998. pp. 279–326.

[57] Fullard J. "The tuning of moth ears," *Experientia* 1988;44(5):423–428.

[58] ter Hofstede HM, Goerlitz HR, Ratcliffe JM, Holderied MW, and Surlykke A. "The simple ears of noctuoid moths are tuned to the calls of their sympatric bat community," *Journal of Experimental Biology* 2013;216(21): 3954–3962.

[59] Fullard JH, Ratcliffe JM, and Jacobs DS. "Ignoring the irrelevant: auditory tolerance of audible but innocuous sounds in the bat-detecting ears of moths," *Naturwissenschaften* 2008;95(3):241–245.

[60] Roeder KD. *Nerve cells and insect behavior*. Cambridge, MA: Harvard University Press; 1998.

[61] Fullard JH, Fenton MB, and Furlonger CL. "Sensory relationships of moths and bats sampled from two Nearctic sites," *Canadian Journal of Zoology* 1983;61(8):1752–1757.

[62] Boyan G and Fullard J, editors. "Functional organization of the afferent auditory pathway in a moth". *Abstracts, Sixth International Meeting on Insect Sound and Vibration*, Denmark; 1987.

[63] Nachtigall PE and Moore PW. *Animal sonar: processes and performance*. New York, NY: Plenum Press; 1988.

[64] Agee H. "Response of flying bollworm moths and other tympanate moths to pulsed ultrasound," *Annals of the Entomological Society of America* 1969;62(4):801–807.

[65] Barber J and Conner W. "Tiger moth responses to a simulated bat attack: timing and duty cycle," *Journal of Experimental Biology* 2006;209(14): 2637–2650.

[66] Goerlitz HR, ter Hofstede HM, Zeale MR, Jones G, and Holderied MW. "An aerial-hawking bat uses stealth echolocation to counter moth hearing," *Current Biology* 2010;20(17):1568–1572.

[67] Roeder KD. "Aspects of the noctuid tympanic nerve response having significance in the avoidance of bats," *Journal of Insect Physiology* 1964; 10(4):529–546.

[68] Agee H. "Response of acoustic sense cell of the bollworm and tobacco budworm to ultrasound," *Journal of Economic Entomology* 1967;60(2):366–369.

[69] Dawson J and Fullard J. "The neuroethology of sound production in tiger moths (Lepidoptera, Arctiidae)," *Journal of Comparative Physiology A* 1995; 176(4):541–549.

[70] Skals N and Surlykke A. "Sound production by abdominal tymbal organs in two moth species: the green silver-line and the scarce silver-line (Noctuoidea: Nolidae: Chloephorinae)," *Journal of Experimental Biology* 1999;202(21): 2937–2949.

[71] Conner WE. "'Un chant d'appel amoureux': acoustic communication in moths," *Journal of Experimental Biology* 1999;202(13):1711–1723.

[72] Ratcliffe JM and Fullard JH. "The adaptive function of tiger moth clicks against echolocating bats: an experimental and synthetic approach," *Journal of Experimental Biology* 2005;208(24):4689–4698.

[73] Miller LA and Surlykke A. "How some insects detect and avoid being eaten by bats: tactics and countertactics of prey and predator," *Bioscience* 2001; 51(7):570–581.

[74] Humphries D and Driver P. "Protean defence by prey animals," *Oecologia* 1970;5(4):285–302.

[75] Bates DL and Fenton MB. "Aposematism or startle? Predators learn their responses to the defenses of prey," *Canadian Journal of Zoology* 1990;68(1): 49–52.

[76] Corcoran AJ, Barber JR, Hristov NI, and Conner WE. "How do tiger moths jam bat sonar?" *Journal of Experimental Biology* 2011;214(14):2416–2425.

[77] Hristov N and Conner W. "Predator–prey interactions: effectiveness of tiger moth chemical defenses against insectivorous bats," *Chemoecology* 2005;15:105–113.

[78] Corcoran AJ, Barber JR, and Conner WE. "Tiger moth jams bat sonar," *Science* 2009;325(5938):325–327.

[79] Fullard JH, Fenton MB, and Simmons JA. "Jamming bat echolocation: the clicks of arctiid moths," *Canadian Journal of Zoology* 1979;57(3):647–649.

[80] Fullard JH, Simmons JA, and Saillant PA. "Jamming bat echolocation: the dogbane tiger moth *Cycnia tenera* times its clicks to the terminal attack calls of the big brown bat *Eptesicus fuscus*," *Journal of Experimental Biology* 1994;194(1):285–298.

[81] Miller LA. "Arctiid moth clicks can degrade the accuracy of range difference discrimination in echolocating big brown bats, *Eptesicus fuscus*," *Journal of Comparative Physiology A* 1991;168(5):571–579.

[82] Tougaard J, Casseday J, and Covey E. "Arctiid moths and bat echolocation: broad-band clicks interfere with neural responses to auditory stimuli in the nuclei of the lateral lemniscus of the big brown bat," *Journal of Comparative Physiology A* 1998;182(2):203–215.

[83] Møhl B and Surlykke A. "Detection of sonar signals in the presence of pulses of masking noise by the echolocating bat, *Eptesicus fuscus*," *Journal of Comparative Physiology A* 1989;165(1):119–124.

[84] Mattocks A. *Chemistry and toxicology of pyrrolizidine alkaloids*. New York, NY: Academic Press; 1986.

[85] von Nickisch-Rosenegk Ev, Schneider D, and Wink M. "Time-course of pyrrolizidine alkaloid processing in the alkaloid exploiting arctiid moth, *Creatonotos transiens*," *Zeitschrift für Naturforschung C* 1990;45(7–8):881–894.

[86] Boppré M. "Lepidoptera and pyrrolizidine alkaloids exemplification of complexity in chemical ecology," *Journal of Chemical Ecology* 1990;16(1): 165–185.

[87] Black DW. *Studies on cardiac glycoside storage in moths*. Miami, FL: University of Miami; 1976.

[88] Dunning DC, Acharya L, Merriman CB, and Ferro LD. "Interactions between bats and arctiid moths," *Canadian journal of Zoology* 1992;70(11): 2218–2223.

[89] Dunning DC and Kruger M. "Aposematic sounds in African moths," *Biotropica* 1995:227–231.

[90] Dowdy NJ and Conner WE. "Acoustic aposematism and evasive action in select chemically defended arctiine (Lepidoptera: Erebidae) species: nonchalant or not?" *PLoS ONE* 2016;11(4):e0152981.

[91] Barber JR and Conner WE. "Acoustic mimicry in a predator–prey interaction," *Proceedings of the National Academy of Sciences USA* 2007;104(22): 9331–9334.

[92] Hristov NI and Conner WE. "Sound strategy: acoustic aposematism in the bat–tiger moth arms race," *Naturwissenschaften* 2005;92(4):164–169.

[93] Barber JR, Chadwell BA, Garrett N, Schmidt-French B, and Conner WE. "Naïve bats discriminate arctiid moth warning sounds but generalize their aposematic meaning," *Journal of Experimental Biology* 2009;212(14): 2141–2148.

[94] Cumming GS. "Mantis movements by night and the interactions of sympatric bats and mantises," *Canadian Journal of Zoology* 1996;74(9):1771–1774.

[95] Yager DD. "Hearing." *The praying mantids*. Baltimore, MD: Johns Hopkins; 1999. pp. 93–113.

[96] Yager D, May M, and Fenton M. "Ultrasound-triggered, flight-gated evasive maneuvers in the praying mantis *Parasphendale agrionina*. I. Free flight," *Journal of Experimental Biology* 1990;152(1):17–39.

[97] Yager DD and May ML. "Ultrasound-triggered, flight-gated evasive maneuvers in the praying mantis *Parasphendale agrionina*. II. Tethered flight," *Journal of Experimental Biology* 1990;152(1):41–58.

[98] Moiseff A, Pollack GS, and Hoy RR. "Steering responses of flying crickets to sound and ultrasound: mate attraction and predator avoidance," *Proceedings of the National Academy of Sciences USA* 1978;75(8):4052–4056.

[99] Montealegre-Z. F, Jonsson T, Robson-Brown KA, Postles M, and Robert D. "Convergent evolution between insect and mammalian audition," *Science* 2012;338(6109):968–971.

[100] Faure P and Hoy R. "The sounds of silence: cessation of singing and song pausing are ultrasound-induced acoustic startle behaviors in the katydid *Neoconocephalus ensiger* (Orthoptera; Tettigoniidae)," *Journal of Comparative Physiology A* 2000;186(2):129–142.

[101] ter Hofstede HM and Fullard JH. "The neuroethology of song cessation in response to gleaning bat calls in two species of katydids, *Neoconocephalus ensiger* and *Amblycorypha oblongifolia*," *Journal of Experimental Biology* 2008;211(15):2431–2441.

[102] Libersat F and Hoy RR. "Ultrasonic startle behavior in bushcrickets (Orthoptera; Tettigoniidae)," *Journal of Comparative Physiology A* 1991; 169(4):507–514.

[103] Yager DD, Cook AP, Pearson DL, and Spangler HG. "A comparative study of ultrasound-triggered behaviour in tiger beetles (Cicindelidae)," *Journal of Zoology* 2000;251(3):355–368.

[104] Forrest T, Farris H, and Hoy R. "Ultrasound acoustic startle response in scarab beetles," *Journal of Experimental Biology* 1995;198(12):2593–2598.

[105] Forrest T, Read M, Farris H, and Hoy R. "A tympanal hearing organ in scarab beetles," *Journal of Experimental Biology* 1997;200(3):601–606.

[106] Miller LA. "Structure of the green lacewing tympanal organ (*Chrysoperla carnea*, Neuroptera)," *Journal of Morphology* 1970;131(4):359–382.

[107] Miller LA. "The behaviour of flying green lacewings, *Chrysoperla carnea*, in the presence of ultrasound," *Journal of Insect Physiology* 1975;21(1): 205–219.

[108] Miller LA and Olesen J. "Avoidance behavior in green lacewings," *Journal of Comparative Physiology* 1979;131(2):113–120.

[109] Rydell J, Jones G, and Waters D. "Echolocating bats and hearing moths: who are the winners?" *Oikos* 1995:419–424.

[110] Jones G. "Scaling of echolocation call parameters in bats," *Journal of Experimental Biology* 1999;202(23):3359–3367.

[111] Fullard J and Dawson J. "The echolocation calls of the spotted bat *Euderma maculatum* are relatively inaudible to moths," *Journal of Experimental Biology* 1997;200(1):129–137.

[112] Painter ML, Chambers CL, Siders M, Doucett RR, Whitaker JJO, and Phillips DL. "Diet of spotted bats (*Euderma maculatum*) in Arizona as indicated by fecal analysis and stable isotopes," *Canadian Journal of Zoology* 2009;87(10):865–875.

[113] Griffin DR. "The importance of atmospheric attenuation for the echolocation of bats (Chiroptera)," *Animal Behaviour* 1971;19(1):55–61.

[114] Neuweiler G. "Auditory adaptations for prey capture in echolocating bats," *Physiological Reviews* 1990;70(3):615–641.

[115] Fenton MB, Faure PA, and Ratcliffe JM. "Evolution of high duty cycle echolocation in bats," *Journal of Experimental Biology* 2012;215(17): 2935–2944.

[116] Schoeman CM and Jacobs DS. "Support for the allotonic frequency hypothesis in an insectivorous bat community," *Oecologia* 2003;134(1): 154–162.

[117] Fenton MB and Bell GP. "Recognition of species of insectivorous bats by their echolocation calls," *Journal of Mammalogy* 1981;62(2):233–243.

[118] Jacobs DS, Barclay RMR, and Schoeman MC. "Foraging and roosting ecology of a rare insectivorous bat species, *Laephotis wintoni* (Thomas, 1901), Vespertilionidae," *Acta Chiropterologica* 2005;7(1):101–109.

[119] Dawkins R. *The extended phenotype: the long reach of the gene.* New York, NY: Oxford University Press; 2016.

[120] Balleri A, Griffiths H, Holderied M, and Baker C, editors. "Bat-inspired multi-harmonic waveforms," *2010 International Waveform Diversity and Design Conference*; Niagara Falls, Canada, Aug. 2010, pp. 86–89.

[121] Corcoran AJ and Conner WE. "Sonar jamming in the field: effectiveness and behavior of a unique prey defense," *Journal of Experimental Biology* 2012;215(24):4278–4287.

Chapter 7

Biologically-inspired coordination of guidance and adaptive radiated waveform for interception and rendezvous problems

Alessio Balleri[1], Alfonso Farina[2] and Alessio Benavoli[3]

7.1 Introduction

The task of intercepting a target and/or rendezvous is an important technical challenge that occurs in many defence operations as well as in civilian applications like robotics, Simultaneous Localisation And Mapping (SLAM) [1], etc. One of the first papers on optimal guidance for interception and rendezvous dates back to 1971 [2]. In that paper, a sensor on the ground delivers optimal guidance to the interceptor on the basis of the estimated trajectories of the interceptor and the target to reach. The radar transmits a suitable waveform which, however, does not change during the task. Another paper [3] years later develops a procedure to adapt the radiated waveform to minimise the estimation error in a tracking case study. This procedure has been recently named fore-active control.

It is known [4] that a bat looking for prey (e.g., a moth or a butterfly), during its search, acquisition, tracking and interception phases and along its trajectory to approach the prey, changes adaptively the radiated waveform of the calls in order to improve the location of the prey. More precisely, the figures from [5,6] show the time-frequency spectrogram of the radiated calls in the successive phases of the interception. It can be argued that the bat develops an optimal rendezvous trajectory together with an adaptive radiated waveform which improves the location capability of the predator.

In this chapter, we take inspiration from the bat and develop an algorithm that guides an airborne radar interceptor towards a target by jointly developing an optimal guidance and automatically adapting and optimising the transmitted waveform on a pulse-to-pulse basis. The algorithm uses a Kalman filter to predict the relative position and speed of the interceptor with respect to the target.

[1]Centre for Electronic Warfare, Information and Cyber, Cranfield University, Defence Academy of the UK, Shrivenham SN6 8LA, UK
[2]FIET, LFIEEE, Selex-ES (retired), Visiting Professor UCL, Rome, Italy
[3]Istituto "Dalle Molle" di Studi sull'Intelligenza Artificiale (IDSIA), Galleria 2, CH-6928, Manno (Lugano), Switzerland

The transmitted waveform is automatically selected based on its ambiguity function and accuracy properties along the approaching path. For each pulse, the interceptor predicts its position and velocity with respect to the target, takes a measurement of range and radial velocity and, with the Kalman filter, refines the relative range and range rate estimates. These are fed into a Linear Quadratic Gaussian (LQG) controller that ensures the interceptor reaches the target automatically and successfully with minimum error and with the minimum guidance energy consumption. We have suitably combined the techniques in [2,3], namely the optimal LQG control law and the fore-active control of the radiated waveform, and this is an original contribution of [7]. The result which we achieve is to emulate what the bat does in its predation: contemporaneously and interactively develop an optimal approaching trajectory and transmit a waveform that adaptively changes during the approaching trajectory phases so that the measurements of the sensor on board of the interceptor are better suited to improve guidance law. Some preliminary results of the proposed technique have been presented in [8].

7.2 Theoretical framework

We study the case of an interceptor and a target both moving with the same type of linear kinematics described by a matrix $\mathbf{F}$ and state equations $\mathbf{x_f}(k)$ and $\mathbf{x_b}(k)$ [9]. The trajectory of both the interceptor and the target are subject to Gaussian random perturbations, $\mathbf{Gw_f}(k-1)$ and $\mathbf{Gw_b}(k-1)$, with zero mean value and covariance matrices $\mathbf{Q_f}$ and $\mathbf{Q_b}$, respectively.

$$\begin{aligned} \mathbf{x_f}(k) &= \mathbf{Fx_f}(k-1) + \mathbf{Gw_f}(k-1) \\ \mathbf{x_b}(k) &= \mathbf{Fx_b}(k-1) + \mathbf{Bu}(k-1) + \mathbf{Gw_b}(k-1) \end{aligned} \tag{7.1}$$

The term $\mathbf{Bu}(k-1)$ is used to model the ability of the interceptor to adapt and control its trajectory at each step. The interceptor is modelled as a controlled system that accepts an input vector $\mathbf{u}(k)$ which is combined linearly with a matrix $\mathbf{B}$ before being applied to the equations describing the target kinematics. We define the difference between the state equations of the interceptor and of the target as the error to reduce to the minimum value at the intercept point

$$\mathbf{e}(k) = \mathbf{x_b}(k) - \mathbf{x_f}(k) = \mathbf{Fe}(k-1) + \mathbf{Bu}(k-1) + \mathbf{Gw_e}(k-1) \tag{7.2}$$

with $\mathbf{Gw_e}(k-1)$ being a Gaussian random processes with zero-mean and covariance matrix $\mathbf{Q}$. At each time k the interceptor transmits a waveform to measure its relative distance and radial velocity with respect to the target and uses the measurements to control its trajectory in order to intercept the target with a limited number of radar transmissions N_T and with the minimum energy consumption of the interceptor. We assume that the measurement $\mathbf{y_e}(k)$ of the distance and velocity relative to each transmission is a linear function of the error $\mathbf{e}(k)$ as [3]

$$\mathbf{y_e}(k) = \mathbf{He}(k) + \nu_e(k;\theta_k) \tag{7.3}$$

where $\mathbf{H}$ is the matrix that maps the error into the measurement and $\nu_e(k;\theta_k)$ is a Gaussian random process with mean value zero and a covariance matrix $\mathbf{N}(\theta_k)$. The covariance matrix of each measurement depends on the accuracy of the transmitted waveform $s(t;\theta_k)$ whose design is fully described by the vector of parameters θ_k that identifies the key waveform properties, such as duration, bandwidth and time-frequency curvature. The mathematical expression of the elements of θ_k and the vector length depend on the waveform design. It has been shown in the literature that $\mathbf{N}(\theta_k)$ corresponds to the Cramér-Rao Lower Bound (CRLB), relative to the task of joint estimation of range and radial velocity between a sensor and a target, when the sensor transmits the signal $s(t;\theta_k)$ [10–13]. The Fisher Information Matrix (FIM) for range and radial velocity in the presence of noise with mean power N_0 can be expressed as

$$\text{FIM} = -\text{SNR}\begin{pmatrix} \frac{4}{c^2}\frac{1}{2}\frac{\partial^2|\chi(\tau,\nu;\theta_k)|^2}{\partial\tau^2} & \frac{4}{c\lambda}\frac{1}{2}\frac{\partial^2|\chi(\tau,\nu;\theta_k)|^2}{\partial\tau\partial\nu} \\ \frac{4}{c\lambda}\frac{1}{2}\frac{\partial^2|\chi(\tau,\nu;\theta_k)|^2}{\partial\nu\partial\tau} & \frac{4}{\lambda^2}\frac{1}{2}\frac{\partial^2|\chi(\tau,\nu;\theta_k)|^2}{\partial\nu^2} \end{pmatrix}\Bigg|_{\tau,\nu=0} \tag{7.4}$$

where c is the speed of propagation, λ is the wavelength, $\text{SNR} = 2E_s/N_0$ is the Signal-to-Noise Ratio and $\chi_k(\tau,\nu)$ is the normalised narrowband Complex Ambiguity Function (CAF) of the signal $\sqrt{E_s}s(t;\theta_k)$ of energy E_s defined as

$$\chi(\tau,\nu;\theta_k) = \int_{-\infty}^{\infty} s(t;\theta_k)s^*(t+\tau;\theta_k)e^{j2\pi\nu t}dt \tag{7.5}$$

The CRLB is obtained as the inverse of the FIM and therefore

$$\mathbf{N}(\theta_k) = [\text{FIM}]^{-1} \tag{7.6}$$

This gives the minimum values of variances and covariances of the measurements of range and range rate. At each time k, the interceptor makes a prediction of the estimation error covariance matrix

$$\mathbf{P}_{k|k-1} = \mathbf{F}\mathbf{P}_{k-1}\mathbf{F}^T + \mathbf{Q} \tag{7.7}$$

and then selects the waveform parameters θ_k so as to minimise the determinant of the residual matrix

$$\mathbf{S}_k = \mathbf{H}\mathbf{P}_{k|k-1}\mathbf{H}^T + \mathbf{N}(\theta_k) \tag{7.8}$$

as described in [3]. The interceptor then produces a pulse, takes a measurement with a waveform of the preselected parameters θ_k and uses the covariance matrix $\mathbf{N}(\theta_k)$ to calculate the Kalman filter gain $\mathbf{K}_k$ as

$$\mathbf{K}_k = \mathbf{P}_{k|k-1}\mathbf{H}^T\mathbf{S}_k^{-1} \tag{7.9}$$

The Kalman gain is then used to calculate the estimation error covariance matrix at the kth step as $\mathbf{P}_k = (\mathbf{I} - \mathbf{K}_k\mathbf{H})\mathbf{P}_{k|k-1}$ and an estimate of the error as

$$\begin{aligned} \hat{\mathbf{e}}(k|k-1) &= \mathbf{F}\hat{\mathbf{e}}(k-1) + \mathbf{B}\mathbf{u}(k-1) \\ \hat{\mathbf{e}}(k) &= \hat{\mathbf{e}}(k|k-1) + \mathbf{K}_k[\mathbf{y}_\mathbf{e}(k) - \mathbf{H}\hat{\mathbf{e}}(k|k-1)] \end{aligned} \tag{7.10}$$

Equation (7.10) shows that the estimate of the error at the kth time step only depends on the control input at the $(k-1)$th time and this will allow us to select the most appropriate control input at the kth time based solely on the estimate of the error.

As previously mentioned, the control task is carried out to ensure the interceptor reaches the target as efficiently as possible. To do this, we define and minimise the cost function

$$J = E\left[\mathbf{e}^T(N_T)\mathbf{M}\mathbf{e}(N_T) + \sum_{k=0}^{N_T} \mathbf{u}^T(k)\mathbf{R}\mathbf{u}(k)\right] \tag{7.11}$$

of the kind of a typical LQG control framework [2]. In (7.11) N_T is the predefined number of transmissions used to intercept the target and $\mathbf{M}$ and $\mathbf{R}$ are two suitable matrices that are applied to the dynamic state error and to the input control signal, respectively. It is worth noting that when $\mathbf{M}$ and $\mathbf{R}$ are identity matrices the cost function is minimised when the error at time N_T is minimised and when the energy of the input signal $\mathbf{u}(k)$ is also minimised. The solution of (7.11) is well known in the literature (e.g., see [2]) and it is such that the control input signal at the time k is a function of the estimate of the error at the time k and of a matrix $\mathbf{L}_k$

$$\mathbf{u}(k) = -\mathbf{L}_k\hat{\mathbf{e}}(k) \tag{7.12}$$

where

$$\mathbf{L}_{k+1} = \left(\mathbf{B}^T\mathbf{U}_{k+1}\mathbf{B} + \mathbf{R}\right)^{-1}\mathbf{B}^T\mathbf{U}_{k+1}\mathbf{F} \tag{7.13}$$

and

$$\mathbf{U}_k = \mathbf{F}^T\left(\mathbf{U}_{k+1} - \mathbf{U}_{k+1}\mathbf{B}(\mathbf{B}^T\mathbf{U}_{k+1}\mathbf{B} + \mathbf{R})^{-1}\mathbf{B}^T\mathbf{U}_{k+1}\right)\mathbf{F}, \text{ with } \mathbf{U}_{N_T} = \mathbf{M} \tag{7.14}$$

7.2.1 Gaussian linear chirp

In this chapter, we limit the study to Linear Frequency Modulated (LFM) chirps with a Gaussian amplitude modulation of the form

$$s(t;\theta_k) = \left(\frac{1}{\pi\lambda_G^2}\right)^{1/4} e^{-t^2/2\lambda_G^2} e^{jb_Gt^2} e^{j2\pi f_0 t} \tag{7.15}$$

whose design depends solely on the parameters λ_G and b_G, that is $\theta_k = [b_G\ \lambda_G]^T$. Selecting and diversifying these parameters results in waveforms with a different time duration $T = 2\lambda_G$ and bandwidth $B = b_G T/\pi$. The use of a Gaussian linear chirp simplifies the analysis because the FIM of range and range rate relative to this class of waveforms is known in the literature (e.g., see [3,14]) and can be expressed as

$$\mathrm{FIM} = \mathrm{SNR}\begin{pmatrix} \frac{4}{c^2}\left(\frac{1}{2\lambda_G^2} + 2\lambda_G^2 b_G^2\right) & \frac{4}{c\lambda}\left(2\pi\lambda_G^2 b_G\right) \\ \frac{4}{c\lambda}\left(2\pi\lambda_G^2 b_G\right) & \frac{4}{\lambda^2}\left(2\pi^2\lambda_G^2\right) \end{pmatrix} \tag{7.16}$$

which leads to a covariance matrix of the estimators of range and range rate

$$\mathbf{N}(\theta_k) = \begin{pmatrix} \frac{c^2\lambda_G^2}{2\mathrm{SNR}} & -\frac{c\lambda\lambda_G^2 b_G}{2\pi\mathrm{SNR}} \\ -\frac{c\lambda\lambda_G^2 b_G}{2\pi\mathrm{SNR}} & \frac{\lambda^2}{4\pi^2\mathrm{SNR}}\left(\frac{1}{2\lambda_G^2} + 2\lambda_G^2 b_G^2\right) \end{pmatrix} \tag{7.17}$$

It can be easily shown that the determinant of the FIM is equal to $\mathrm{SNR}^2\,\frac{16\pi^2}{c^2\lambda^2}$ and does not depend on the parameters b_G and λ_G [3]. This will significantly simplify the calculations of the optimal θ_k in the next section.

7.3 Two-dimensional case study

For simplicity, we study the case of a static target and an interceptor moving on a two-dimensional plane. The state variables are expressed in polar coordinates and consist of the range between the target and the interceptor ρ, the radial velocity $\dot{\rho}$ and the relative angle θ. Although, in practice, problems of tracking in 2D and 3D are treated in Cartesian coordinates with the use of non-linear equations, using polar coordinates allows the treatment of the problem tackled in this chapter with linear equations. The use of linear equation is of particular importance as it guarantees the convergence of the LGQ control. A noisy perturbation is applied to the component representing the acceleration and to the angular velocity of the interceptor. To represent this scenario, we define the matrices **F** and **G** as

$$\mathbf{F} = \begin{pmatrix} 1 & T_s & 0 \\ 0 & 1 & 0 \\ 0 & 0 & 1 \end{pmatrix} \tag{7.18}$$

and

$$\mathbf{G} = \begin{pmatrix} 0 & 0 \\ 1 & 0 \\ 0 & 1 \end{pmatrix} \tag{7.19}$$

The resulting interceptor state equations in the absence of control are

$$\begin{cases} \rho(k) = \rho(k-1) + \dot{\rho}(k-1)T_s \\ \dot{\rho}(k) = \dot{\rho}(k-1) + a \\ \theta(k) = \theta(k-1) + b \end{cases} \tag{7.20}$$

where T_s is the radar scan period and a and b are two independent Gaussian processes with zero mean value and variance σ_a^2 and σ_b^2, respectively.[1]

The sensor measures the distance between the interceptor and the target, their relative radial velocity and the relative angle. The matrix $\mathbf{H}$ is defined as

$$\mathbf{H} = \begin{pmatrix} 1 & 0 & 0 \\ 0 & 1 & 0 \\ 0 & 0 & 1 \end{pmatrix} \tag{7.21}$$

The covariance matrix of the measurement is obtained by modifying (7.17) so as to account for the measurement of the angle as

$$\hat{\mathbf{N}}(\theta_k) = \begin{pmatrix} \dfrac{c^2\lambda_G^2}{2\text{SNR}} & -\dfrac{c\lambda\lambda_G^2 b_G}{2\pi\text{SNR}} & 0 \\ -\dfrac{c\lambda\lambda_G^2 b_G}{2\pi\text{SNR}} & \dfrac{\lambda^2}{4\pi^2\text{SNR}}\left(\dfrac{1}{2\lambda_G^2} + 2\lambda_G^2 b_G^2\right) & 0 \\ 0 & 0 & \sigma_\theta^2 \end{pmatrix} \tag{7.22}$$

under the assumption that the measurement of the angle is statistically independent of the measurements of the range and the radial velocity. The waveform parameters of the Gaussian linear chirp $\theta_k = [b_G \quad \lambda_G]^T$ at the kth time step can be estimated from the element p_{ij} of the matrix $\mathbf{P}_{k|k-1}$ so as to minimise the matrix $\mathbf{S}_k$. The derivation of these equations follows and it is an extension of the solutions provided in [3] for the case when the sensor measures only the range and the range rate. The calculations are straightforward and result from the non-dependency of the determinant of the covariance matrix of the measurements on the parameters of the Gaussian chirp.

They are carried out by noting that because $\mathbf{H}$ is an identity matrix the matrix $\mathbf{S}_k$ is of the form:

$$\mathbf{S}_k = \begin{pmatrix} p_{1,1} + n_{1,1} & p_{1,2} + n_{1,2} & p_{1,3} \\ p_{2,1} + n_{2,1} & p_{2,2} + n_{2,2} & p_{2,3} \\ p_{3,1} & p_{3,2} & p_{3,3} + n_{3,3} \end{pmatrix} \tag{7.23}$$

where n_{ij} indicates the element (i, j) of $\hat{\mathbf{N}}(\theta_k)$.

[1] In this chapter, we consider the case of one pulse per scan and hence the scan period corresponds to the Pulse Repetition Interval (PRI) of the radar.

The determinant of $\mathbf{S}_k$ is equal to

$$\begin{aligned}\det(\mathbf{S}_k) = (p_{3,3} + n_{3,3})\left[(p_{1,1} + n_{1,1})(p_{2,2} + n_{2,2}) - (p_{1,2} + n_{1,2})^2\right] \\ + p_{3,1}\left[p_{2,3}(p_{1,2} + n_{1,2}) - p_{1,3}(p_{2,2} + n_{2,2})\right] \\ - p_{3,2}\left[p_{2,3}(p_{1,1} + n_{1,1})\right] - p_{1,3}(p_{2,1} + n_{2,1})]\end{aligned} \tag{7.24}$$

and can be expressed as function of the determinant of $\mathbf{P}_{k|k-1}$ and the determinant of $\mathbf{N}(\theta_k)$ as

$$\begin{aligned}\det(\mathbf{S}_k) = p\left[\det(\mathbf{P}_{k|k-1}) + \det(\mathbf{N}(\theta_k)) + p_{1,1}n_{2,2} + p_{2,2}n_{1,1} - 2p_{1,2}n_{1,2}\right] \\ + p_{3,1}\left[p_{2,3}(p_{1,2} + n_{1,2}) - p_{1,3}(p_{2,2} + n_{2,2})\right] \\ - p_{3,2}\left[p_{2,3}(p_{1,1} + n_{1,1})\right] - p_{1,3}(p_{2,1} + n_{2,1})]\end{aligned} \tag{7.25}$$

We derive (7.25) with respect to the parameters of the chirp λ_G and b_G, where the quantities $\det(\mathbf{P}_{k|k-1})$ and $\det(\mathbf{N}(\theta_k))$ do not depend on the chirp parameters, to find that the determinant of $\mathbf{S}_k$ is minimised for

$$b_G = \frac{\gamma w_0}{4\beta} \tag{7.26}$$

and

$$\lambda_G = \left(\frac{\beta}{\delta}\right)^{0.25} \tag{7.27}$$

with

$$\begin{cases} \hat{p} = p_{3,3} + \sigma_\theta^2 \\ \alpha = \hat{p}p_{2,2} - p_{3,2}p_{2,3} \\ \beta = \hat{p}p_{1,1} - p_{3,1}p_{1,3} \\ \gamma = p_{3,2}p_{1,3} - 2\hat{p}p_{1,2} + p_{3,1}p_{2,3} \\ \delta = w_0^2\alpha - 2\gamma w_0 b_G + 4b_G^2\beta \end{cases} \tag{7.28}$$

7.4 Simulation results

Simulations are carried out for the case of an interceptor that for $k = 0$ is at 10 km from the target moving with a relative radial velocity of -20 m/s and with a relative angle of 0° ($\mathbf{e}(0) = [10^4 \; -200]^{\mathrm{T}}$). The filter is initialised with $\hat{\mathbf{e}}(0)$ being a realisation of a Gaussian random variable with mean value $\mathbf{e}(0)$ and covariance matrix $\mathbf{P}_0$. This is a diagonal matrix with all the elements on the diagonal equal to 10^6. The covariance matrix $\mathbf{P}_0$ is very large in order to guarantee the filter accepts the first set of measurements which typically, for a single radar pulse, are characterised by a very large covariance matrix due to the low accuracy of a single chirp. The SNR at time zero is set to 20 dB and the parameter σ_θ is fixed to 0.002 as a representative beamwidth of a 2 m aperture at 10 GHz. Simulations are carried out for two different cases. Firstly, we simulate the case when no constraints are applied to the pulse duration and pulse bandwidth. The algorithm, in this case, is left free to select any chirp parameters so as to minimise the determinant of the matrix $\mathbf{S}_k$. Simulations are then carried out for the case when constraints are applied to both the pulse width and the pulse bandwidth at each step in order to guarantee a minimum possible value of the radar peak power, that all target echoes arrive after transmissions are finished (that is to avoid eclipsing losses) and to guarantee a suitable value of the range resolution. The control is set to reach the target within $N_T = 300$ transmissions.

Figure 7.1 shows the range-range rate diagram for the case with no constraints on the chirp parameters. Each dot in the diagram corresponds to a pulse

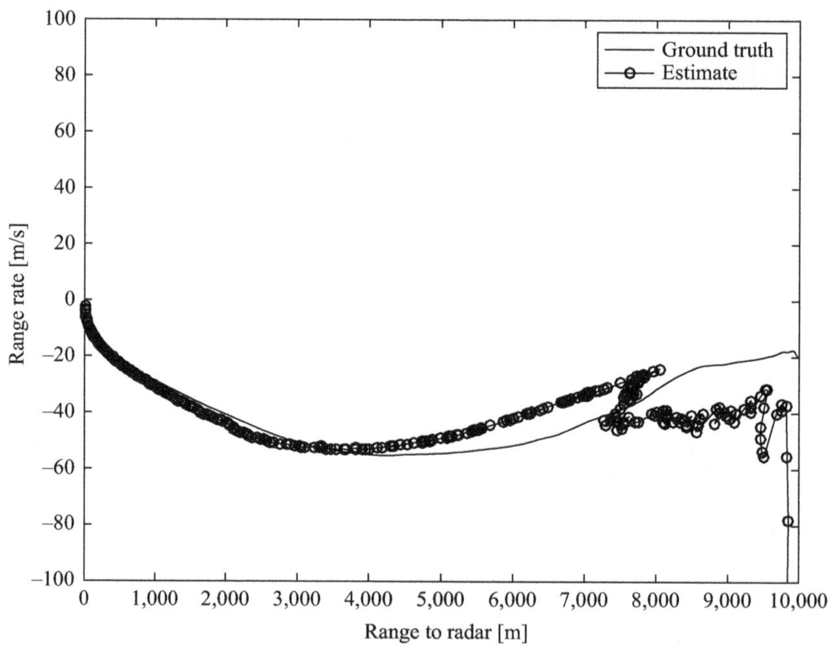

Figure 7.1 Range–range rate diagram for the case when no constraints are applied to the chirp parameters

transmission. Results show that, as expected, the interceptor reaches the target at the end of the sequence of N_T pulses. The error between the estimated range and range rate is large at the beginning of the sequence due to the large covariance matrix of the error $\mathbf{P}_k$. Figures 7.2 and 7.3 show the values of the pulse width and pulse bandwidth resulting from minimising the determinant of the matrix $\mathbf{S}_k$. The parameters are such that the accuracy of both range and radial velocity are very large and the filter takes a few steps to reduce the error. This is the reason why the elements of the covariance matrix of the error $\mathbf{P}_0$ were chosen large. If these were small, the filter would discard the measurements and trust the predictions but this could lead to even bigger errors. Results also show that the algorithm converges to chirp parameters which are not typical of common radar systems. The resulting bandwidth is very small throughout the sequence leading to a very low range resolution and the pulse width is very long compared to the typical duration of a radar waveform which is typically only a few μs. Comparing the pulse width with the longer pulse width possible to avoid eclipsing in Figure 7.2 shows that eclipsing would occur almost at each transmission. In other words only a system deploying two antennas, one for the transmitter and one for the receiver, would be capable of operating with such parameters. Figure 7.4 shows the AF of the first, 75th, 225th

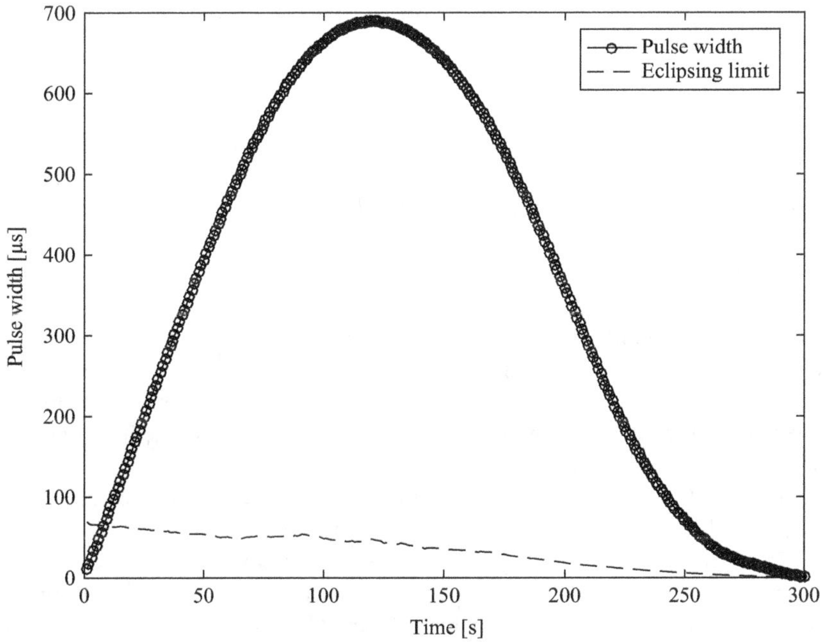

Figure 7.2 Pulse width of each transmission for the case when no constraints are applied to the chirp parameters

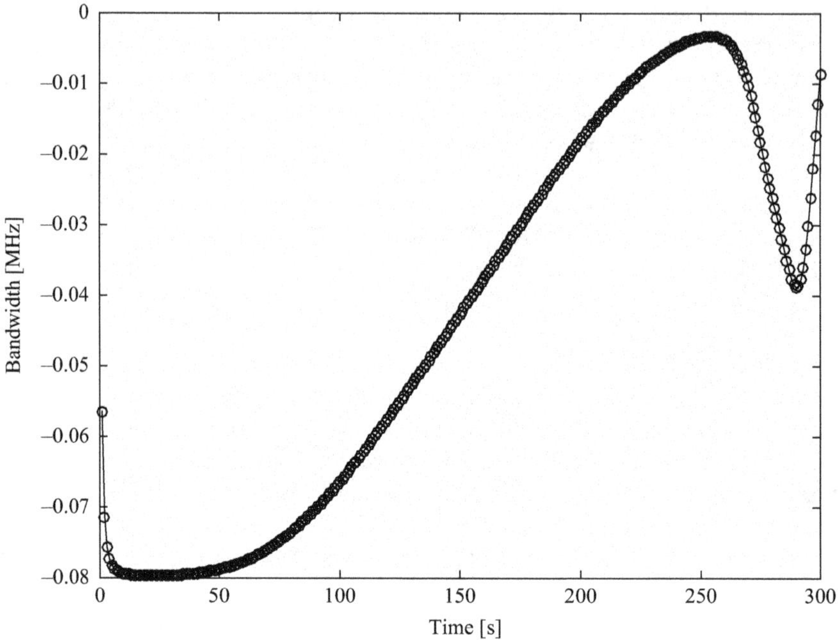

Figure 7.3 Pulse bandwidth of each transmission for the case when no constraints are applied to the chirp parameters

and 300th transmitted waveforms. Results show that the wedge of the AF rotates anticlockwise along the trajectory. When the interceptor approaches the target, the pulse duration becomes shorter, the range resolution improves and the Doppler resolution decreases.

Figure 7.5 shows the range-range rate diagram for the case with constraints on both the pulse width and the bandwidth of the chirp. The pulse width cannot assume values below 1 μs and cannot assume values that would generate eclipsing. The eclipsing limit is calculated by using the estimate of the range between the target and the interceptor at each step k as $2\rho_k/c$. The bandwidth cannot reach values below 5 MHz, corresponding to a minimum range resolution equal to 30 m. Results show that the LQG control converges as expected and that the interceptor reaches the target within the predefined N_T transmissions. The error at the beginning of the sequence is significantly lower than the case when no constraints are applied to the chirp parameters. This is because, due to the applied constraints, the chirp parameters assume values that result in a lower measurement covariance matrix starting from the very first transmissions. Results in Figure 7.6 show that the pulse duration assumes values between 1 μs up to about 30 μs. The bandwidth remains constant throughout the sequence at 5 MHz.

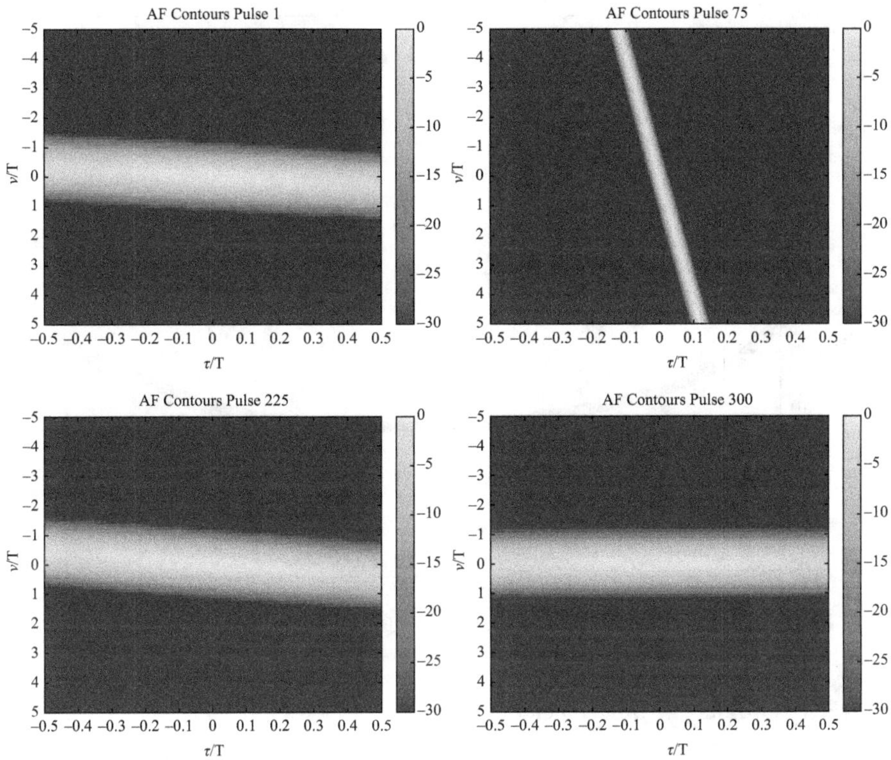

Figure 7.4 Contours of the AF of the first, 75th, 225th and 300th transmitted waveforms for the case when no constraints are applied to the chirp parameters

Figure 7.7 shows the AF of the first, 75th, 225th and 300th transmitted waveforms. Results show that the wedge of the AF rotates anticlockwise along the trajectory although the total angle of rotation is smaller with respect to the case with no constraints on the chirp parameters. When the interceptor approaches the target, the bandwidth does not change significantly and the range resolution remains constant. However, as the pulse duration becomes shorter the Doppler resolution decreases. The waveform is Doppler tolerant throughout the mission,[2] that is, the output of the matched-filter remains high in the presence of a Doppler mismatch. To assess the advantages of using the proposed algorithm, Figure 7.8 shows the determinant of the matrix $\mathbf{S}_k$ for the adaptive case with and without constraints on the chirp parameters and for the case when the chirp parameters are

[2]Doppler tolerance is a characteristic of linear chirps when the narrowband approximation is satisfied [14].

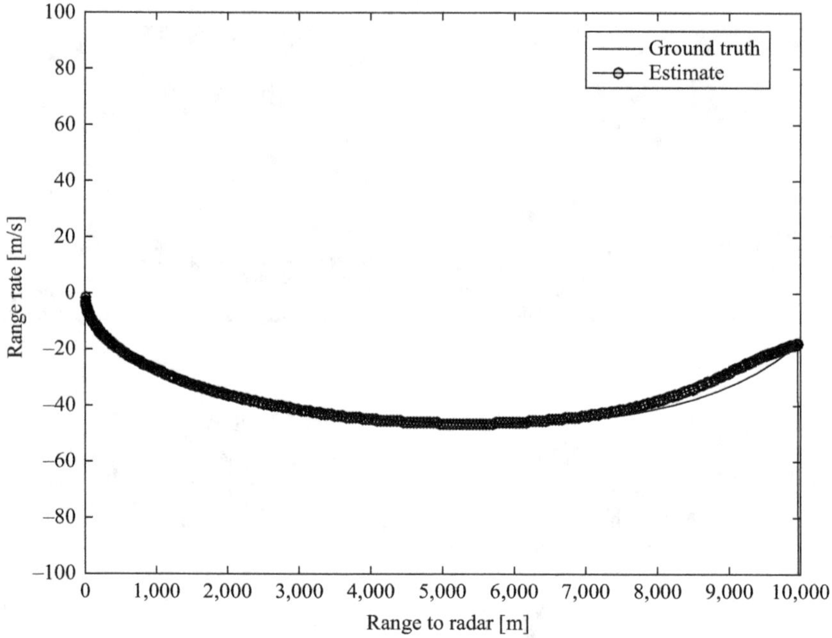

Figure 7.5 Range–range rate diagram for the case with constraints applied to the chirp parameters

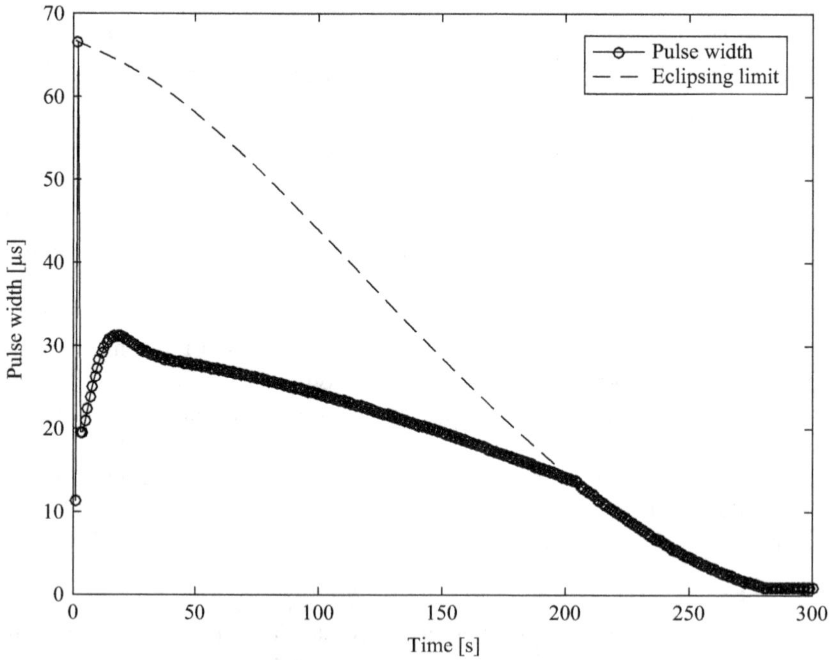

Figure 7.6 Pulse width of each transmission for the case with constraints applied to the chirp parameters

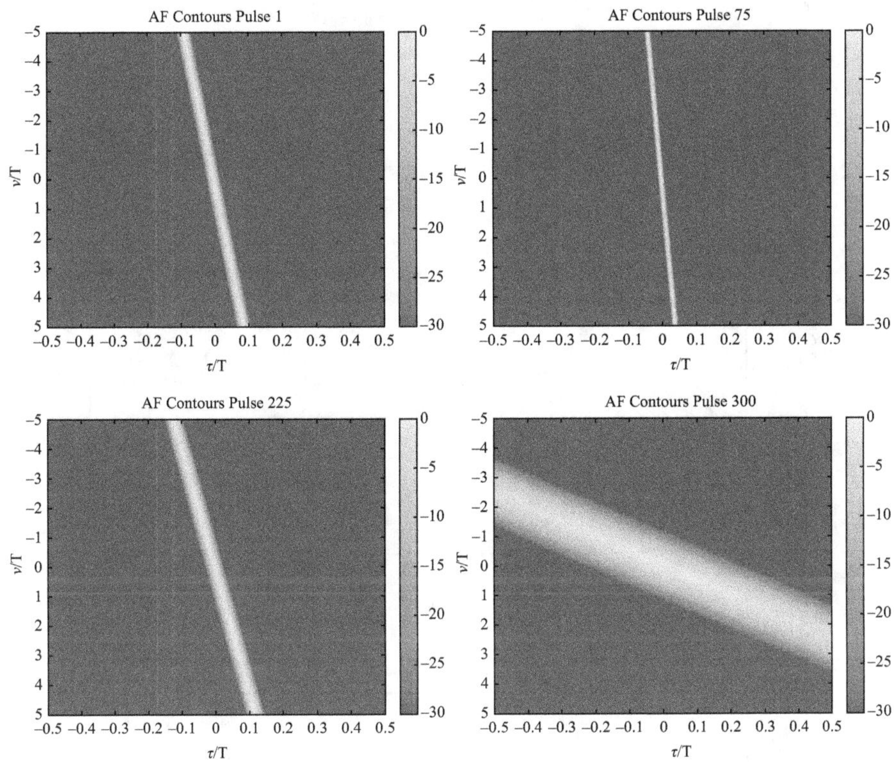

Figure 7.7 Contours of the AF of the first, 75th, 225th and 300th transmitted waveforms for the case with constraints applied to the chirp parameters

kept constant to a pulse width of 1 μs and a bandwidth of 5 MHz. For all three cases the value of the determinant of $\mathbf{S}_k$ drops significantly from pulse-to-pulse. Because the proposed algorithm is designed to minimise the determinant of $\mathbf{S}_k$, as expected, results show that the determinant assumes lower values at each step for the adaptive case than the case with constant waveform. The results in the figure have been obtained with 100 Monte Carlo trials. It is important to observe that minimising the determinant of $\mathbf{S}_k$ does not necessarily guarantee a minimum of the trace of the error covariance matrix $\mathbf{P}_k$, that is, it does not correspond to a minimum variance of the estimate of the state vector. This was evident by a comparison between Figures 7.1 and 7.5 where the error at the beginning of the transmitting sequence for the adaptive case with no constraints, that is the case that converges to the minimum of the determinant of $\mathbf{S}_k$ at every step, was much larger than that of the adaptive case with constraints. This is corroborated by Figures 7.9–7.11 that show the mean value and variance of all errors for the three cases. Results show

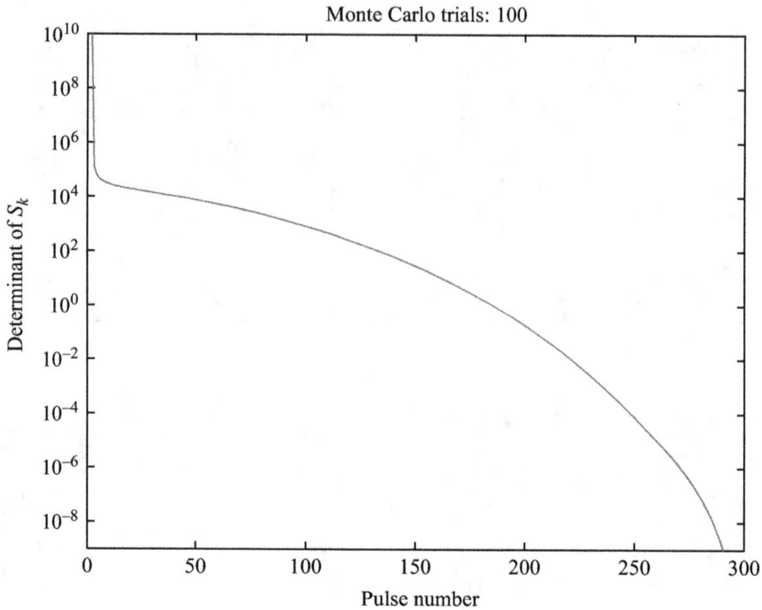

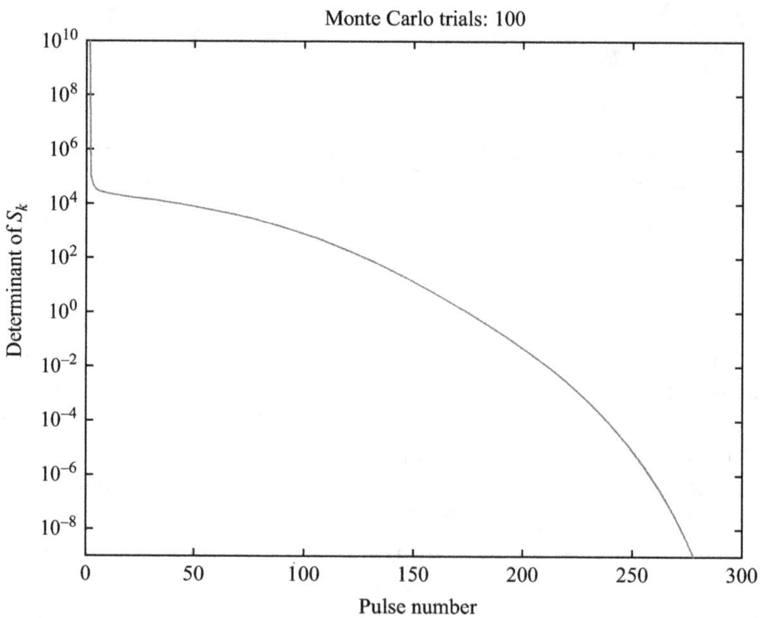

Figure 7.8 Determinant of the matrix $\mathbf{S}_k$ as a function of time

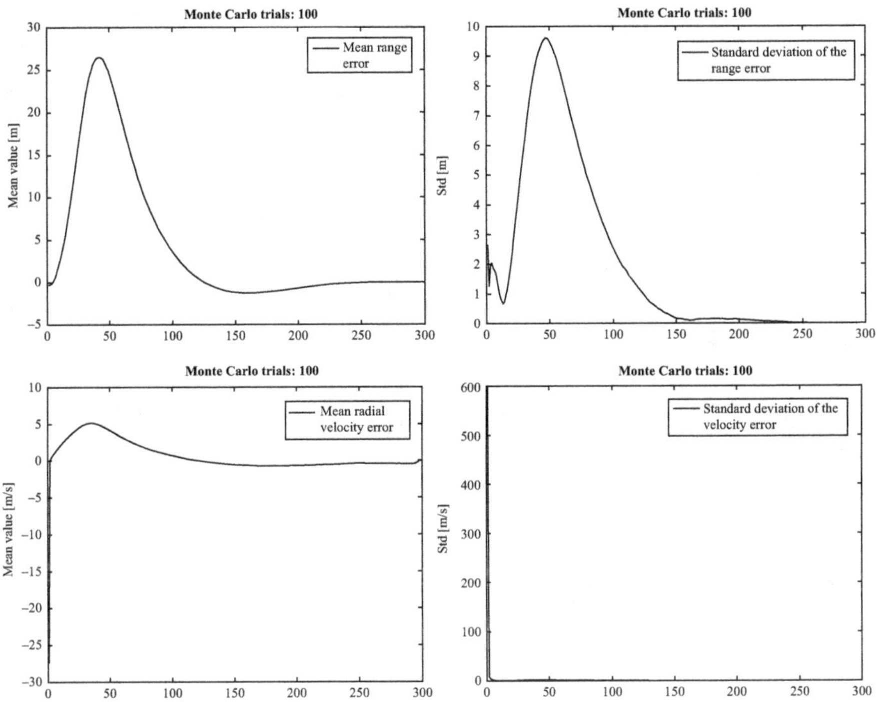

Figure 7.9 Mean value and standard deviation of the error estimates of range and range rate for a constant waveform with pulse width 1 μs and bandwidth 5 MHz throughout the sequence

that the estimation error for the case with constant chirp parameters is lower than that of both cases with adaptive waveform.

7.5 Conclusion

In this chapter, we have presented an algorithm that allows an interceptor aircraft with an on-board radar to adapt its trajectory in order to intercept a target and to automatically optimise the transmitted waveform on a pulse-to-pulse basis. To achieve this, we have suitably combined two techniques, namely the optimal Linear Quadratic Gaussian (LQG) control law and the fore-active control of the radiated waveform. The case study takes inspiration from existing predator-prey relationships in nature, such as that of a bat which captures a moth or a butterfly.

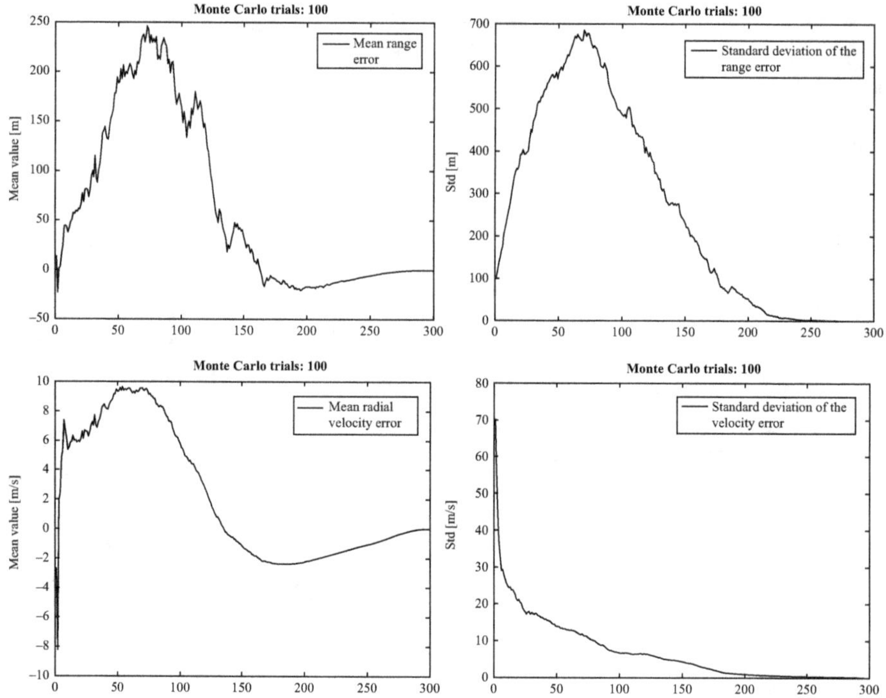

Figure 7.10 Mean value and standard deviation of the error estimates of range and range rate for the case when no constraints are applied to the chirp parameters

The bat during its search, acquisition, tracking and interception of the insect adjusts both its trajectory to approach the prey and adaptively changes the radiated waveform of the echolocation calls in order to improve the localisation of the prey. Simulation results show that the interceptor can successfully reach the target within the predefined number of transmissions and automatically adapt the waveform during the mission. Future work will look at using different types of waveform designs and different optimisation criteria. We plan to implement the tracking algorithm in Cartesian coordinates. This means to consider a non-linear measurement equation and hence a non-linear tracking filter. LQG control can no longer be used in this context, because of the non-linearity. Moreover, the separation principle between filtering and control no longer holds and, therefore, we cannot optimise the waveform and control cost separately. To address this issue, we plan to derive an optimal waveform and control law by means of a stochastic Model Predictive Control based approach, implemented using a receding horizon strategy [15,16]. The case of the bat will be further studied by expanding the algorithm presented in this chapter to the wideband waveform case.

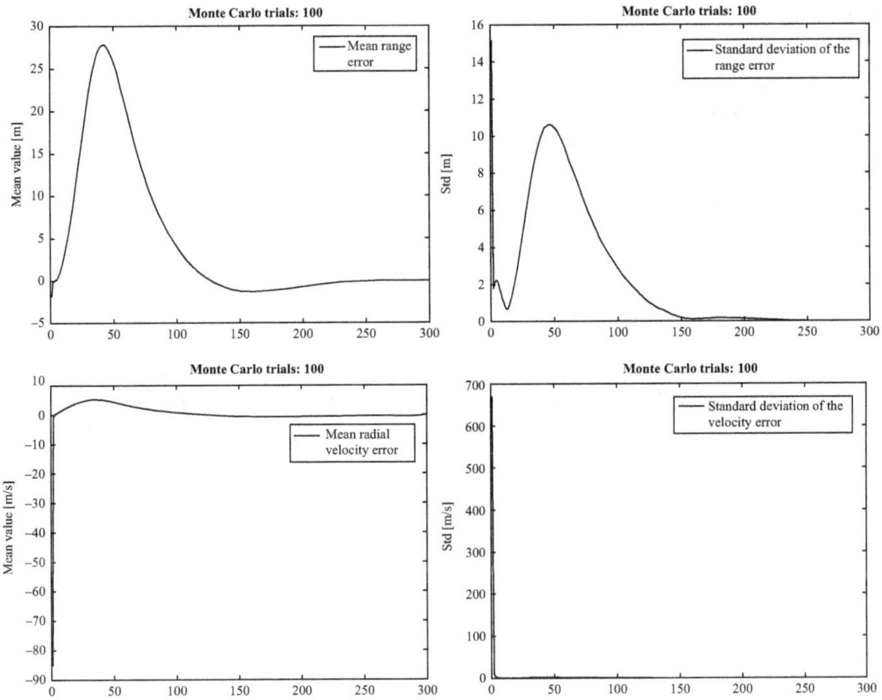

Figure 7.11 Mean value and standard deviation of the error estimates of range and range rate for the case with constraints applied to the chirp parameters

References

[1] M. W. M. G. Dissanayake, P. Newman, S. Clark, H. F. Durrant-Whyte, and M. Csorba, "A solution to the simultaneous localization and map building (SLAM) problem," *IEEE Transactions on Robotics and Automation*, vol. 17, no. 3, pp. 229–241, Jun. 2001.

[2] M. Athans, "On optimal allocation and guidance laws for linear interception and rendezvous problems," *IEEE Transactions on Aerospace and Electronic Systems*, vol. AES-7, no. 5, pp. 843–853, Sep. 1971.

[3] D. Kershaw and R. Evans, "Optimal waveform selection for tracking systems," *IEEE Transactions on Information Theory*, vol. 40, no. 5, pp. 1536–1550, Sep. 1994.

[4] C. Baker, H. Griffiths, and A. Balleri, "Biologically inspired waveform diversity (in Waveform Design and Diversity for Advanced Radar Systems)," *Institution of Engineering and Technology, Series on Radar, Sonar, Navigation and Avionics*, pp. 149–172, 2012.

[5] P. Flandrin, "Time-frequency processing of bat sonar signals," in *Animal Sonar: Processes and Performance*. Boston, MA: Springer US, pp. 797–802, 1988.

[6] F. Hlawatsch and G. F. Boudreaux-Bartels, "Linear and quadratic time–frequency signal representations," *IEEE Signal Processing Magazine*, vol. 9, no. 2, pp. 21–67, Apr. 1992.

[7] A. Balleri, A. Farina, and A. Benavoli, "Coordination of optimal guidance law and adaptive radiated waveform for interception and rendezvous problems," *IET Radar, Sonar and Navigation* Dec. 2016. doi:10.1049/iet-rsn.2016.0547.

[8] A. Farina, "Cognitive radar signal processing," *Keynote Speech, IET International Radar Conference*, Hangzhou, China, 14–16 Oct. 2015.

[9] R. Kalman, "Mathematical description of linear dynamical systems," *Journal of the Society for Industrial and Applied Mathematics, Series A: Control*, vol. 1, no. 2, pp. 152–192, 1963.

[10] H. L. Van Trees, *Detection, Estimation and Modulation Theory, Part III. Radar-Sonar Signal Processing and Gaussian Signals in Noise*. New York, NY: Wiley, 2001.

[11] C. E. Cook and M. Bernfeld, *Radar Signals: An Introduction to Theory and Application*. London: Artech House, 1987.

[12] E. Kelly, "The radar measurement of range, velocity and acceleration," *IRE Transactions on Military Electronics*, vol. MIL-5, no. 2, pp. 51–57, Apr. 1961.

[13] A. Dogandzic and A. Nehorai, "Cramer–Rao bounds for estimating range, velocity, and direction with an active array," *IEEE Transactions on Signal Processing*, vol. 49, no. 6, pp. 1122–1137, Jun. 2001.

[14] A. Balleri and A. Farina, "Ambiguity function and accuracy of the hyperbolic chirp: comparison with the linear chirp," *IET Radar, Sonar & Navigation*, vol. 11, no. 1, pp. 142–153, 2017.

[15] D. de la Penad, A. Bemporad, and T. Alamo, "Stochastic programming applied to model predictive control," in *Proceedings of the 44th IEEE Conference on Decision and Control*, Seville, Spain, pp. 1361–1366, 12–15 Dec. 2005.

[16] D. Van Hessem and O. Bosgra, "A conic reformulation of model predictive control including bounded and stochastic disturbances under state and input constraints," in *Proceedings of the 41st IEEE Conference on Decision and Control*, vol. 4, Las Vegas, NV, pp. 4643–4648, 10–13 Dec. 2002.

Chapter 8

Cognitive sensor/processor system framework for target tracking

Kristine L. Bell[1], Chris Baker[2], Graeme E. Smith[3], Joel T. Johnson[3] and Muralidhar Rangaswamy[4]

8.1 Introduction

In this chapter we examine the emerging topic of artificial cognition and its application to target tracking. Most sensor/processor systems employ a feed-forward processing chain in which they first perform some low-level processing of received sensor data (such as obtaining target detections) and then pass the processed data on to some higher-level processor (such as a tracker or classifier), which extracts information (such as target kinematic parameters and target type) to achieve a system objective. Tracking and classification can be improved using adaptation between the information extracted from the sensor/processor and the design and transmission of subsequent illuminating waveforms. As such the application of artificial cognition to sensor/processor systems offers much promise for improved sensing and processing [1–7].

Cognitive systems take inspiration from the neurobiological processes of human cognition [8,9] so the sensor may interpret the environment and adjust its system parameters to obtain the most relevant information from the scene. The heart of cognition is the *perception–action cycle* that both informs and is informed by *memory* [9]. Cognition requires stimulation by *sensors*. In the human this is via hearing, touch, smell, vision, and taste. The nervous system *processes* the sensed stimuli and converts them into a *perception* of the world. We are able to take informed *action* by interpreting our perception of the world and making decisions. As a consequence, the nervous system sends signals that activate our muscles, thus

[1]Metron, Inc., 1818 Library Street, Suite 600, Reston, VA 20190, USA
[2]40 St Peters Road, Great Malvern, Worcestershire, WR14 1 QS, UK
[3]Department of Electrical and Computer Engineering and ElectroScience Lab, The Ohio State University, 1330 Kinnear Road, Columbus, OH 43212, USA
[4]Air Force Research Lab/RYAP, 2241 Avionics Circle, Building 620, Wright-Patterson Air Force Base, OH 45433-7320, USA

enabling the desired action to take place. Informed *decision-making* implies the establishment of choices and the selection of one according to a desired goal. *Prediction* is necessary for decision-making and requires an *understanding* of the situation and the effect actions may have on it. *Attention* is closely related to perception and may be thought of as the requirement to allocate and direct the sensing resources towards relevant information. *Artificial cognition* in a sensor/processor system attempts to mimic the perception–action cycle to make the best use of system resources for the situation at hand.

Haykin introduced the concept of cognitive radar in [1]; however, the motivation and many of the underlying ideas grew out of the fields of knowledge-aided signal processing [5–7,10–13], agile waveform design [14–21], sensor network management [22–31], sonar ping control [32–34], biomimetic signal processing [35–40], and echoic flow [41,42]. In [2,3], Haykin makes the distinction between traditional feed-forward radar, fully adaptive radar, and truly cognitive radar. While a fully adaptive radar [5,6] may employ feedback and use prior knowledge stored in memory, a cognitive radar predicts the consequences of actions, performs explicit decision-making, learns from the environment, and uses memory to store the learned knowledge [3]. Haykin's approach uses Bayesian filtering as the basis for the artificial perception–action cycle. It provides the processing, perception, and memory inherent in a cognitive system. The Bayesian filter is augmented with a decision-making controller that uses perception from the filter and prediction of future outcomes to determine the next action taken by the sensor. The Bayesian approach has also been used in [17–34,43] for a variety of applications.

Using a first principles approach, we generalize and formalize the work in [1–3,17–34] and develop a general cognitive sensor/processor framework that can be applied to a variety of system objectives, or tasks. We assume the reader has familiarity with detection and estimation theory and Bayesian filtering and quote fundamental results from [44–48] in our development. The reader may consult these references for full derivations of the fundamental material.

The basic framework consists of four components: the scene (target and environment), sensor, processor, and controller. The framework includes the feedback mechanism and optimization criterion used to obtain the next set of sensor data to optimize the performance of the processor. The framework is then specialized for the task of single target tracking. We further develop the tracking framework by partitioning the processor into a detector and a tracker, allowing feedback from the tracker to perform guided detection. This is a key feature of our framework that allows us to cognitively control both the sensor and processor for improved performance.

In order to apply the framework to a specific problem, we must specify the *application model* and the *implementation details*. The application model includes the target motion model, the sensor measurement model, the sensor parameters being controlled and their costs or constraints, and the controller optimization criteria for determining the next set of sensor parameters. The implementation details include the tracker type (e.g., Kalman filter or particle filter) and the method for solving the controller optimization problem. By separating the general principles from the specific application model and implementation details, our formulation provides a general framework that can be applied to a wide variety of tracking problems.

Examples are provided for two applications: (i) a distributed sensor system in which system resources are allocated to optimize tracking estimation performance and (ii) a pulse-Doppler radar system in which the pulse repetition frequency (PRF) is adjusted to optimize tracking estimation performance, while keeping the target from being Doppler-aliased and away from the zero-Doppler clutter. Results are shown for the second application on experimentally collected data using a software defined radar (SDR) system. The examples demonstrate that cognitive systems can achieve significant performance improvements over static, feed-forward systems.

8.2 Framework

8.2.1 Cognitive sensor/processor system framework

There are three components in a typical feed-forward sensor/processor system: (i) the scene, which includes the target and environment, (ii) the sensor that observes the scene and generally consists of a transmitter and receiver, and (iii) the processor that converts the observed data into information about the target to achieve a system objective. A cognitive sensor/processor system includes a fourth component: (iv) the controller that determines the next actions taken by the sensor and processor based on feedback (perception) from the processor, as shown in Figure 8.1.

We now develop the mathematical model that describes the cognitive sensor/processor system. We assume that the objective of the system is to estimate the

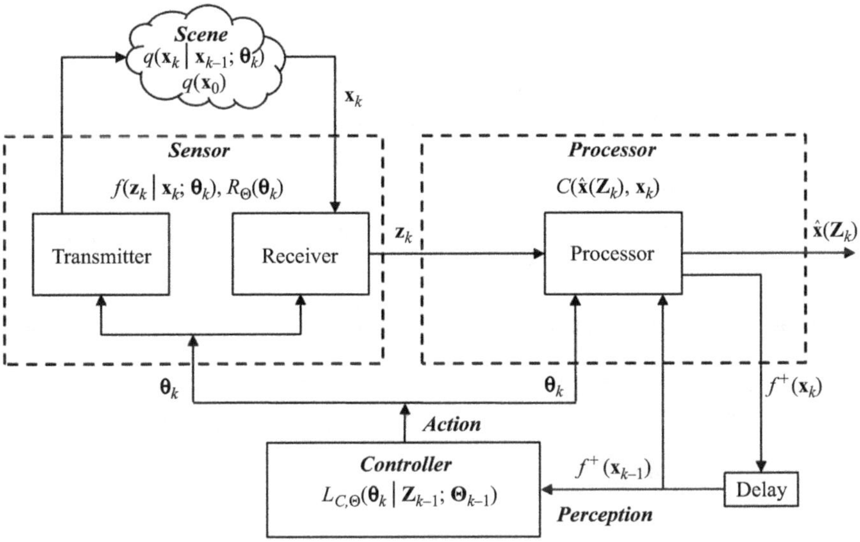

Figure 8.1 Cognitive sensor/processor system

state of a target in the scene. The target state at time t_k is denoted as $\mathbf{x}_k$. The sensor observes the scene and produces a measurement vector $\mathbf{z}_k$ that depends on the target state $\mathbf{x}_k$ and the sensor parameters $\boldsymbol{\theta}_k$. We assume that the estimate of the target state at time t_k is a function of the observations up to time t_k, which in turn depend on the sensor parameters up to time t_k, which we denote as $\mathbf{Z}_k \equiv \{\mathbf{z}_1, \mathbf{z}_2, \ldots, \mathbf{z}_k\}$ and $\boldsymbol{\Theta}_k \equiv \{\boldsymbol{\theta}_1, \boldsymbol{\theta}_2, \ldots, \boldsymbol{\theta}_k\}$, respectively.

We assume a Markov motion model with initial target state probability density function (PDF) $q(\mathbf{x}_0)$ and transition PDF $q(\mathbf{x}_k|\mathbf{x}_{k-1}; \boldsymbol{\theta}_k)$, which represents the probability that a target in state $\mathbf{x}_{k-1}$ will evolve to state $\mathbf{x}_k$. The transition PDF may depend on the sensor parameters. This will occur, for example, when the time difference $\Delta t_k = t_k - t_{k-1}$ depends on the choice of sensor parameters. The measurement model is described by the conditional PDF, or likelihood function, $f(\mathbf{z}_k|\mathbf{x}_k; \boldsymbol{\theta}_k)$. The cost of obtaining an observation and any constraints on the sensor parameters are modelled by the sensor cost function $R_\Theta(\boldsymbol{\theta}_k)$. The processor processes the data and produces an estimate of the target state $\hat{\mathbf{x}}_k(\mathbf{Z}_k)$ by minimizing the expected value of the processor cost function $C(\hat{\mathbf{x}}_k(\mathbf{Z}_k), \mathbf{x}_k)$.

The controller decides on the next value for the sensor parameters $\boldsymbol{\theta}_k$ by minimizing a loss function $L_{C,\Theta}(\cdot)$ that balances the performance of the processor via the processor cost function $C(\cdot,\cdot)$ and the cost of using the sensor via the sensor cost function $R_\Theta(\cdot)$.

For the Markov motion model, the conditional or posterior PDF of $\mathbf{x}_k$ given $\mathbf{Z}_k$ may be obtained from the Bayes–Markov recursion [44,45]:

$$f^-(\mathbf{x}_k) \equiv f(\mathbf{x}_k|\mathbf{Z}_{k-1}; \boldsymbol{\Theta}_k) = \int q(\mathbf{x}_k|\mathbf{x}_{k-1}; \boldsymbol{\theta}_k) f^+(\mathbf{x}_{k-1}) d\mathbf{x}_{k-1} \tag{8.1}$$

$$f^+(\mathbf{x}_k) \equiv f(\mathbf{x}_k|\mathbf{Z}_k; \boldsymbol{\Theta}_k) = \frac{f(\mathbf{z}_k|\mathbf{x}_k; \boldsymbol{\theta}_k) f^-(\mathbf{x}_k)}{\int f(\mathbf{z}_k|\mathbf{x}_k; \boldsymbol{\theta}_k) f^-(\mathbf{x}_k) d\mathbf{x}_k} \tag{8.2}$$

where $f^-(\mathbf{x}_k)$ is the motion-updated predicted PDF and $f^+(\mathbf{x}_k)$ is the information-updated posterior PDF. The recursion is initialized with

$$f^+(\mathbf{x}_0) = q(\mathbf{x}_0) \tag{8.3}$$

The *conditional Bayes risk* is the expected value of the processor cost function with respect to the conditional PDF of $\mathbf{x}_k$ given $\mathbf{Z}_k$

$$R_C^+(\mathbf{Z}_k; \boldsymbol{\Theta}_k) = E_k^+\{C(\hat{\mathbf{x}}(\mathbf{Z}_k), \mathbf{x}_k)\} \tag{8.4}$$

where $E_k^+\{\cdot\}$ denotes expectation with respect to $f^+(\mathbf{x}_k)$. The estimator is found by minimizing the conditional Bayes risk [46,47]:

$$\hat{\mathbf{x}}(\mathbf{Z}_k) = \arg \min_{\hat{\mathbf{x}}(\mathbf{Z}_k)} R_C^+(\mathbf{Z}_k; \boldsymbol{\Theta}_k) \tag{8.5}$$

In the controller, we assume that we have received the observations up to time t_{k-1} and want to find the next set of sensor parameters to optimize the performance of the state estimator that will include the next observation $\mathbf{z}_k$ as well as the previous observations $\mathbf{Z}_{k-1}$. We define the *joint conditional PDF* of $\mathbf{x}_k$ and $\mathbf{z}_k$ conditioned on $\mathbf{Z}_{k-1}$ as:

$$\begin{aligned} f^{\uparrow}(\mathbf{x}_k,\mathbf{z}_k) &\equiv f(\mathbf{x}_k,\mathbf{z}_k|\mathbf{Z}_{k-1};\boldsymbol{\Theta}_k) \\ &= f(\mathbf{z}_k|\mathbf{x}_k;\boldsymbol{\theta}_k)f(\mathbf{x}_k|\mathbf{Z}_{k-1};\boldsymbol{\Theta}_k) \\ &= f(\mathbf{z}_k|\mathbf{x}_k;\boldsymbol{\theta}_k)f^{-}(\mathbf{x}_k) \end{aligned} \tag{8.6}$$

We define the *predicted conditional Bayes risk* for the estimator $\hat{\mathbf{x}}_k(\mathbf{Z}_k)$ by taking the expectation of the processor cost function with respect to the joint conditional PDF $f^{\uparrow}(\mathbf{x}_k,\mathbf{z}_k)$,

$$R_C^{\uparrow}(\boldsymbol{\theta}_k|\mathbf{Z}_{k-1};\boldsymbol{\Theta}_{k-1}) \equiv E_k^{\uparrow}\{C(\hat{\mathbf{x}}(\mathbf{Z}_k),\mathbf{x}_k)\} \tag{8.7}$$

It is important to emphasize that the predicted conditional Bayes risk is a function of the known past observations $\mathbf{Z}_{k-1}$ but not the unknown next observation $\mathbf{z}_k$ since it has been averaged over both $\mathbf{z}_k$ and $\mathbf{x}_k$. It is also a function of all the sensor parameters in $\boldsymbol{\Theta}_k$; however, we separate the dependence on the unknown next sensor parameter $\boldsymbol{\theta}_k$ from the known past sensor parameters $\boldsymbol{\Theta}_{k-1}$ so that we may optimize over $\boldsymbol{\theta}_k$.

The next value of $\boldsymbol{\theta}_k$ is chosen to minimize a loss function that balances the predicted conditional Bayes risk and the sensor cost,

$$L_{C,\Theta}(\boldsymbol{\theta}_k|\mathbf{Z}_{k-1};\boldsymbol{\Theta}_{k-1}) = L\left\{R_C^{\uparrow}(\boldsymbol{\theta}_k|\mathbf{Z}_{k-1};\boldsymbol{\Theta}_{k-1}),R_{\Theta}(\boldsymbol{\theta}_k)\right\} \tag{8.8}$$

The controller optimization problem is then given by:

$$\boldsymbol{\theta}_k = \arg\min_{\boldsymbol{\theta}} L_{C,\Theta}(\boldsymbol{\theta}|\mathbf{Z}_{k-1};\boldsymbol{\Theta}_{k-1}) \tag{8.9}$$

The cognitive sensor/processor system framework described by Figure 8.1 and (8.1)–(8.9) is very general and can be applied to many problems. In the next section, we specialize the framework for systems whose objective is single target tracking.

8.2.2 Cognitive single target tracking

To specialize the framework for single target tracking, we specify the processor cost function and derive the corresponding state estimator and predicted conditional Bayes risk function used by the controller.

For vector parameter estimation, a commonly used cost function is the sum of the squared estimation errors. A more general form, which allows for linear combinations of errors, is given by:[1]

$$C(\hat{\mathbf{x}}(\mathbf{Z}_k),\mathbf{x}_k) = \mathrm{tr}\left\{\mathbf{C}^T[\hat{\mathbf{x}}(\mathbf{Z}_k)-\mathbf{x}_k][\hat{\mathbf{x}}(\mathbf{Z}_k)-\mathbf{x}_k]^T\mathbf{C}\right\} \tag{8.10}$$

[1]When $\mathbf{C}=\mathbf{I}$, this reduces to the sum of the squared errors.

For the cost function in (8.10), the solution to (8.5) is the minimum mean-square error (MMSE) estimator, which is the conditional mean [46,47]:

$$\hat{\mathbf{x}}(\mathbf{Z}_k) = \boldsymbol{\mu}_k^+ \equiv E_k^+\{\mathbf{x}_k\} \tag{8.11}$$

The predicted conditional Bayes risk is given by:

$$R_C^\uparrow(\boldsymbol{\theta}_k|\mathbf{Z}_{k-1};\boldsymbol{\Theta}_{k-1}) = \text{tr}\left\{\mathbf{C}^T\mathbf{S}_k^\uparrow(\boldsymbol{\theta}_k|\mathbf{Z}_{k-1};\boldsymbol{\Theta}_{k-1})\mathbf{C}\right\} \tag{8.12}$$

where

$$\mathbf{S}_k^\uparrow(\boldsymbol{\theta}_k|\mathbf{Z}_{k-1};\boldsymbol{\Theta}_{k-1}) \equiv E_k^\uparrow\left\{[\hat{\mathbf{x}}(\mathbf{Z}_k) - \mathbf{x}_k][\hat{\mathbf{x}}(\mathbf{Z}_k) - \mathbf{x}_k]^T\right\} \tag{8.13}$$

is the predicted conditional mean square error (MSE) matrix.

An alternative estimator that may be easier to calculate is the maximum a posteriori (MAP) estimator, which is the maximum of the posterior PDF. Regardless of the estimation strategy, the MSE of the estimator is the desired performance metric. In most cases, it is not possible to evaluate the MSE analytically or numerically. However, the *Bayesian Cramér–Rao lower bound* (BCRLB), which is the inverse of the *Bayesian information matrix* (BIM), provides a (matrix) lower bound on the MSE matrix of any estimator [47,48] and is usually analytically tractable. It is frequently used as a tool for system analysis in place of the MSE matrix. An overview of the BCRLB and BIM is given in Appendix A.

For tracking applications, application of the BCRLB theory yields the posterior Cramér–Rao lower bound (PCRLB) [48,49]. The PCRLB provides a lower bound on the global MSE that has been averaged over $\mathbf{x}_k$ and $\mathbf{Z}_k$, thus it characterizes tracker performance for all possible data that might have been received. Here we develop a *predicted conditional Cramér–Rao lower bound* (PC-CRLB) to bound the predicted conditional MSE (PC-MSE) matrix in (8.13), which is averaged over the joint PDF of $\mathbf{x}_k$ and $\mathbf{z}_k$ conditioned on $\mathbf{Z}_{k-1}$. The PC-CRLB differs from the PCRLB in that it characterizes performance conditioned on the actual data that has been received.

Following Appendix A, we define the *predicted conditional Bayesian information matrix* (PC-BIM) as:

$$\mathbf{B}_k^\uparrow(\boldsymbol{\theta}_k|\mathbf{Z}_{k-1};\boldsymbol{\Theta}_{k-1}) \equiv -E_k^\uparrow\left\{\Delta_{\mathbf{x}_k}^{\mathbf{x}_k}\ln f^\uparrow(\mathbf{x}_k,\mathbf{z}_k)\right\} \tag{8.14}$$

where

$$\Delta_{\mathbf{x}}^{\mathbf{x}}g(\mathbf{x}) \equiv \nabla_{\mathbf{x}}[\nabla_{\mathbf{x}}g(\mathbf{x})]^T \tag{8.15}$$

denotes the matrix of second-order partial derivatives of a function $g(\mathbf{x})$ with respect to the components of $\mathbf{x}$ and $\nabla_{\mathbf{x}}g(\mathbf{x})$ denotes the gradient,

$$\nabla_{\mathbf{x}}g(\mathbf{x}) \equiv \left[\frac{\partial}{\partial x_1}g(\mathbf{x}) \quad \frac{\partial}{\partial x_2}g(\mathbf{x}) \quad \cdots \quad \frac{\partial}{\partial x_N}g(\mathbf{x})\right]^T \tag{8.16}$$

The PC-CRLB is the inverse of the PC-BIM and has the property:

$$R_C^{\uparrow}(\boldsymbol{\theta}_k|\mathbf{Z}_{k-1};\boldsymbol{\Theta}_{k-1}) \geq \mathrm{tr}\left\{\mathbf{C}^T\mathbf{B}_k^{\uparrow}(\boldsymbol{\theta}_k|\mathbf{Z}_{k-1};\boldsymbol{\Theta}_{k-1})^{-1}\mathbf{C}\right\} \tag{8.17}$$

As shown in Appendix A, the PC-BIM may be expressed as the sum of a prior term and a data term as follows [47,48]:

$$\mathbf{B}_k^{\uparrow}(\boldsymbol{\theta}_k|\mathbf{Z}_{k-1};\boldsymbol{\Theta}_{k-1}) \equiv \mathbf{B}_k^{-}(\boldsymbol{\theta}_k|\mathbf{Z}_{k-1};\boldsymbol{\Theta}_{k-1}) + \mathbf{J}_k^{-}(\boldsymbol{\theta}_k|\mathbf{Z}_{k-1};\boldsymbol{\Theta}_{k-1}) \tag{8.18}$$

From (8.6), the predicted PDF $f^{-}(\mathbf{x}_k)$ serves as the prior PDF of $\mathbf{x}_k$ in the calculation of the PC-BIM, therefore we call the prior term in (8.18) the *predicted information matrix* (PIM). It is defined as:

$$\mathbf{B}_k^{-}(\boldsymbol{\theta}_k|\mathbf{Z}_{k-1};\boldsymbol{\Theta}_{k-1}) \equiv -E_k^{-}\left\{\Delta_{\mathbf{x}_k}^{\mathbf{x}_k}\ln f^{-}(\mathbf{x}_k)\right\} \tag{8.19}$$

where the expectation is with respect to $f^{-}(\mathbf{x}_k)$. Evaluation of the PIM analytically is difficult in many cases. One case where it can be evaluated is for a Gaussian PDF, where it is equal to the inverse of the predicted covariance matrix. Using the intuition gained from the Gaussian case, we approximate the PIM by the inverse of the predicted covariance matrix, which is straightforward to evaluate numerically. In the Kalman filter family of trackers, the predicted PDF is presumed to be Gaussian and the predicted covariance matrix is available from the recursion. In particle filter or grid-based trackers, the predicted covariance matrix can be calculated numerically from the set of particles or grid points and weights that represent the predicted PDF. The approximation is:

$$\mathbf{B}_k^{-}(\boldsymbol{\theta}_k|\mathbf{Z}_{k-1};\boldsymbol{\Theta}_{k-1}) \cong \boldsymbol{\Sigma}_k^{-}(\boldsymbol{\theta}_k|\mathbf{Z}_{k-1};\boldsymbol{\Theta}_{k-1})^{-1} \tag{8.20}$$

where

$$\boldsymbol{\Sigma}_k^{-}(\boldsymbol{\theta}_k|\mathbf{Z}_{k-1};\boldsymbol{\Theta}_{k-1}) \equiv E_k^{-}\left\{\left[\mathbf{x}_k - \boldsymbol{\mu}_k^{-}\right]\left[\mathbf{x}_k - \boldsymbol{\mu}_k^{-}\right]^T\right\} \tag{8.21}$$

and

$$\boldsymbol{\mu}_k^{-} \equiv E_k^{-}\{\mathbf{x}_k\} \tag{8.22}$$

The data term in (8.18) is the expected value of the Fisher information matrix (FIM) with respect to the predicted PDF $f^{-}(\mathbf{x}_k)$. The *expected Fisher information matrix* (EFIM) is given by:

$$\mathbf{J}_k^{-}(\boldsymbol{\theta}_k|\mathbf{Z}_{k-1};\boldsymbol{\Theta}_{k-1}) = E_k^{-}\{\mathbf{J}_{\mathbf{x}}(\mathbf{x}_k;\boldsymbol{\theta}_k)\} \tag{8.23}$$

where $\mathbf{J}(\mathbf{x}_k;\boldsymbol{\theta}_k)$ is the standard FIM [47,48]:

$$\mathbf{J}_{\mathbf{x}}(\mathbf{x}_k;\boldsymbol{\theta}_k) \equiv -E_{\mathbf{z}_k|\mathbf{x}_k;\boldsymbol{\theta}_k}\left\{\Delta_{\mathbf{x}_k}^{\mathbf{x}_k}\ln f(\mathbf{z}_k|\mathbf{x}_k;\boldsymbol{\theta}_k)\right\} \tag{8.24}$$

and the expectation is with respect to $f(\mathbf{z}_k|\mathbf{x}_k;\boldsymbol{\theta}_k)$. We assume that we have an analytical expression for the likelihood function $f(\mathbf{z}_k|\mathbf{x}_k;\boldsymbol{\theta}_k)$ and can evaluate the FIM $\mathbf{J}(\mathbf{x}_k;\boldsymbol{\theta}_k)$. The expectation in (8.23) can be accomplished analytically or numerically.

The expressions in (8.1)–(8.3), (8.11), and (8.17)–(8.24) provide the Bayes–Markov tracking recursion, the state estimate, and the predicted conditional Bayes risk expressions for a cognitive sensor/processor system whose objective is single target tracking. The cognitive single target tracking system recursion is summarized in Table 8.1. It provides a generalization and formalism to the cognitive radar tracking formulations in [1–3,17–34]. By separating the general principles from the specific application and implementation details, our formulation provides a general framework that can be applied to a wide variety of tracking problems.

An important component of our formulation is the PC-BIM (or its inverse, the PC-CRLB) used in the controller. The PC-CRLB is a generalization of the posterior expected covariance matrix used in [3] and more suitable than the PCRLB used

Table 8.1 Cognitive Single Target Tracking Recursion

Initialization	
1	$f^{+}(\mathbf{x}_0) = q(\mathbf{x}_0)$
Controller Optimization	
2	$f^{-}(\mathbf{x}_k;\boldsymbol{\theta}) = \int q(\mathbf{x}_k|\mathbf{x}_{k-1};\boldsymbol{\theta})f^{+}(\mathbf{x}_{k-1})d\mathbf{x}_{k-1}$
3	$\mathbf{B}_k^{-}(\boldsymbol{\theta}|\mathbf{Z}_{k-1};\boldsymbol{\Theta}_{k-1}) \cong \boldsymbol{\Sigma}_k^{-}(\boldsymbol{\theta}|\mathbf{Z}_{k-1};\boldsymbol{\Theta}_{k-1})^{-1}$
4	$\mathbf{J}_{\mathbf{x}}(\mathbf{x}_k;\boldsymbol{\theta}) \equiv -E_{\mathbf{z}_k|\mathbf{x}_k;\boldsymbol{\theta}}\left\{\Delta_{\mathbf{x}_k}^{\mathbf{x}_k} \ln f(\mathbf{z}_k|\mathbf{x}_k;\boldsymbol{\theta})\right\}$
5	$\mathbf{J}_k^{-}(\boldsymbol{\theta}|\mathbf{Z}_{k-1};\boldsymbol{\Theta}_{k-1}) = E_k^{-}\{\mathbf{J}_{\mathbf{x}}(\mathbf{x}_k;\boldsymbol{\theta})\}$
6	$\mathbf{B}_k^{\uparrow}(\boldsymbol{\theta}|\mathbf{Z}_{k-1};\boldsymbol{\Theta}_{k-1}) \equiv \mathbf{B}_k^{-}(\boldsymbol{\theta}|\mathbf{Z}_{k-1};\boldsymbol{\Theta}_{k-1}) + \mathbf{J}_k^{-}(\boldsymbol{\theta}|\mathbf{Z}_{k-1};\boldsymbol{\Theta}_{k-1})$
7	$\boldsymbol{\theta}_k = \arg\min_{\boldsymbol{\theta}} L\left\{\mathbf{B}_k^{\uparrow}(\boldsymbol{\theta}|\mathbf{Z}_{k-1};\boldsymbol{\Theta}_{k-1}), R_{\Theta}(\boldsymbol{\theta})\right\}$
Motion Update	
8	$f^{-}(\mathbf{x}_k) = \int q(\mathbf{x}_k|\mathbf{x}_{k-1};\boldsymbol{\theta}_k)f^{+}(\mathbf{x}_{k-1})d\mathbf{x}_{k-1}$
Measurement	
9	Obtain measurement $\mathbf{z}_k$ according to $\boldsymbol{\theta}_k$
Information Update	
10	$f^{+}(\mathbf{x}_k) = \dfrac{f(\mathbf{z}_k|\mathbf{x}_k;\boldsymbol{\theta}_k)f^{-}(\mathbf{x}_k)}{\int f(\mathbf{z}_k|\mathbf{x}_k;\boldsymbol{\theta}_k)f^{-}(\mathbf{x}_k)d\mathbf{x}_k}$
Track Estimate	
11	Obtain $\hat{\mathbf{x}}(\mathbf{Z}_k)$ from mean or maximum of $f^{+}(\mathbf{x}_k)$

in [1]. The PCRLB characterizes tracker performance that has been averaged over all possible data that might have been received rather than performance that is conditioned on the actual data that has been received. The PCRLB is the proper tool for characterizing the global MSE performance of the tracker in a static system, but the PC-CRLB is more appropriate for cognitive systems because it more accurately reflects the MSE performance of the current realization of the tracker. The current literature is not clear on this point. While most formulations recognize that the tracker performance is characterized by the PC-MSE matrix, the PCRLB is widely used to bound the performance. Exceptions are [20] in which the PCRLB is modified to calculate an EFIM with respect to the predicted density, and [22], in which a 'predictive PCRLB' is developed that is essentially the same as the PC-CRLB developed here. A 'conditional PCRLB' is also developed in [50] which has an elegant recursive form; however, it is a joint bound on the past and current target states, and is therefore a weaker bound than the PC-CRLB and predictive PCRLB. However, we note that in some problems, such as the linear Gaussian model, this point is moot since the PC-MSE matrix, posterior covariance matrix of the tracker, and the various bounds do not depend on the data and therefore are all the same.

The recursions in Table 8.1 are applicable to detection-based and track-before-detect (TBD) tracking systems. In a detection-based system, the sensor performs 'hard detection,' i.e., there is some signal processing of the sensor data that converts the data to a detection surface, which is then thresholded to produce a 'detection' in the form of a measurement in the natural measurement space of the sensor (e.g., angle, range, and/or Doppler). The observation $\mathbf{z}_k$ is the measurement obtained from hard detection processing. In a TBD system, $\mathbf{z}_k$ is the sensor data, and the information update involves computing the likelihood function of the sensor data with respect to the target state vector. This can be computationally intensive but can yield significant performance improvements over a detection-based system. In the next subsection, we extend the recursions to maximum a posteriori penalty function (MAP-PF) tracking systems, in which the processor includes both the detector and tracker; however, the detector performs 'guided' hard detection processing using feedback from the tracker.

8.2.3 Cognitive MAP-PF single target tracking

The MAP-PF methodology [35,51–56], applies to problems in which the likelihood function depends on the target state vector only through a known, possibly nonlinear mapping to the 'natural parameters' of the sensor. For example, in radar systems, the natural parameters of the sensor are typically the angle, range, and/or Doppler frequency, while the state parameters are typically the target position and velocity. The MAP-PF methodology offers reduced computational complexity over TBD systems, while maintaining the performance advantage over detection-based systems. The cognitive MAP-PF tracking system, in which

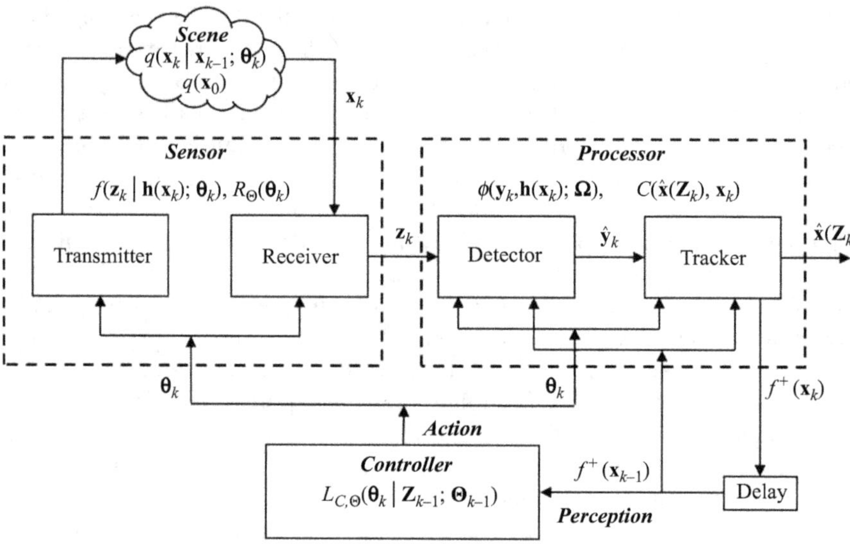

Figure 8.2 Cognitive MAP-PF sensor/processor tracking system

the processor includes the detector and tracker, is shown in Figure 8.2. It is particularly suitable for cognitive systems, as it already contains feedback within the tracking processor.

MAP-PF is a multi-target tracking methodology developed in [51–56] and described in Chapter 6 of [35]. In this approach, the multi-target track estimation problem is formulated directly from the sensor data $\mathbf{z}_k$ using the MAP estimation criterion. The penalty function method of non-linear programming [57] is used to obtain a tractable solution. The result is a two-step estimation process similar to traditional feed-forward detection-based systems, except the processes are coupled via the penalty function and the data association step of traditional multi-target tracking approaches is eliminated. In the detection process, the penalty function uses the current target states to guide the detector to the relevant region of the detection surface. In the track estimation process, the penalty function determines the influence of the detector measurements on the final track estimates by adaptively adjusting the measurement error variance using the FIM. In [58], the MAP-PF tracker was shown to achieve the PCRLB while the standard feed-forward version of the same detector and tracker did not.

Let $\mathbf{y}_k$ denote the natural parameters. They are related to the state parameters by the non-linear mapping

$$\mathbf{y}_k = \mathbf{h}(\mathbf{x}_k) \tag{8.25}$$

Let $f(\mathbf{z}_k|\mathbf{y}_k;\boldsymbol{\theta}_k)$ denote the likelihood function with respect to the natural parameters and let $\mathbf{J}_{\mathbf{y}}(\mathbf{y}_k;\boldsymbol{\theta}_k)$ denote the FIM of the natural parameters, which is defined as:

$$\mathbf{J}_{\mathbf{y}}(\mathbf{y}_k;\boldsymbol{\theta}_k) \equiv -E_{\mathbf{z}_k|\mathbf{y}_k;\boldsymbol{\theta}_k}\left\{\Delta_{\mathbf{y}_k}^{\mathbf{y}_k} \ln f(\mathbf{z}_k|\mathbf{y}_k;\boldsymbol{\theta}_k)\right\} \tag{8.26}$$

The likelihood function and FIM with respect to the state parameters are given by [47]:

$$f(\mathbf{z}_k|\mathbf{x}_k;\boldsymbol{\theta}_k) = f(\mathbf{z}_k|\mathbf{y}_k = \mathbf{h}(\mathbf{x}_k);\boldsymbol{\theta}_k) \tag{8.27}$$

$$\mathbf{J}_{\mathbf{x}}(\mathbf{x}_k;\boldsymbol{\theta}_k) = \mathbf{H}(\mathbf{x}_k)^T\mathbf{J}_{\mathbf{y}}(\mathbf{h}(\mathbf{x}_k);\boldsymbol{\theta}_k)\mathbf{H}(\mathbf{x}_k) \tag{8.28}$$

where $\mathbf{H}(\mathbf{x}_k)$ is the Jacobian matrix, defined as

$$\mathbf{H}(\mathbf{x}_k) \equiv \left[\nabla_{\mathbf{x}}\,\mathbf{h}(\mathbf{x})^T\right]^T\Big|_{\mathbf{x}=\mathbf{x}_k} \tag{8.29}$$

Following [35], the MAP-PF motion update is the same as in (8.1) and the MAP-PF information update has the form:

$$f^+(\mathbf{x}_k) \equiv f(\mathbf{x}_k|\mathbf{Z}_k;\boldsymbol{\Theta}_k) = \frac{f(\mathbf{z}_k|\mathbf{h}(\mathbf{x}_k);\boldsymbol{\theta}_k)f^-(\mathbf{x}_k)}{\int f(\mathbf{z}_k|\mathbf{h}(\mathbf{x}_k);\boldsymbol{\theta}_k)f^-(\mathbf{x}_k)d\mathbf{x}_k} \tag{8.30}$$

The MAP-PF algorithm breaks up the information update into three steps using a penalty function. The penalty function $\phi(\mathbf{y},\mathbf{h}(\mathbf{x}))$ has the property that it is equal to zero when $\mathbf{y} = \mathbf{h}(\mathbf{x})$ and becomes smaller (more negative) as the distance between $\mathbf{y}$ and $\mathbf{h}(\mathbf{x})$ increases. For example, a quadratic penalty function is:

$$\phi(\mathbf{y},\mathbf{h}(\mathbf{x});\Omega) \equiv -\frac{1}{2}[\mathbf{y}-\mathbf{h}(\mathbf{x})]^T\Omega^{-1}[\mathbf{y}-\mathbf{h}(\mathbf{x})] \tag{8.31}$$

where Ω is a matrix chosen to weight the components of the penalty function in some desirable manner. The three-step MAP-PF information update is given by [35]:

$$\hat{\mathbf{x}}_k^- = \arg\max_{\mathbf{x}_k} f^-(\mathbf{x}_k) \tag{8.32}$$

$$\hat{\mathbf{y}}_k = \arg\max_{\mathbf{y}} \ln f(\mathbf{z}_k|\mathbf{y};\boldsymbol{\theta}_k) + \phi(\mathbf{y},\mathbf{h}(\hat{\mathbf{x}}_k^-);\Omega) \tag{8.33}$$

$$f^+(\mathbf{x}_k) = \frac{\exp\{\phi(\hat{\mathbf{y}}_k,\mathbf{h}(\mathbf{x}_k);\Omega)\}f^-(\mathbf{x}_k)}{\int \exp\{\phi(\hat{\mathbf{y}}_k,\mathbf{h}(\mathbf{x}_k);\Omega)\}f^-(\mathbf{x}_k)d\mathbf{x}_k} \tag{8.34}$$

In the first step, the MAP estimate of the predicted PDF is found. Depending on the implementation, it may be easier to find the MMSE estimate, which is the mean of the predicted PDF, instead.

In the second step, the optimization problem in (8.33) is a penalized maximum likelihood (ML) problem. If the second term in (8.33) had the form $\ln f(\mathbf{y})$, it would be a MAP estimation problem. Thus the penalty function can be interpreted as a prior term in a MAP estimation problem. In a traditional feed-forward detection-based tracking system, the optimal detector would solve the standard ML problem (i.e., (8.33) without the penalty function) to get the detector measurement. In MAP-PF, the penalty function restricts the detector estimate to be in the vicinity of where the tracker predicts it to be, hence MAP-PF is performing 'guided' detection. By specifying a quadratic penalty function, we are implicitly modelling the prior distribution of $\mathbf{y}_k$ as Gaussian with mean $\mathbf{h}(\hat{\mathbf{x}}_k^-)$ and covariance matrix Ω. We have some flexibility in choosing Ω, and a logical choice would be the covariance matrix of the predicted PDF of $\mathbf{y}_k$ obtained from a transformation of the predicted PDF $f^-(\mathbf{x}_k)$. Using a locally linear approximation of the function $\mathbf{h}(\mathbf{x})$ at the point $\hat{\mathbf{x}}_k^-$, we choose Ω to be the predicted covariance matrix of $\mathbf{y}_k$, which is approximately given by:

$$\Omega_{MAP,k} = \boldsymbol{\Sigma}_{\mathbf{y}_k}^- \cong \mathbf{H}(\hat{\mathbf{x}}_k^-)\boldsymbol{\Sigma}_k^-\mathbf{H}(\hat{\mathbf{x}}_k^-)^T \tag{8.35}$$

The third step in (8.34) looks like a standard information update with $\hat{\mathbf{y}}_k$ acting as the measurement vector and the exponential of the penalty function, $\exp\{\phi(\hat{\mathbf{y}}_k, \mathbf{h}(\mathbf{x}_k); \Omega)\}$, acting as the measurement likelihood function $f(\hat{\mathbf{y}}_k|\mathbf{x}_k; \boldsymbol{\theta}_k)$. Here the quadratic penalty function is implicitly modelling $\hat{\mathbf{y}}_k$ as Gaussian with mean $\mathbf{h}(\mathbf{x}_k)$ and covariance matrix Ω. As in [35,51–56], we choose Ω to be the inverse of the FIM of the natural parameters, $\mathbf{J}_\mathbf{y}(\mathbf{y}_k; \boldsymbol{\theta}_k)$. Calculation of the FIM often requires knowledge of the true value of $\mathbf{y}_k$; however, we can obtain a reasonably accurate approximation to the FIM by replacing $\mathbf{y}_k$ with an estimate. The transformation of the predicted state estimate $\mathbf{h}(\hat{\mathbf{x}}_k^-)$ is a less volatile estimate than the current measurement $\hat{\mathbf{y}}_k$, therefore we evaluate the FIM at $\mathbf{h}(\hat{\mathbf{x}}_k^-)$. Thus, for the information update we choose

$$\Omega_{I,k}^{-1} = \mathbf{J}_\mathbf{y}(\mathbf{h}(\hat{\mathbf{x}}_k^-); \boldsymbol{\theta}_k) \tag{8.36}$$

The cognitive single target MAP-PF recursion is summarized in Table 8.2.

8.2.4 Summary

We have developed a general framework for a cognitive sensor/processor tracking system. The framework is based on the *perception–action cycle* and includes *sensing* in the sensor; *processing* in the detector and tracker; *perception* in the conversion of sensor data to the posterior PDF of the state vector; *memory* of all the past data in the posterior PDF; *attention* in the penalty function of the guided adaptive detector, which focuses the detector on the relevant region of the detection surface; *prediction* in the PC-BIM, which predicts the performance of the next

Table 8.2 Cognitive single target MAP-PF tracking recursion

Initialization	
1	$f^+(\mathbf{x}_0) = q(\mathbf{x}_0)$
Controller Optimization	
2	$f^-(\mathbf{x}_k;\boldsymbol{\theta}) = \int q(\mathbf{x}_k\|\mathbf{x}_{k-1};\boldsymbol{\theta})f^+(\mathbf{x}_{k-1})d\mathbf{x}_{k-1}$
3	$\mathbf{B}_k^-(\boldsymbol{\theta}\|\mathbf{Z}_{k-1};\boldsymbol{\Theta}_{k-1}) \cong \boldsymbol{\Sigma}_k^-(\boldsymbol{\theta}\|\mathbf{Z}_{k-1};\boldsymbol{\Theta}_{k-1})^{-1}$
4	$\mathbf{J}_\mathbf{y}(\mathbf{y}_k;\boldsymbol{\theta}) \equiv -E_{\mathbf{z}_k\|\mathbf{y}_k;\boldsymbol{\theta}}\left\{\Delta_{\mathbf{y}_k}^{\mathbf{y}_k} \ln f(\mathbf{z}_k\|\mathbf{y}_k;\boldsymbol{\theta})\right\}$
5	$\mathbf{J}_\mathbf{x}(\mathbf{x}_k;\boldsymbol{\theta}) = \mathbf{H}(\mathbf{x}_k)^T\mathbf{J}_\mathbf{y}(\mathbf{h}(\mathbf{x}_k);\boldsymbol{\theta})\mathbf{H}(\mathbf{x}_k)$
6	$\mathbf{J}_k^-(\boldsymbol{\theta}\|\mathbf{Z}_{k-1};\boldsymbol{\Theta}_{k-1}) = E_k^-\{\mathbf{J}_\mathbf{x}(\mathbf{x}_k;\boldsymbol{\theta})\}$
7	$\mathbf{B}_k^{\uparrow}(\boldsymbol{\theta}\|\mathbf{Z}_{k-1};\boldsymbol{\Theta}_{k-1}) \equiv \mathbf{B}_k^-(\boldsymbol{\theta}\|\mathbf{Z}_{k-1};\boldsymbol{\Theta}_{k-1}) + \mathbf{J}_k^-(\boldsymbol{\theta}\|\mathbf{Z}_{k-1};\boldsymbol{\Theta}_{k-1})$
8	$\boldsymbol{\theta}_k = \arg\min_{\boldsymbol{\theta}} L\left\{\mathbf{B}_k^{\uparrow}(\boldsymbol{\theta}\|\mathbf{Z}_{k-1};\boldsymbol{\Theta}_{k-1}), R_\Theta(\boldsymbol{\theta})\right\}$
Motion Update	
9	$f^-(\mathbf{x}_k) = \int q(\mathbf{x}_k\|\mathbf{x}_{k-1};\boldsymbol{\theta}_k)f^+(\mathbf{x}_{k-1})d\mathbf{x}_{k-1}$
Measurement	
10	Obtain measurement $\mathbf{z}_k$ according to $\boldsymbol{\theta}_k$
11	Obtain $\hat{\mathbf{x}}_k^-$ from mean or maximum of $f^-(\mathbf{x}_k)$
12	$\Omega_{MAP,k} = \mathbf{H}(\hat{\mathbf{x}}_k^-)\boldsymbol{\Sigma}_k^-\mathbf{H}(\hat{\mathbf{x}}_k^-)^T$
13	$\hat{\mathbf{y}}_k = \arg\max_{\mathbf{y}} \ln f(\mathbf{z}_k\|\mathbf{y};\boldsymbol{\theta}_k) + \phi(\mathbf{y},\mathbf{h}(\hat{\mathbf{x}}_k^-);\Omega_{MAP,k})$
Information Update	
14	$\Omega_{I,k}^{-1} = \mathbf{J}_\mathbf{y}(\mathbf{h}(\hat{\mathbf{x}}_k^-);\boldsymbol{\theta}_k)$
15	$f^+(\mathbf{x}_k) = \dfrac{\exp\{\phi(\hat{\mathbf{y}}_k,\mathbf{h}(\mathbf{x}_k);\Omega_{I,k})\}f^-(\mathbf{x}_k)}{\int \exp\{\phi(\hat{\mathbf{y}}_k,\mathbf{h}(\mathbf{x}_k);\Omega_{I,k})\}f^-(\mathbf{x}_k)d\mathbf{x}_k}$
Track Estimate	
16	Obtain $\hat{\mathbf{x}}(\mathbf{Z}_k)$ from mean or maximum of $f^+(\mathbf{x}_k)$

measurement; and *decision-making* in the controller, which decides on the next values for the sensor parameters based on the predicted performance.

For a specific application, we would then need to specify the components of the state vector, the motion and measurement models, the sensor parameters being controlled, and the form of the controller loss function. Finally, we would need to specify the implementation details that include the type of tracker used to implement the Bayes–Markov recursion and the method for solving the controller optimization problem. We now give two examples.

8.3 Distributed sensor example

In this section, we demonstrate how the general framework can be applied to a distributed sensor management problem [59–61]. We first define the model for this problem, then discuss the implementation, and finally provide some simulation results.

8.3.1 Model

In this example, the target state is the two-dimensional position in the xy-plane,

$$\mathbf{x}_k = \begin{bmatrix} x_k \\ y_k \end{bmatrix} \tag{8.37}$$

The sensor consists of N independent sensors that produce noisy target angle-of-arrival estimates. Let (x_n, y_n) denote the position of the nth sensor. The angle and range to the target are:

$$\phi_n(\mathbf{x}_k) = \arctan\left(\frac{y_k - y_n}{x_k - x_n}\right) \tag{8.38}$$

$$r_n(\mathbf{x}_k) = \sqrt{(x_k - x_n)^2 + (y_k - y_n)^2} \tag{8.39}$$

A typical scenario is shown in Figure 8.3.

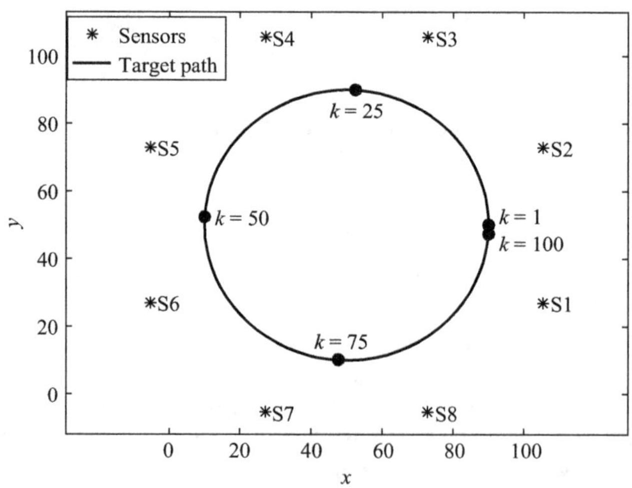

Figure 8.3 Sensor placement and target path

We assume that there is a fixed amount of observation time that can be allocated among the sensors. Let $\theta_{n,k}$ denote the fraction of observation time on the nth sensor. The sensor parameter vector is defined as

$$\boldsymbol{\theta}_k \equiv \begin{bmatrix} \theta_{1,k} \\ \vdots \\ \theta_{N,k} \end{bmatrix} \tag{8.40}$$

The sensor cost function reflects the constraints on the sensor parameters: each component must be nonnegative and their sum must be less than or equal to one. The cost is zero when the constraints are met and infinite otherwise:

$$R_\Theta(\boldsymbol{\theta}_k) = \begin{cases} 0 & \theta_{n,k} \geq 0; n = 1, \ldots, N, \ \sum_{n=1}^{N} \theta_{n,k} \leq 1 \\ \infty & \text{otherwise} \end{cases} \tag{8.41}$$

The motion model consists of a prior PDF $q(\mathbf{x}_0)$ and a transition PDF $q(\mathbf{x}_k|\mathbf{x}_{k-1})$ that does not depend on the sensor parameters. We assume an initial target state distribution of the form $\mathbf{x}_0 \sim N(\boldsymbol{\mu}_0, \boldsymbol{\Sigma}_0)$, therefore

$$q(\mathbf{x}_0) = \frac{1}{2\pi\sqrt{|\boldsymbol{\Sigma}_0|}} \exp\left\{ -\frac{1}{2} [\mathbf{x}_0 - \boldsymbol{\mu}_0]^T \boldsymbol{\Sigma}_0^{-1} [\mathbf{x}_0 - \boldsymbol{\mu}_0] \right\} \tag{8.42}$$

We assume a 'nearly constant' motion model

$$\mathbf{x}_k = \mathbf{x}_{k-1} + \mathbf{v}_k \tag{8.43}$$

where $\mathbf{v}_k \sim N(\mathbf{0}, \mathbf{Q})$ and $\mathbf{Q} = \sigma_x^2 \mathbf{I}$, so that the transition PDF is

$$q(\mathbf{x}_k|\mathbf{x}_{k-1}) = \frac{1}{2\pi\sigma_x^2} \exp\left\{ -\frac{1}{2\sigma_x^2} \|\mathbf{x}_k - \mathbf{x}_{k-1}\|^2 \right\} \tag{8.44}$$

In this model, transition PDF does not depend on $\boldsymbol{\theta}_k$, so we have dropped it from the notation.

The measurement model is characterized by the likelihood function. When $\theta_{n,k} > 0$, the observation at the nth sensor is the noisy angle estimate

$$z_{n,k} = \phi_n(\mathbf{x}_k) + w_{n,k} \tag{8.45}$$

where $w_{n,k}$ is zero-mean additive white Gaussian noise whose variance is inversely proportional to the observation time on that sensor and whose value is truncated so that $z_{n,k}$ is restricted to the 180° (π radians) sector $[\phi_{n,0}, \phi_{n,0} + \pi)$ looking into the field of view from that sensor. The likelihood function is the truncated Gaussian PDF:

$$\begin{aligned} f_n(z_{n,k}|\mathbf{x}_k; \theta_{n,k}) &= c_{n,k} \exp\left\{ -\frac{1}{2\sigma_n^2/\theta_{n,k}} (z_{n,k} - \phi_n(\mathbf{x}_k))^2 \right\} \\ z_{n,k} &\in [\phi_{n,0}, \phi_{n,0} + \pi) \end{aligned} \tag{8.46}$$

where $c_{n,k}$ is the normalization constant,

$$c_{n,k}^{-1} = \int_{\phi_{n,0}}^{\phi_{n,0}+\pi} \exp\left\{-\frac{1}{2\sigma_n^2/\theta_{n,k}}\left(z_{n,k} - \phi_n(\mathbf{x}_k)\right)^2\right\} dz \tag{8.47}$$

When $\theta_{n,k} = 0$, no observation is made and the truncated Gaussian likelihood function becomes the uniform likelihood function:

$$f_n\left(z_{n,k}|\mathbf{x}_k; \theta_{n,k} = 0\right) = \frac{1}{\pi}; \quad z_{n,k} \in \left[\phi_{n,0}, \phi_{n,0} + \pi\right) \tag{8.48}$$

Let $\mathbf{z}_k$ denote the N-dimensional observation vector over all sensors,

$$\mathbf{z}_k \equiv \begin{bmatrix} z_{1,k} \\ \vdots \\ z_{N,k} \end{bmatrix} \tag{8.49}$$

The joint likelihood function of the observation vector $\mathbf{z}_k$ is the product of the individual likelihood functions:

$$f(\mathbf{z}_k|\mathbf{x}_k; \boldsymbol{\theta}_k) = \prod_{n=1}^{N} f_n\left(z_{n,k}|\mathbf{x}_k; \theta_{n,k}\right) \tag{8.50}$$

The FIM can be evaluated from the likelihood function. It is the sum of the FIMs for each sensor individually

$$\mathbf{J}(\mathbf{x}_k; \boldsymbol{\theta}_k) = \sum_{n=1}^{N} \mathbf{J}_n\left(\mathbf{x}_k; \theta_{n,k}\right) \tag{8.51}$$

where $\mathbf{J}_n\left(\mathbf{x}_k; \theta_{n,k}\right)$ is the FIM for the nth sensor. It has the form:

$$\begin{aligned} \mathbf{J}_n\left(\mathbf{x}_k; \theta_{n,k}\right) &= -E_{z_{n,k}|\mathbf{x}_k;\theta_{n,k}}\left\{\Delta_{\mathbf{x}_k}^{\mathbf{x}_k} \ln f_n\left(z_{n,k}|\mathbf{x}_k; \theta_{n,k}\right)\right\} \\ &= \frac{\theta_{n,k}}{\sigma_n^2 r_n(\mathbf{x}_k)^2} \mathbf{u}_n(\mathbf{x}_k)\mathbf{u}_n(\mathbf{x}_k)^T \end{aligned} \tag{8.52}$$

where $\mathbf{u}_n(\mathbf{x}_k)$ is the unit vector perpendicular to the vector pointing from the sensor to the target,

$$\mathbf{u}_n(\mathbf{x}_k) = \frac{1}{r_n(\mathbf{x}_k)} \begin{bmatrix} -(y_k - y_n) \\ (x_k - x_n) \end{bmatrix} \tag{8.53}$$

The nth sensor FIM in (8.52) can be expressed as the product of the sensor parameter $\theta_{n,k}$ and a unit sensor FIM, $\mathbf{J}_n(\mathbf{x}_k)$, defined as the sensor FIM when $\theta_{n,k} = 1$,

$$\mathbf{J}_n(\mathbf{x}_k) \equiv \frac{1}{\sigma_n^2 r_n(\mathbf{x}_k)^2} \mathbf{u}_n(\mathbf{x}_k)\mathbf{u}_n(\mathbf{x}_k)^T \tag{8.54}$$

The total FIM can then be expressed as a weighted sum of the unit sensor FIMs

$$\mathbf{J}(\mathbf{x}_k;\boldsymbol{\theta}_k) = \sum_{n=1}^{N} \theta_{n,k}\mathbf{J}_n(\mathbf{x}_k) \tag{8.55}$$

Consequently, the EFIM can also be expressed as a weighted sum of the unit sensor EFIMs

$$\mathbf{J}_k^-(\boldsymbol{\theta}_k|\mathbf{Z}_{k-1};\boldsymbol{\Theta}_{k-1}) = \sum_{n=1}^{N} \theta_{n,k}\mathbf{J}_{n,k}^-(\mathbf{Z}_{k-1};\boldsymbol{\Theta}_{k-1}) \tag{8.56}$$

where the unit sensor EFIM is defined as

$$\mathbf{J}_{n,k}^-(\mathbf{Z}_{k-1};\boldsymbol{\Theta}_{k-1}) \equiv E^-\{\mathbf{J}_n(\mathbf{x}_k)\} \tag{8.57}$$

The PC-BIM is therefore the PIM plus a weighted sum of the unit sensor EFIMS

$$\mathbf{B}_k^\uparrow(\boldsymbol{\theta}_k|\mathbf{Z}_{k-1};\boldsymbol{\Theta}_{k-1}) = \mathbf{B}_k^-(\mathbf{Z}_{k-1};\boldsymbol{\Theta}_{k-1}) + \sum_{n=1}^{N} \theta_{n,k}\mathbf{J}_{n,k}^-(\mathbf{Z}_{k-1};\boldsymbol{\Theta}_{k-1}) \tag{8.58}$$

The controller optimization problem is to minimize the trace of the PC-CRLB subject to the constraints imposed by the sensor cost function:

$$\begin{aligned}\boldsymbol{\theta}_k &= \arg\min_{\boldsymbol{\theta}} \operatorname{tr}\left\{\mathbf{B}_k^\uparrow(\boldsymbol{\theta}|\mathbf{Z}_{k-1};\boldsymbol{\Theta}_{k-1})^{-1}\right\} \\ &\quad \text{s.t. } \theta_n \geq 0; \quad n = 1,\ldots,N \sum_{n=1}^{N} \theta_n \leq 1\end{aligned} \tag{8.59}$$

8.3.2 Implementation

The implementation details include the tracking algorithm used to implement the Bayes–Markov recursions and the method for solving the controller optimization problem. The tracker is implemented on a discrete 2D grid and expectations and integrals in the motion and information updates are implemented as discrete sums. The controller optimization problem in (8.59) does not have a convenient closed form solution due to the non-negativity constraints, therefore we develop an iterative solution. Full details of the implementation are given in [59].

8.3.3 Simulation results

The locations of the $N = 8$ sensors and the target path are shown in Figure 8.3. The parameter space is the square [0,100] × [0,100]. The angles that define the field of view for the sensors are $\phi_{1,0} = \phi_{2,0} = \pi/2$, $\phi_{3,0} = \phi_{4,0} = -\pi$, $\phi_{5,0} = \phi_{6,0} = -\pi/2$, $\phi_{7,0} = \phi_{8,0} = 0$. The unit measurement variance of the sensors is $\sigma_n^2 = (0.04\pi)^2$. The prior PDF has mean $\boldsymbol{\mu}_0 = [50, 50]^T$ and covariance matrix $\boldsymbol{\Sigma}_0 = (30)^2\mathbf{I}$.

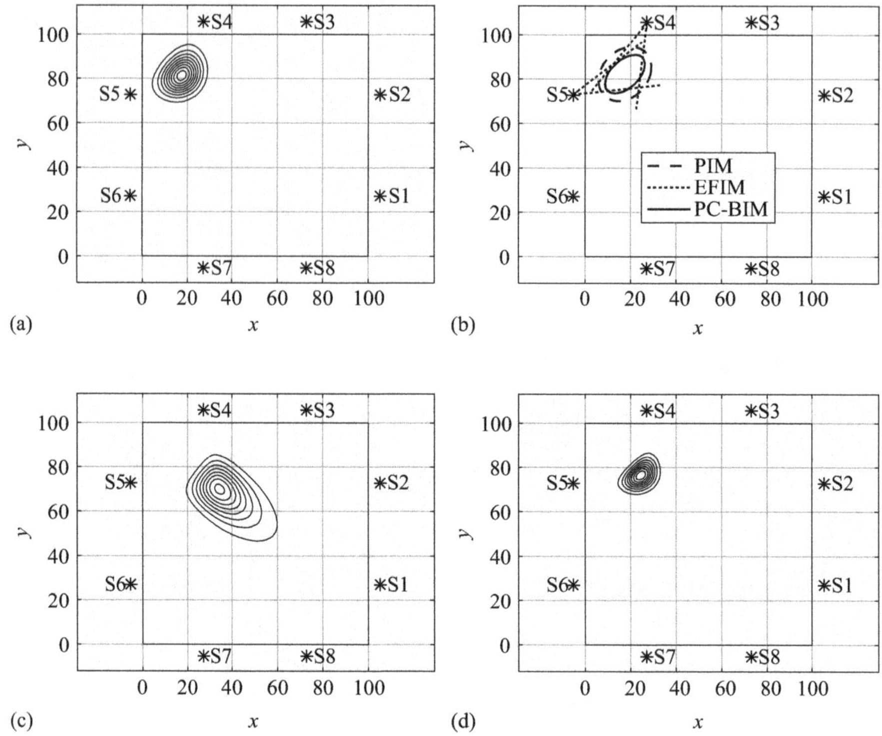

Figure 8.4 Cognitive tracking operation at $k = 39$. (a) Predicted PDF $f(\mathbf{x}_k|\mathbf{Z}_{k-1})$, (b) PC-BIM, $\theta_5 = 0.53$, $\theta_4 = 0.47$, (c) LF $f(z_k|\mathbf{x}_k)$, $\theta_5 = 0.53$, $\theta_4 = 0.47$ and (d) posterior PDF $f(\mathbf{x}_k|\mathbf{Z}_k)$

The transition covariance matrix is $\mathbf{Q} = (4)^2\mathbf{I}$. There are 101 grid points in each dimension with a spacing of one unit for a total of $M = 10{,}201$ grid points. We set the components of $\boldsymbol{\theta}_0$ to be equal, i.e., $\theta_{n,0} = 1/8; n = 1, \ldots, 8$.

Figure 8.4 illustrates the controller optimization at time $k = 39$. Prior to obtaining the next sample, the predicted PDF is shown in Figure 8.4(a). The PIM is equal to the inverse of the predicted covariance matrix and is represented by the 2σ (dashed line) error ellipse shown in Figure 8.4(b). The controller allocates resources to sensors 4 and 5 with $\theta_4 = 0.47$ and $\theta_5 = 0.53$. This produces the 2σ resolution sectors (dotted lines) determined by the sensor EFIMs, and the 2σ (solid line) error ellipse of the PC-BIM. The sensor measurements are obtained and the resulting likelihood function is shown in Figure 8.4(c). The posterior PDF is shown in Figure 8.4(d).

Figure 8.5 shows the optimal allocation of sensor resources over time. Before any data is received, the optimal allocation is equal across all eight sensors. After a few iterations, the controller is able to extract enough information from the data to focus resources on sensors 1 and 2, which are closest to the target. As the target moves past sensors 2–8, the allocation shifts to an approximately equal allocation between the closest two sensors.

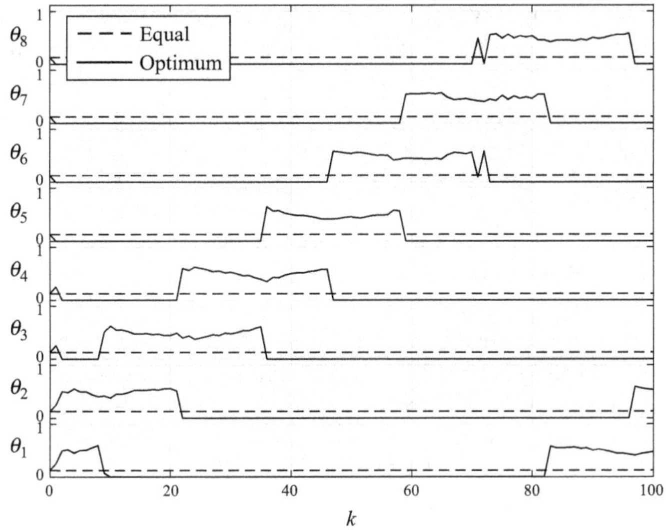

Figure 8.5 Optimal sensor parameters vs time

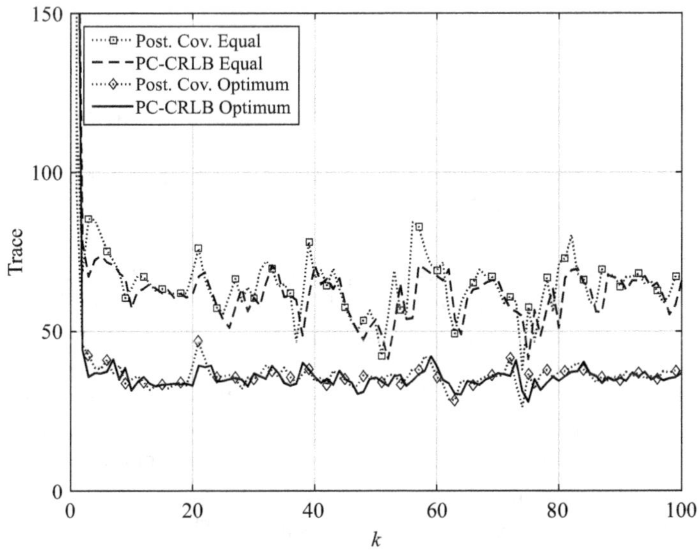

Figure 8.6 Trace of PC-CRLB vs time

The PC-CRLB represents the theoretical performance before a measurement is taken. Figure 8.6 shows the trace of the PC-CRLB over time as well as the actual trace of the posterior covariance matrix after the data is received. They are very nearly the same and decrease rapidly as the first few measurements are taken and then remain fairly constant. We also show the traces of the PC-CRLB and posterior covariance matrix when the sensor allocation is not optimized, but remains constant throughout the track. Both are higher for the static system, thus the cognitive system is making better use of the system resources and achieves better tracking performance. This simple example demonstrates the potential for significant performance gains from using a cognitive sensor/processor system.

8.4 Software defined radar example

We now demonstrate the performance of a pulse-Doppler radar system in which the PRF is adjusted to optimize tracking performance, while keeping the target from being Doppler-aliased and away from the zero-Doppler clutter. Results are shown on experimentally collected data using an SDR system [62–69] available at the ElectroScience Laboratory of The Ohio State University. Some of these results have been reported in a conference paper [70].

The experimental configuration for the data collection is described in [66–68]. The system operates at frequency $f_c = 8$ GHz and transmits linear frequency modulation (LFM) pulses with bandwidth $B = 500$ MHz, therefore the range resolution is 0.3 m. The human target walks back and forth over a 5 m range and exhibits varying velocities and a fluctuating radar cross section (RCS) as it moves. The objective is to track the range and velocity of the target based on radar measurements in range/Doppler (RD) space.

12,200 pulses of data were collected with a fixed PRF of 800 Hz over 15.25 s. Cognitive processing is performed ‘after the fact’ on the stored data, with the PRF being adjusted artificially by downsampling the pulses by a factor of one to four to achieve PRFs of 800, 400, 267, and 200 Hz.

To visualize the data, we performed Hamming-windowed Doppler processing to produce an RD surface on overlapping 128-pulse coherent processing intervals (CPIs) spaced 0.1 s apart. There is strong clutter in the zero-Doppler bin and a few surrounding bins, therefore we perform ‘clutter-nulling’ by simply discarding the data in these bins. Range and Doppler cuts at the peak RD response vs time are shown in Figures 8.7 and 8.8. Ranges are shown with a minimum reference range, R_0, subtracted. The signal-to-noise ratio (SNR) is higher when the target is closer in range and lower when it is farther away. The peak value in Doppler corresponds to the motion of the torso and there is significant micro-Doppler structure due to the motion of the arms and legs of the human subject. The peak Doppler is largest when the target is at mid-range and passes through zero when the target is at the end ranges and turning around. The peak Doppler has a maximum value of about 135 Hz, thus we have room to lower the PRF without causing Doppler aliasing.

For the cognitive processing, each non-overlapping scan consists of a CPI with $N_p = 128$ pulses at the designated PRF, thus the scan interval varies with the PRF,

$$\Delta t_k = \frac{N_p}{PRF_k} \tag{8.60}$$

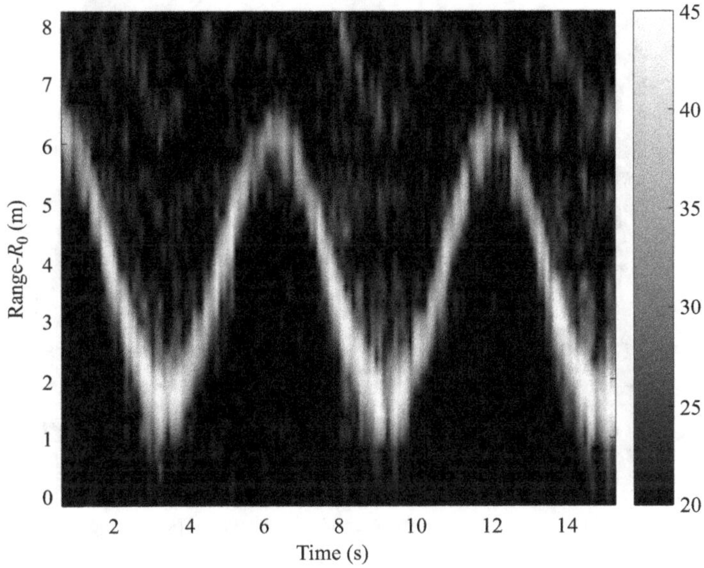

Figure 8.7 RD surface (dB) vs range and time

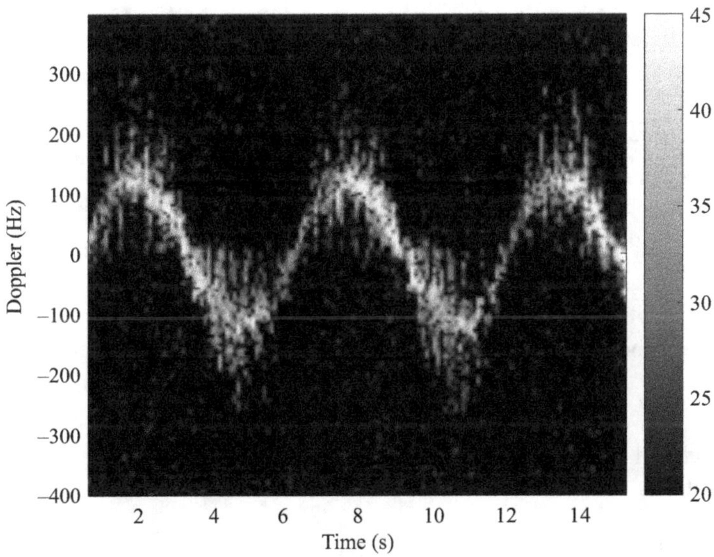

Figure 8.8 RD surface (dB) vs Doppler and time

A typical RD surface for 128 pulses at a PRF of 400 Hz is shown in Figure 8.9. There is strong clutter in the zero-Doppler bin and a few surrounding bins. The target is spread in both range and Doppler, as we see some aliasing of the micro-Doppler returns. We are only interested in tracking the single strongest peak and

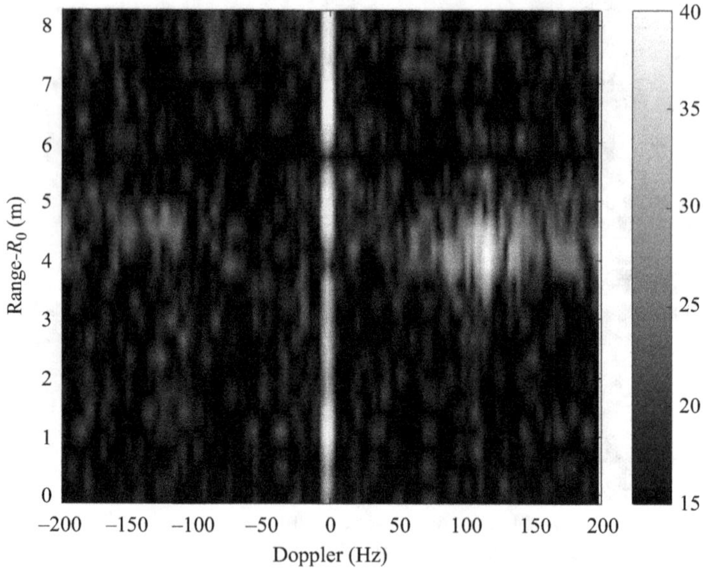

Figure 8.9 Typical RD surface (dB)

the penalty function will be critical in suppressing distracting micro-Doppler and clutter returns.

8.4.1 Model

The target kinematic parameters that we wish to estimate and track are the range r_k and velocity v_k. Let $\boldsymbol{\rho}_k$ denote the kinematic parameter vector at time t_k,

$$\boldsymbol{\rho}_k \equiv [r_k \quad v_k]^T \tag{8.61}$$

The SDR observes the target via the natural radar parameters of range r_k and Doppler frequency f_k. Let $\boldsymbol{\eta}_k$ denote the vector of natural radar parameters,

$$\boldsymbol{\eta}_k \equiv [r_k \quad f_k]^T \tag{8.62}$$

The natural parameters are related to the kinematic parameters by the linear transformation

$$\boldsymbol{\eta}_k = \mathbf{T}\boldsymbol{\rho}_k \tag{8.63}$$

where

$$\mathbf{T} = \begin{bmatrix} 1 & 0 \\ 0 & -\dfrac{2f_c}{c} \end{bmatrix} \tag{8.64}$$

where c is the speed of light.

To calculate the FIM, we need to know the post-processing SNR, therefore we track the SNR (in decibels (dB)) along with the kinematic parameters. Let s_k denote the post-processing SNR in dB:

$$s_k \equiv 10\log_{10}(SNR_k) \tag{8.65}$$

We augment the kinematic and natural parameters with s_k to obtain:

$$\mathbf{x}_k \equiv \begin{bmatrix} r_k & v_k & s_k \end{bmatrix}^T \tag{8.66}$$

$$\mathbf{y}_k \equiv \begin{bmatrix} r_k & f_k & s_k \end{bmatrix}^T \tag{8.67}$$

The natural parameter vector is related to the target state vector by the mapping

$$\mathbf{h}(\mathbf{x}_k) \equiv \mathbf{H}\mathbf{x}_k \tag{8.68}$$

where

$$\mathbf{H} = \begin{bmatrix} 1 & 0 & 0 \\ 0 & -2f_c/c & 0 \\ 0 & 0 & 1 \end{bmatrix} \tag{8.69}$$

The sensor parameter θ is the PRF. The sensor cost function restricts the allowable PRFs to a finite set. The cost is zero when the PRF is one of the pre-defined values and infinite otherwise:

$$R_\Theta(\theta) = \begin{cases} 0 & \theta \in \{200, 267, 400, 800\} \\ \infty & \text{otherwise} \end{cases} \tag{8.70}$$

We assume a nearly constant velocity motion model with additive white Gaussian noise of the form:

$$\mathbf{x}_{k+1} = \mathbf{F}(\Delta t_k)\mathbf{x}_k + \mathbf{v}_k \tag{8.71}$$

where

$$\mathbf{F}(\Delta t_k) = \begin{bmatrix} 1 & \Delta t_k & 0 \\ 0 & 1 & 0 \\ 0 & 0 & 1 \end{bmatrix} \tag{8.72}$$

and $\mathbf{v}_k \sim N(\mathbf{0}, \mathbf{Q}(\Delta t_k))$, with $\mathbf{Q}(\Delta t_k)$ determined through empirical data analysis to have the form [71]:

$$\mathbf{Q}(\Delta t_k) = \begin{bmatrix} 4\Delta t_k^2 & 0 & 0 \\ 0 & 5\sqrt{\Delta t_k} & 0 \\ 0 & 0 & 64\Delta t_k \end{bmatrix} \tag{8.73}$$

We assume an initial target state distribution of the form $\mathbf{x}_0 \sim N(\boldsymbol{\mu}_0, \boldsymbol{\Sigma}_0)$. We assume that the range is in the interval $[R_{\min}, R_{\max}]$ and the velocity is in the interval $(c/2f_c)[-PRF_0/2, PRF_0/2]$, where PRF_0 is the initial PRF. We centre the distribution in the middle of the range interval and at zero velocity. We assume a low value for the SNR, determined through empirical data analysis. The prior mean and covariance matrix are given by [71]:

$$\boldsymbol{\mu}_0 = \begin{bmatrix} 0.5(R_{\min} + R_{\max}) & 0 & 19 \end{bmatrix}^T \tag{8.74}$$

$$\boldsymbol{\Sigma}_0 = \begin{bmatrix} \dfrac{(R_{\max} - R_{\min})^2}{12} & 0 & 0 \\ 0 & \dfrac{(PRF_0)^2}{12}\left(\dfrac{c}{2f_c}\right)^2 & 0 \\ 0 & 0 & (2.65)^2 \end{bmatrix} \tag{8.75}$$

Let $RD(\boldsymbol{\eta}; \mathbf{z}_k, \theta_k)$ denote the RD surface obtained from processing $\mathbf{z}_k$, which was collected with $\mathrm{PRF} = \theta_k$. Empirical data analysis gives us the following approximations to the likelihood function and FIM [71]:

$$\ln f(\mathbf{z}_k|\mathbf{y}; \theta_k) \approx 10 \log_{10} RD(\boldsymbol{\eta}; \mathbf{z}_k, \theta_k) \tag{8.76}$$

$$\mathbf{J}_{\mathbf{y}}(s_k; \theta_k) = \begin{bmatrix} 0.16 \cdot 10^{s_k/10} & 0 & 0 \\ 0 & 41\dfrac{10^{s_k/10}}{\theta_k^2} & 0 \\ 0 & 0 & 0.142 \end{bmatrix} \tag{8.77}$$

Track estimation performance is characterized by the root MSE (RMSE). Let ξ_r and ξ_v denote the RMSE in range and velocity, respectively. PRF affects the SDR system performance in several conflicting ways:

- As PRF decreases, Δt_k increases and motion model uncertainty in $\mathbf{Q}(\Delta t_k)$ increases. This makes both range and velocity tracking resolution poorer (ξ_r and ξ_v larger).
- As PRF decreases, the Doppler bin width (PRF/128) in the RD surface decreases. This improves the Doppler measurement resolution (as reflected in the Doppler entry of the FIM) and makes velocity tracking resolution better (ξ_v smaller).
- The zero-Doppler clutter occupies a few Doppler bins regardless of PRF. As PRF decreases and Doppler bin width decreases, the target Doppler bin, which has index I_D, moves away from the zero-Doppler clutter.
- As PRF decreases, there will be Doppler aliasing if the absolute value of target Doppler is larger than PRF/2.

The controller optimization problem finds maximum PRF (minimum scan time) that keeps the target out of the zero-Doppler clutter and achieves a velocity resolution goal, without Doppler aliasing. Prevention of Doppler aliasing takes priority and we specify the clutter and resolution goals of the form:

$$\begin{aligned} I_D &\geq I_{\min} \\ \xi_v &\leq \xi_{v,\,\max} \end{aligned} \tag{8.78}$$

8.4.2 Implementation

We use a Kalman filter for the tracking method. The MAP-PF guided detector obtains range and Doppler estimates from the peak of the RD surface. We estimate the noise from the median of the RD surface at the estimated range, and compute the SNR (in dB) as the value at the peak of the RD surface minus the noise (in dB).

In the controller optimization step, we start with the largest PRF and, as long as we are not Doppler aliased, keep decreasing until the resolution goal and target distance from clutter goals are met. For each PRF, we compute a motion update and calculate the PC-CRLB. From the motion update, we form a 1.5σ confidence interval on the Doppler. If any portion of the confidence interval is aliased, then we stop and keep the previous PRF. It there is no aliasing, then we check the resolution goal by comparing the standard deviation of the velocity component of the PC-CRLB to our goal value. We also check to see how far away the predicted Doppler estimate is from the zero-Doppler clutter. If both goals are achieved, then we stop, otherwise we decrease the PRF and continue. Implementation details are given in [70,71].

8.4.3 Collected data results

Figures 8.10 and 8.11 show screen shots of the display that is produced as the tracker runs. The display consists of nine panels as follows:

- Upper left: RD surface at the current scan.
- Middle left: Clutter-nulled RD surface with penalty function at the current scan. The dashed ellipse is the 2σ ellipse of the likelihood penalty function centred on the predicted range and Doppler. The solid ellipse is the 2σ ellipse of the measurement FIM centred on the measurement obtained from the peak of the penalized RD surface.
- Lower left: 2σ ellipses of the predicted and posterior densities for the range and velocity.
- Middle three panels: Range, velocity, and SNR tracks with 2σ confidence intervals, and the measurement provided to the tracker. The track is initiated when the SNR measurement exceeds 15 dB.
- Upper right: Predicted confidence interval maximum and mean target Doppler bins with goals.
- Middle right: Velocity RMSE vs time with performance goal.
- Lower right: PRF vs time.

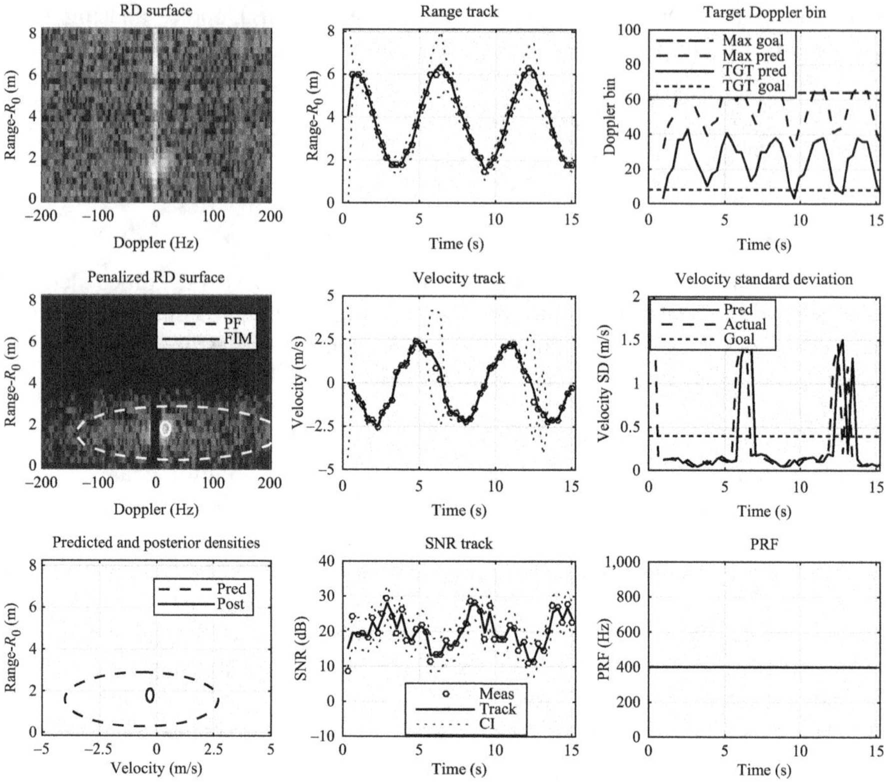

Figure 8.10 Static tracking results

Figure 8.10 shows the results of static processing with a PRF of 400 Hz and Figure 8.11 shows the results of cognitive processing. In both, the penalized RD surface guides the detector in the MAP-PF tracker to the right place and eliminates clutter returns that might cause false detections. The strictness of the penalty function varies according to conditions reported by the tracker. This is a critical step that contributes to the overall performance of both the static and cognitive systems, which are able to maintain track on the target.

In the static system, the target Doppler remains out of the clutter and the boundary of the Doppler confidence interval remains below the aliasing limit on most trials. The velocity resolution goal is achieved on most scans, except when the target is at the far ranges and the SNR is low. In the cognitive system, the optimal PRF varies between 267 and 800 Hz. There is no Doppler aliasing and the target is kept away from the clutter. The velocity resolution goal is achieved on more scans, in particular during the low SNR period around 6 s.

This example demonstrates that we can adjust the PRF to avoid clutter and Doppler aliasing, as well as to achieve a desired track resolution.

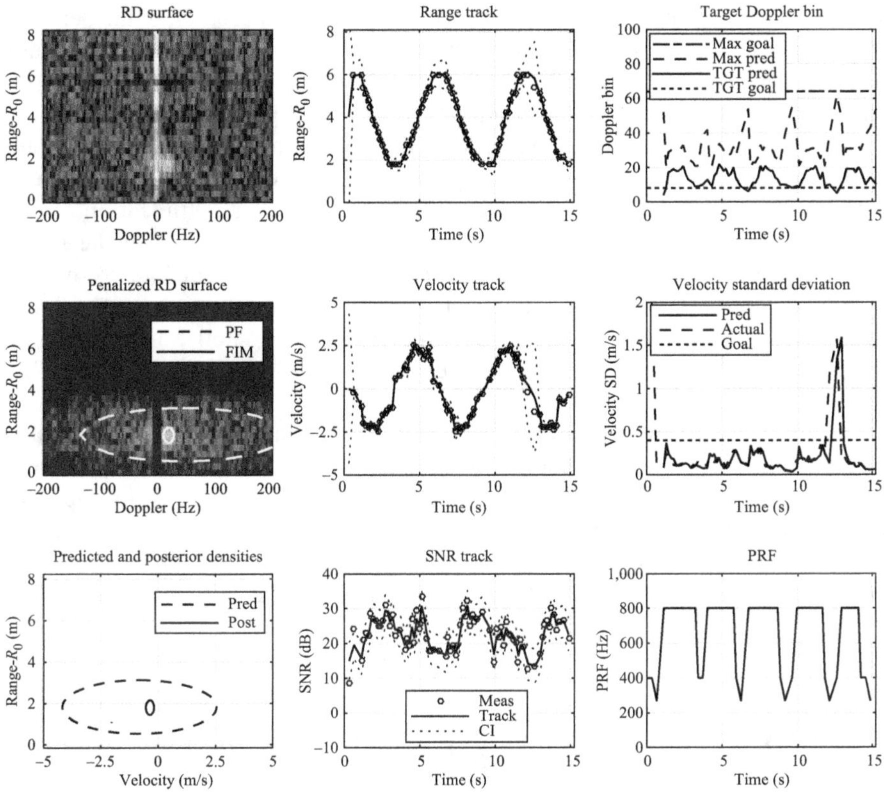

Figure 8.11 Cognitive tracking results

8.5 Conclusions

In this chapter, we developed a general cognitive sensor/processor framework that can be applied to a variety of system objectives, or tasks. The basic framework consists of four components: the scene (target and environment), the sensor, the processor, and the controller. The framework includes the feedback mechanism and optimization criterion used to obtain the next set of sensor data to optimize the performance of the processor. The framework was then specialized for the task of single target tracking. We further developed the tracking framework by partitioning the processor into a detector and a tracker, allowing feedback from the tracker to perform guided detection. This is a key feature of our framework that allows us to cognitively control both the sensor and processor for improved performance.

The framework is based on the *perception–action cycle* and includes *sensing* in the transmitter and receiver; *processing* in the detector and tracker; *perception* in the conversion of sensor data to the posterior PDF of the state vector; *memory* of all the past data in the posterior PDF; *attention* in the penalty function of the guided adaptive detector, which focuses the detector on the relevant region of the detection surface; *prediction* in the PC-BIM, which predicts the performance of the next measurement; and *decision-making* in the controller, which decides on the next values for the sensor parameters based on the predicted performance.

For a specific application, we specify the application model which includes the target motion and sensor measurement models, the radar parameters being controlled and their costs or constraints, and the controller optimization criteria for determining the next set of sensor parameters. We also specify the implementation details that include the tracking implementation and the method for solving the controller optimization problem. By separating the general principles from the specific application model and implementation details, our formulation provides a general framework that can be applied to a wide variety of tracking problem.

Examples were provided for two applications: (i) a distributed sensor system in which system resources are allocated to optimize tracking estimation performance and (ii) a pulse-Doppler radar system in which the PRF was adjusted to optimize tracking estimation performance, while keeping the target from being Doppler-aliased and away from the zero-Doppler clutter. Results were shown for the second application on experimentally collected data using an SDR system. The examples demonstrate that cognitive systems can achieve significant performance improvements over static, feed-forward systems.

Appendix A: Bayesian Cramér–Rao lower Bound

Following [47,48], suppose we wish to estimate a random vector $\mathbf{x}$ based on an observation vector $\mathbf{z}$. Let $f(\mathbf{x})$, $f(\mathbf{z}|\mathbf{x})$, and $f(\mathbf{x},\mathbf{z}) = f(\mathbf{z}|\mathbf{x})f(\mathbf{x})$ denote the prior PDF of $\mathbf{x}$, the conditional PDF of $\mathbf{z}$ given $\mathbf{x}$, and the joint PDF of $\mathbf{x}$ and $\mathbf{z}$, respectively. Let $\hat{\mathbf{x}}(\mathbf{z})$ denote an estimate of $\mathbf{x}$. The MSE matrix is defined as

$$\mathbf{S} \equiv E_{\mathbf{x},\mathbf{z}}\left\{[\hat{\mathbf{x}}(\mathbf{z}) - \mathbf{x}][\hat{\mathbf{x}}(\mathbf{z}) - \mathbf{x}]^T\right\} \tag{A.1}$$

where the expectation is taken with respect to the joint PDF $f(\mathbf{x},\mathbf{z})$.

The *Bayesian Cramér–Rao lower bound* (BCRLB) provides a lower bound[2] on the MSE matrix of the form:

$$\mathbf{S} \geq \mathbf{B}^{-1} \tag{A.2}$$

where $\mathbf{B}$ is the *Bayesian information matrix* (BIM) and $\mathbf{B}^{-1}$ is the BCRLB matrix. The BIM is defined as:

$$\mathbf{B} \equiv -E_{\mathbf{x},\mathbf{z}}\left\{\Delta_{\mathbf{x}}^{\mathbf{x}} \ln f(\mathbf{x},\mathbf{z})\right\} \tag{A.3}$$

[2]A matrix lower bound of the form $\mathbf{B} \geq \mathbf{A}$ means that $\mathbf{B} - \mathbf{A}$ is a positive semidefinite matrix, or equivalently that $\mathrm{tr}\{\mathbf{C}^T\mathbf{B}\mathbf{C}\} \geq \mathrm{tr}\{\mathbf{C}^T\mathbf{A}\mathbf{C}\}$ for any non-zero matrix $\mathbf{C}$.

The BIM can be expressed as the sum of two terms:

$$\begin{aligned}\mathbf{B} &\equiv -E_{\mathbf{x},\mathbf{z}}\{\Delta_{\mathbf{x}}^{\mathbf{x}} \ln f(\mathbf{x}, \mathbf{z})\} \\ &= -E_{\mathbf{x},\mathbf{z}}\{\Delta_{\mathbf{x}}^{\mathbf{x}} \ln f(\mathbf{x})\} - E_{\mathbf{x},\mathbf{z}}\{\Delta_{\mathbf{x}}^{\mathbf{x}} \ln f(\mathbf{z}|\mathbf{x})\} \\ &= -E_{\mathbf{x}}\{\Delta_{\mathbf{x}}^{\mathbf{x}} \ln f(\mathbf{x})\} - E_{\mathbf{x}}\{E_{\mathbf{z}|\mathbf{x}}\{\Delta_{\mathbf{x}}^{\mathbf{x}} \ln f(\mathbf{z}|\mathbf{x})\}\} \\ &= \mathbf{B}_p + \mathbf{B}_D\end{aligned} \tag{A.4}$$

where

$$\mathbf{B}_P \equiv -E_{\mathbf{x}}\{\Delta_{\mathbf{x}}^{\mathbf{x}} \ln f(\mathbf{x})\} \tag{A.5}$$

and

$$\mathbf{B}_D \equiv E_{\mathbf{x}}\{-E_{\mathbf{z}|\mathbf{x}}\{\Delta_{\mathbf{x}}^{\mathbf{x}} \ln f(\mathbf{z}|\mathbf{x})\}\} \tag{A.6}$$

The term $\mathbf{B}_P$ represents prior information and the term $\mathbf{B}_D$ represents information in the data.

The inner term in (A.6) is the standard Fisher information matrix (FIM), defined as:

$$\mathbf{J}(\mathbf{x}) \equiv -E_{\mathbf{z}|\mathbf{x}}\{\Delta_{\mathbf{x}}^{\mathbf{x}} \ln f(\mathbf{z}|\mathbf{x})\} \tag{A.7}$$

Thus the data term can be written as:

$$\mathbf{B}_D = E_{\mathbf{x}}\{\mathbf{J}(\mathbf{x})\} \tag{A.8}$$

References

[1] S. Haykin, "Cognitive radar: a way of the future," *IEEE Signal Processing Magazine*, vol. 23, no. 1, pp. 30–40, Jan. 2006.

[2] S. Haykin, Y. Xue, and M. P. Setoodeh, "Cognitive radar: step toward bridging the gap between neuroscience and engineering," *Proceedings of the IEEE*, vol. 100, no. 11, pp. 3102–3130, Nov. 2012.

[3] S. Haykin, *Cognitive Dynamic Systems (Perception–Action Cycle, Radar, and Radio)*, Cambridge, UK: Cambridge University Press, 2012.

[4] S. Haykin, Ed. "Cognitive dynamic systems," *Proceedings of the IEEE (special issue)*, vol. 102, no. 4, pp. 414–416, Apr. 2014.

[5] F. Gini and M. Rangaswamy, Eds., *Knowledge Based Radar Detection, Tracking, and Classification*, Hoboken, NJ: Wiley, 2008.

[6] J. R. Guerci, *Cognitive Radar: The Knowledge Aided Fully Adaptive Approach*, Reading, MA: Artech House, 2008.

[7] J. Guerci, R. Guerci, M. Rangaswamy, J. Bergin, and M. Wicks, "Cognitive fully adaptive radar (CoFAR)," in *Proceedings of the 2014 IEEE Radar Conference*, Cincinnati, OH, May 2014.

[8] B. Kolb and I. Q. Whishaw, *An Introduction to Brain and Behavior*, 2nd ed. New York, NY: Worth Publishers, 2006.

[9] J. M. Fuster, *Cortex and Mind: Unifying Cognition*, Oxford, UK: Oxford University Press, 2003.

[10] S. L. C. Miranda, C. J. Baker, K. D. Woodbridge, and H. D. Griffiths, "Knowledge based resource management for multifunction radar," *IEEE Signal Processing Magazine*, vol. 23, no. 1, pp. 66–76, 2006.

[11] S. L. C. Miranda, C. J. Baker, K. D. Woodbridge, and H. D. Griffiths, "Fuzzy logic approach for prioritization of radar tasks and sectors of surveillance in multifunction radar," *IET Radar, Sonar and Navigation*, vol. 1, no. 2, pp. 131–141, 2007.

[12] M. Vespe, C. J. Baker, and H. D. Griffiths, "Automatic target recognition using multi-diversity radar," *IET Radar, Sonar and Navigation*, vol. 1, no. 6, pp. 470–478, Dec. 2007.

[13] S. Miranda, C. J. Baker, K. Woodbridge, and H. D. Griffiths, "Intelligent radar resource management," *Knowledge Based Radar Detection, Tracking and Classification*, F. Gini and M. Rangaswamy, Eds., Hoboken, NJ: Wiley, 2008.

[14] A. Aubry, A. De Maio, M. Piezzo, A. Farina, and M. Wicks, "Cognitive design of the receive filter and transmitted phase code in reverberating environment," *IET Radar, Sonar and Navigation*, vol. 6, no. 9, pp. 822–833, 2012.

[15] P. Stinco, M. Greco, and F. Gini, "Spectrum sensing and sharing for cognitive radars," *IET Radar, Sonar and Navigation*, vol. 10, no. 3, pp. 595–602, 2016.

[16] P. Stinco, M. Greco, F. Gini, and B. Himed "Cognitive radars in spectrally dense environments," *IEEE Aerospace and Electronic Systems Magazine*, vol. 31, no. 10, pp. 20–27, Oct. 2016.

[17] D. J. Kershaw and R. J. Evans, "Optimal waveform selection for tracking systems," *IEEE Transactions on Information Theory*, vol. 40, no. 5, pp. 1536–1550, Sep. 1994.

[18] D. Fuhrmann, "Active-testing surveillance systems, or, playing twenty questions with radar," in *Proceedings of the 11th Annual Adaptive Sensor and Array Processing (ASAP 2003) Workshop*, MIT Lincoln Laboratory, Lexington, MA, Mar. 2003.

[19] S. P. Sira, A. Papandreou-Suppappola, and D. Morrell, "Dynamic configuration of time-varying waveforms for agile sensing and tracking in clutter," *IEEE Transactions on Signal Processing*, vol. 55, no. 7, pp. 3207–3217, Jul. 2007.

[20] M. Hurtado, T. Zhao, and A. Nehorai, "Adaptive polarized waveform design for target tracking based on sequential Bayesian inference," *IEEE Transactions on Signal Processing*, vol. 56, no. 13, pp. 1120–1133, Mar. 2008.

[21] S. P. Sira, Y. Li, A. Papandreou-Suppappola, D. Morrell, D. Cochran, and M. Rangaswamy "Waveform-agile sensing for tracking," *IEEE Signal Processing Magazine*, vol. 26, no. 1, pp. 53–64, Jan. 2009.

[22] M. L. Hernandez, T. Kirubarajan, and Y. Bar-Shalom, "Multisensor resource deployment using posterior Cramér–Rao bounds," *IEEE Transactions on Aerospace and Electronic Systems*, vol. 40, no. 2, pp. 399–416, Apr. 2004.

[23] C. Kreucher, A. O. Hero, K. Kastella, and D. Chang, "Efficient methods of non-myopic sensor management for multitarget tracking," in *Proceedings of the 43rd IEEE Conference Decision and Control*, Atlantis, Bahamas, pp. 722–727, Dec. 2004.

[24] C. Kreucher, K. Kastella, and A. O. Hero, "Sensor management using an active sensing approach," *Signal Processing*, vol. 85, no. 3, pp. 607–624, Mar. 2005.

[25] C. Kreucher, A. O. Hero, and K. Kastella, "A comparison of task driven and information driven sensor management for target tracking," in *Proceedings of the 44th IEEE Conference on Decision and Control*, Seville, Spain, pp. 4004–4009, Dec. 2005.

[26] C. M. Kreucher, A. O. Hero, K. Kastella, and B. Shapo, "Information-based sensor management for simultaneous multitarget tracking and identification," in *Proceedings of the 13th Annual Conference on Adaptive Sensor Array Processing*, MIT Lincoln Laboratory, Lexington, MA, Jun. 2005.

[27] C. Kreucher, A. Hero, K. Kastella, and M. Morelande, "An information-based approach to sensor management in large dynamic networks," *Proceedings of the IEEE*, vol. 95, no. 5, pp. 978–999, May 2007.

[28] R. Tharmarasa, T. Kirubarajan, and M. L. Hernandez, "Large-scale optimal sensor array management for multitarget tracking," *IEEE Transactions on Systems, Man, and Cybernetics C: Applications and Reviews*, vol. 37, no. 5, pp. 803–814, Sep. 2007.

[29] E. K. P. Chong, C. M. Kreucher, and A. O. Hero, "Monte-Carlo-based partially observable Markov decision process approximations for adaptive sensing," in *Proceedings of the Ninth International Workshop on Discrete Event Systems*, Goteborg, Sweden, pp. 173–180, May 2008.

[30] P. Chavali and A. Nehorai, "Scheduling and power allocation in a cognitive radar network for multiple-target tracking," *IEEE Transactions on Signal Processing*, vol. 60, no. 2, pp. 715–729, Feb. 2012.

[31] R. A. Romero and N. A. Goodman, "Cognitive radar network: cooperative adaptive beamsteering for integrated search-and-track application," *IEEE Transactions on Aerospace and Electronic Systems*, vol. 49, no. 2, pp. 915–931, Apr. 2013.

[32] A. Saksena and I.-J. Wang, "Dynamic ping optimization for surveillance in multistatic sonar buoy networks with energy constraints," in *Proceedings of the 47th IEEE Conference Decision and Control*, Cancun, Mexico, pp. 1109–1114, Dec. 2008.

[33] C. Y. Wakayama and D. J. Grimmett, "Adaptive ping control for track-holding in multistatic active sonar networks,"in *Proceedings of the 13th International Conference on Information Fusion*, Edinburgh, UK, Jul. 2010.

[34] C. Y. Wakayama, D. J. Grimmett, and Z. B. Zabinsky, "Forecasting probability of target presence for ping control in multistatic sonar networks using detection and tracking models," in *Proceedings of the 14th International Conference on Information Fusion*, Chicago, IL, Jul. 2011.

[35] M. W. Holdereid, C. J. Baker, M. Vespe, M., and G. Jones, "Understanding signal design during the pursuit of aerial insects by echo locating bats: tools and applications," *Integrative and Comparative Biology*, vol. 48, pp. 78–84, May 2008.

[36] C. J. Baker and H. D. Griffiths, "Biologically inspired waveform diversity," *Waveform Design and Diversity for Advanced Radar Systems*, F. Gini, A. De Maio, and L. Patton, Eds., Stevenage, UK: IET Publishing, Aug. 2012.

[37] A. Balleri, H. D. Griffiths, C. J. Baker, K. Woodbridge, and M. W. Holderied, "Analysis of acoustic echoes from a bat-pollinated plant species: insight into strategies for radar and sonar target classification," *IET Radar, Sonar System Navigation*, vol. 6, no. 6, pp. 536–544, Jul. 2012.

[38] C. J. Baker and G. E. Smith, "Aspects of cognition and echolocation," in *Proceedings of the 2012 IEEE Antennas and Propagation Society International Symposium (APSURSI)*, Chicago, IL, pp. 1–2, 8–14 July 2012.

[39] C. J. Baker, G. E. Smith, A. Balleri, M. Holderied, and H. D. Griffiths, "Biomimetic echolocation with application to radar and sonar sensing," *Proceedings of the IEEE*, vol. 102, no. 4, pp. 447–458, Apr. 2014.

[40] C. J. Baker, G. E. Smith, A. Balleri, M. Holderied, and H. D. Griffiths, "Sensing, cognition, and engineering application [further thoughts]," *Proceedings of the IEEE*, vol. 102, no. 4, p. 459, Apr. 2014.

[41] G. E. Smith and C. J. Baker, "Echoic flow for radar and sonar," *IET Electronics Letters*, vol. 48, no. 18, pp. 1160–1161, Aug. 2012.

[42] G. E. Smith and C. J. Baker, "Echoic flow for autonomous navigation," in *Proceedings of Radar 2012, The International Conference on Radar Systems*, Glasgow, UK, 2012.

[43] A. Charlish and F. Hoffmann, "Anticipation in cognitive radar using stochastic control," in *Proceedings of the 2015 IEEE Radar Conference*, Arlington, VA, pp. 751–756, May 2015.

[44] L. D. Stone, R. L. Streit, T. L. Corwin, and K. L. Bell, *Bayesian Multiple Target Tracking*, 2nd ed. Norwood, MA: Artech House, 2014.

[45] B. Ristic, S. Arulampalam, and N. Gordon, *Beyond the Kalman Filter: Particle Filters for Tracking Applications*, Boston, MA: Artech House, 2004.

[46] Y. Bar-Shalom, X. R. Li, and T. Kirubarajan, *Estimation with Applications to Tracking and Navigation*, New York, NY: Wiley, 2001.

[47] H. L. Van Trees, K. L. Bell, and Z. Tian, *Detection, Estimation, and Modulation Theory, Part I*, 2nd ed. Hoboken, NJ: Wiley, 2013.

[48] H. L. Van Trees and K. L Bell, Eds., *Bayesian Bounds for Parameter Estimation and Nonlinear Filtering/Tracking*, Piscataway, NJ: Wiley-IEEE Press, 2007.

[49] P. Tichavsky, C. H. Muravchik, and A. Nehorai, "Posterior Cramér–Rao bounds for discrete-time nonlinear filtering," *IEEE Transactions on Signal Processing*, vol. 46, no. 5, pp. 1386–1396, May 1998.

[50] L. Zuo, R. Niu, and P. K. Varshney, "Conditional posterior Cramér–Rao lower bounds for nonlinear sequential Bayesian estimation," *IEEE Transactions on Signal Processing*, vol. 59, no. 1, pp. 1–14, Jan. 2011.

[51] R. E Zarnich, K. L. Bell, and H. L. Van Trees, "A unified method for measurement and tracking of multiple contacts from sensor array data," *IEEE Transactions on Signal Processing*, vol. 49, no. 12, pp. 2950–2961, Dec. 2001.

[52] K. L. Bell, "MAP-PF position tracking with a network of sensor arrays," in *Proceedings of the 2005 IEEE International Conference on Acoustics, Speech, and Signal Processing (ICASSP 2005)*, Philadelphia, PA, vol. IV, pp. 849–852, Mar. 2005.

[53] K. L. Bell and R. Pitre, "MAP-PF 3D position tracking using multiple sensor arrays," in *Proceedings of the Fifth IEEE Sensor Array and Multichannel Signal Processing Workshop (SAM 2008)*, Darmstadt, Germany, pp. 238–242, Jul. 2008.

[54] K. L. Bell, R. E. Zarnich, and R. Wasyk, "MAP-PF wideband multitarget and colored noise tracking," in *Proceedings of the 2010 IEEE International Conference on Acoustics, Speech, and Signal Processing (ICASSP 2010)*, Dallas, TX, pp. 2710–2713, Mar. 2010.

[55] K. L. Bell, "MAP-PF multi-mode tracking for over-the-horizon radar," in *Proceedings of the 2012 IEEE Radar Conference*, Atlanta, GA, pp. 326–331, May 2012.

[56] K. L. Bell and R. E. Zarnich, "MAP-PF multitarget tracking with propagation modeling uncertainties," in *Proceedings of the 47th Asilomar Conference Signals, Systems, and Computers*, Pacific Grove, CA, Nov. 2013.

[57] W. I. Zangwill, *Nonlinear Programming: A Unified Approach*, Englewood Cliffs, NJ: Prentice-Hall, 1969.

[58] K. L. Bell and H. L. Van Trees, "Posterior Cramér–Rao bound for tracking target bearing," in *Proceedings of the 13th Annual Workshop on Adaptive Sensor Array Processing (ASAP 2005)*, MIT Lincoln Lab, Lexington, MA, Jun. 2005.

[59] K. L. Bell, C. J. Baker, G. E. Smith, J. T. Johnson, and M. Rangaswamy, "Fully adaptive radar for target tracking – part I: single target tracking," in *Proceedings of the 2014 IEEE Radar Conference*, Cincinnati, OH, May 2014.

[60] K. L. Bell, C. J. Baker, G. E. Smith, J. T. Johnson, and M. Rangaswamy, "Fully adaptive radar for target tracking – part II: target detection and track initiation," in *Proceedings of the 2014 IEEE Radar Conference*, Cincinnati, OH, May 2014.

[61] K. L. Bell, C. J. Baker, G. E. Smith, J. T. Johnson, and M. Rangaswamy, "Cognitive radar framework for target detection and tracking," *IEEE Journal of Selected Topics in Signal Processing*, vol. 9, no. 8, pp. 1427–1439, Dec. 2015.

[62] M. Frankford, N. Majurec, and J. T. Johnson, "Software-defined radar for MIMO and adaptive waveform applications," in *Proceedings of the 2010 IEEE Radar Conference*, Washington, DC, pp. 724–728, May 2008.

[63] M. Frankford, J. T. Johnson, and E. Ertin, "Including spatial correlations in the statistical MIMO radar target model," *IEEE Signal Processing Letters*, vol. 17, no. 6, pp. 575–578, Jun. 2010.

[64] M. Frankford, K. B. Stewart, N. Majurec, and J. T. Johnson, "Numerical and experimental studies of target detection with MIMO radar," *IEEE Transactions on Aerospace and Electronic Systems*, vol. 50, no. 2, pp. 1569–1577, Apr. 2014.

[65] K. Stewart, M. Frankford, J. T. Johnson, and E. Ertin, "MIMO radar target measurements," in *Proceedings of the 45th Asilomar Conference on Signals, Systems, and Computers*, Pacific Grove, CA, pp. 1067–1071, Nov. 2011.

[66] J. Park, "Multi-frequency radar signatures of human motion: measurements and models," Ph.D. Dissertation, The Ohio State University, 2012.

[67] J. Park, J. T. Johnson, N. Majurec, K. Stewart, G. E. Smith, and L. Westbrook, "Simulation and analysis of polarimetric radar signatures of human gaits," *IEEE Transactions on Aerospace and Electronic Systems*, vol. 50, no. 3, pp. 2164–2175, Jul. 2014.

[68] J. Park, J. T. Johnson, N. Majurec, *et al.*, "Software defined radar studies of human motion signatures," in *Proceedings of the 2012 IEEE Radar Conference*, Atlanta, GA, pp. 596–601, May 2012.

[69] K. B. Stewart, M. T. Frankford, N. Majurec, and J. T. Johnson, "Software defined radar for cognitive applications," *URSI National Radio Science Meeting*, Chicago, IL, 2012.

[70] K. L. Bell, J. T. Johnson, G. E. Smith, C. J. Baker, and M. Rangaswamy, "Cognitive radar for target tracking using a software defined radar system," in *Proceedings of the 2015 IEEE Radar Conference*, Arlington, VA, May 2015.

[71] K. L. Bell, C. J. Baker, G. E. Smith, and J. T. Johnson, "Fully adaptive radar," AFRL Tech. Report AFRL-RY-WP-TR-2014-0072, Mar. 2014.

Chapter 9

The biosonar of the Mediterranean Bottlenose dolphins: analysis and modelling of echolocation signals

Maria Greco[1] *and Fulvio Gini*[1]

9.1 Introduction

This chapter is devoted to the analysis and modelling of echolocation signals emitted by dolphins, in particular in the family of the *Tursiops truncatus* (Bottlenose dolphin) living in the Tuscany Archipelago Park. The analysis covers both audio and ultrasonic frequency bands.

Dolphins have a rich vocal repertoire that has been categorized into three classes:

- broadband, short-duration clicks, called *sonar clicks*, used in echolocation for orientation, perception, and navigation;
- wideband pulsed sounds, called *burst pulses*, used in social contexts;
- narrowband frequency-modulated *whistles* also used in social contexts.

Dolphins use a range of frequencies extending between 1 and 150 kHz. Communication signals (burst pulses and whistles) have a range of frequencies from 1 to 25 kHz. Generally sonar signals have a range of frequencies from 25 to 150 kHz. Dolphins can emit various different types of sounds at the same time and independently.

Bottlenose dolphins have a remarkable range of hearing extending from less than 1 kHz to more than 120 kHz and a range of frequency-dependent sensitivity of nearly 100 dB μPa. They have excellent frequency discrimination capability and are capable of determining changes in frequency as small as 0.2%–0.4% [1]. This degree of discrimination is comparable to that observed in humans, but it is preserved across a much broader range of frequencies. The broad range of hearing and sensitivity and excellent frequency discrimination has likely evolved as part of the biological sonar system (echolocation) used by dolphins for exploitation of a visually limited marine environment. Dolphins respond to pure tone signals in a similar manner as humans. Therefore, the spectral filtering property of the dolphin ear can be modelled by

[1]Dipartimento di Ingegneria dell'Informazione, Università di Pisa, via G. Caruso 14, I-56122 Pisa, Italy

means of a bank of contiguous constant-Q filters, as for humans. Other hearing characteristics that are similar for dolphins and humans include frequency discrimination and sound localization capabilities in 3D space [2].

Marine mammals do not use their mouths and throats to generate the sound – vocal chords rely on air. In dolphins, sound is produced below the nasal plug and then focused by combination of reflection off the skull and passage through a lens mechanism formed by the melon, a mass of fatty tissue in the forehead [2,3]. The acoustic vibrations are then radiated from the bone of the rostrum into the blubber and seawater.

The acoustic field in the immediate vicinity of a dolphin head has no sharp null in the diagram of near field and of beam. This is because short broadband pulses do not show effects of the constructive and destructive interference from multipath. The system of transmission of these pulses has the same radiation characteristics of a directional antenna with 3 dB beam patterns of approximately 10° on the vertical and horizontal planes, as shown in Figure 9.1. The beam is highly dependent on frequency, becoming narrower as the frequency increases. The directivity index of the transmitted beam pattern is approximately 26 dB in Bottlenose dolphins [2].

Moreover, the emitted signal has different shapes according to the position of the animal with respect to the hydrophone. With an array of hydrophones, these different characteristics have been evidenced [2]. On the vertical plane (perpendicular to the head of the dolphin) the signal in the time domain became progressively distorted with respect to the signal on the major axis at +5° as well as in the horizontal plane. The signals were not symmetrical about the beam axis, which is expected since the structure of the skull is not symmetrical about the midline of the animal [2,5].

Echolocation experiments with captive dolphins began about five decades ago to determine the capabilities of odontocetes to detect, discriminate, localize, and recognize targets. Mainly conducted with object that are familiar to humans more than to the marine mammals, these experiments have proved that dolphins can

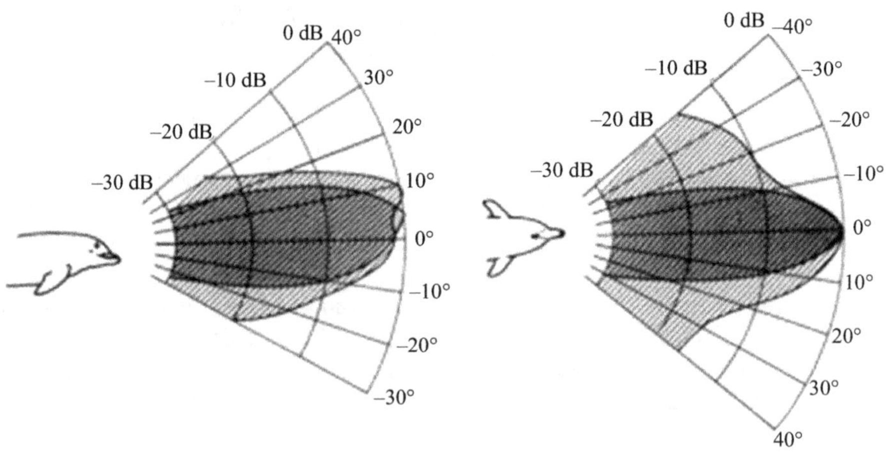

Figure 9.1 Bottlenose dolphin beampattern [4]

easily detect and discriminate objects with a range resolution of about 3 cm [2,6], even in an environment strongly limited by noise and reverberation, with a maximum range of about 100–150 m. The short coverage of the dolphin biosonar is not a problem, because in the wild they are not interested in preys that are further than 150 m. About the echolocation receiver, there is evidence that the dolphin's lower jaw is a component of it [7] and the equally spaced rows of teeth (the Tursiops lower jaw has two rows of about 22 teeth, spaced at approximately 1 cm and diverging at an angle of around 12°) can form a receiving array where the teeth act as resonant receivers arranged in geometries appropriate for beamforming.

9.2 Data acquisition

In our experiments, the chain of data acquisition and recording was composed of a hydrophone, an amplification stage, and a digital card on a laptop. In our recording, we first used a simple digital card with audio band (0–16 kHz) and then we acquired by National Instruments the digital card DAQCard-6062E, with a maximum sampling frequency of 500 kHz.[1]

9.2.1 The hydrophone

The interface between the acquisition system and the underwater world is represented by the hydrophone, an underwater microphone that converts a sound pressure in a proportional difference of tension. In Figure 9.2 the custom-built hydrophone used during our experiments is shown. Its body, dark coloured, is a ceramic toroid sensible to the pressure. It works in the frequency range [0 Hz–180 kHz] and it is almost omnidirectional. This characteristic can increase the possibility of recording sounds, but unfortunately, it can also prevent from localizing their direction of arrival.

The hydrophone is dragged by the boat through a cable connected to the amplifier. This cable is 20 m long and it allows the hydrophone to stay generally 2 m

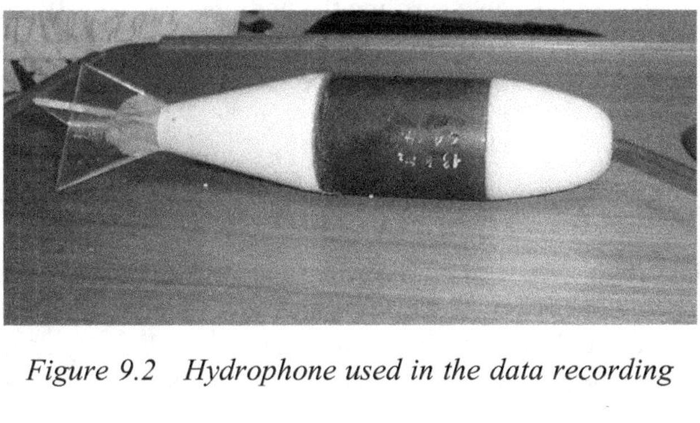

Figure 9.2 Hydrophone used in the data recording

[1]Data acquisition was made in collaboration with the CETUS Research Center of Viareggio (LU) that since 1997 monitors and studies the cetaceans living in the Tuscany Archipelago.

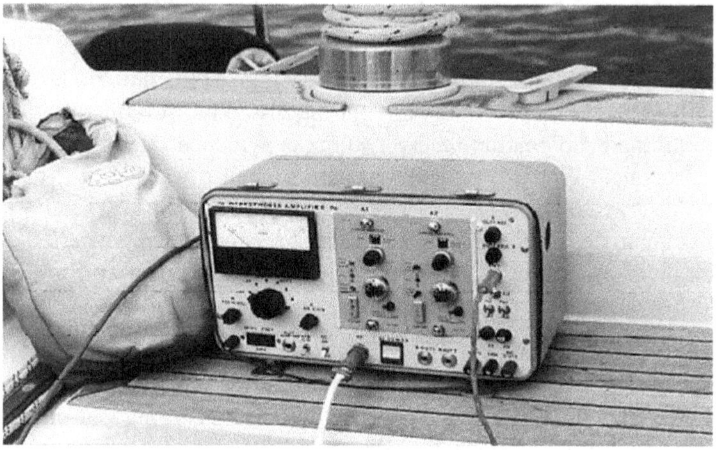

Figure 9.3 Amplifier used in the data recording

below the surface, inside the thermocline that however results almost irrelevant both in summer, and in winter. The cable is screened, to avoid combinations with external signals, and shows a parasite power that is removed from the input stage of the amplifier. The cable vibrations also produce low frequency noise which is then removed by the amplifier.

9.2.2 The amplifier

The amplification stage is shown in Figure 9.3. It is composed of two cascaded charge amplifiers. The input impedance of the amplifier is about 10 MΩ, and it has a band-pass behaviour from 0 Hz up to 180 kHz. The amplifier also allows manual regulation of the gain so that the optimal level of the signal can be maintained during the recording. There is also an active high-pass (HP) filter in the amplifier that removes the components of noise due to the boat engine, to the rinsing of the sea, to the vibrations of the cable carrying the hydrophone. The HP filter has a pole at 400 Hz with band of transition that decays 20 dB/dec. More details on the technical characteristics of the amplifier and of the hydrophone can be found in [8].

9.2.3 Digital card

During the first recording days we used a simple digital card with audio band (0–16 kHz), then we acquired by National Instruments the digital card DAQCard-6062E (see Figure 9.4). This card allows recording even at ultrasonic band because its maximum sampling frequency is 500 kHz, then it is possible to catch signals until 250 kHz. In our files dolphin echolocation signals were digitally sampled at a rate of 360 kHz, providing a Nyquist frequency for all recordings of 180 kHz, that is, the bandwidth of the hydrophone. Recordings were obtained from free-ranging Bottlenose dolphins in the Mediterranean Sea, along the cost in front of Tuscany on 10 occasions.

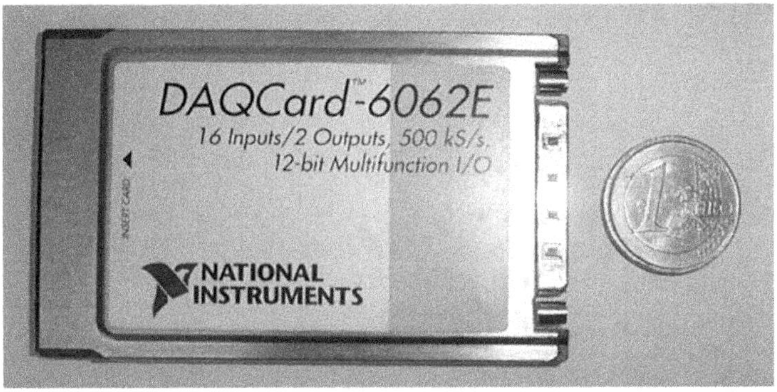

Figure 9.4 Digital card used in the data recording

9.3 Biosonar model

The term **sonar** is the acronym for SOund Navigation and Ranging and it was coined during the World War II. It refers to the principle of detection and localization of objects in submarine environment, through emission of sonorous pulses and the processing of the returning echoes from the same objects. With the term "echolocation" being indicated the orientation ability using the transmission of ultrasonic pulses and the reception of the echoes. The words sonar clicks, echolocation clicks, and biosonar are used to describe the activity of guideline, of navigation, and of localization of the animal that emit acoustic energy and analyse the received echo. The first unequivocal demonstration of the use of the biosonar from dolphin dates back to 1960. Kenneth and Norris placed rubber suction cups over the eyes of a Tursiop to eliminate its use of vision. The dolphin swam normally, emitting ultrasonic pulses and avoiding obstacles, including pipes suspended vertically to form a maze [9].

Dolphins use pulse trains as biosonar. A click train is plotted in Figure 9.5.

The number of clicks and the temporal interval between successive clicks depend on several factors such as, e.g., the distance from the target, the environmental conditions, and the expectation of the animal on the presence/absence of the prey. When the dolphin is in motion, the time that elapses between clicks often changes. A train of clicks can contain from just a few clicks to hundreds of clicks. If the pulses repeat rapidly, say every 5 ms, we perceive them indifferently as a continuous tone [2]. Generally, the dolphin sends a click and waits for the return echo before sending the successive click. The time elapsing between the reception of the echo and the emission of a new click (lag time) depends on the distance from the target. From several studies [2,10], it turns out that the mean lag time (LT) is 15 ms with targets distant from 0.4 to 4 m, 2.5 ms at less than 0.4 m and 20 ms from 4 to 40 m. From several experiments, it is possible to assert that the dolphin can adapt the spectral content of the biosonar to the context in which they work in order to obtain the maximum efficiency [2] and the emitted pulses have duration that is different from a family to the other, in the range from ten to one hundred microseconds [11,12].

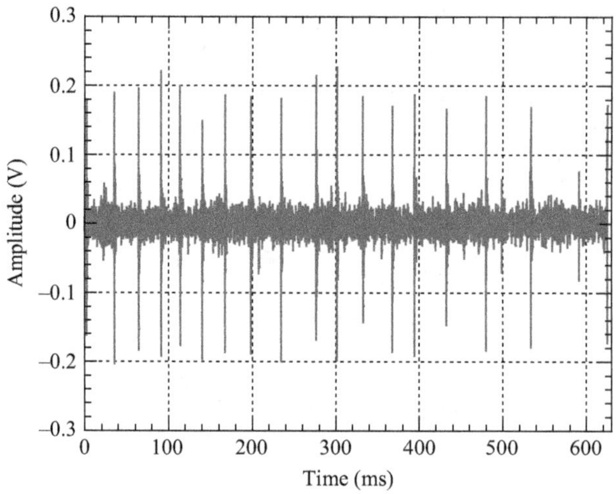

Figure 9.5 Sonar click train

The high resolution of biosonar and the ability to process the echoes allows the dolphin to distinguish geometric figures, three-dimensional objects and to estimate the organic/inorganic composition of whichever object [2]. Moreover, in their hunt for a prey, they are able to move around their target, changing the angle of view and continuing their tracking even when the power scattered by the target is 20–30 dB below its maximum, looking for the "most informative" angle, not always for the most powerful, as described in [4]. Therefore, the important parameter for the dolphin may be the information level, rather than the echo level. The capability of the dolphins to change their trajectory and to adapt it based on what they "sense" about the target is fundamental for the efficiency of their biosonar and could be borrowed by the technological sonars/radars through multiple aspect tracking/discrimination algorithms [13].

The biosonar signal has a peak-to-peak SPL (Sound Pressure Level at a reference range of 1 m and a reference pressure of 1 μPa) that varies between 120 and 230 dB. The levels of SPL change considerably from family to family. The clicks of high level (>210 dB) introduce peaks of frequency at high frequency (hundreds of kHz). Au *et al.* in fact postulated in [2,10] that the high frequencies are by-product of producing high intensity clicks. In other words, dolphins can only emit high-level clicks (greater than 210 dB) if they use high frequencies. Dolphins maybe can emit high frequency clicks at low amplitudes, but cannot produce low frequency clicks at high amplitudes. Moreover, the dolphins can vary the amplitude of the emitted clicks in relation to the environmental conditions and to the distance of the target [14].

Frequency peaks are located between 5 and 150 kHz. In open sea, the dolphins emit biosonar at high frequency with high level. In captivity, they produce echolocation clicks with peak frequencies an octave inferior and levels smaller of 15–30 dB. This is because in open sea noise is much higher [15] and the targets can be much farther than in a captivity environment, therefore a correct echolocation click can only happen through high frequency and high level. In captivity and in highly

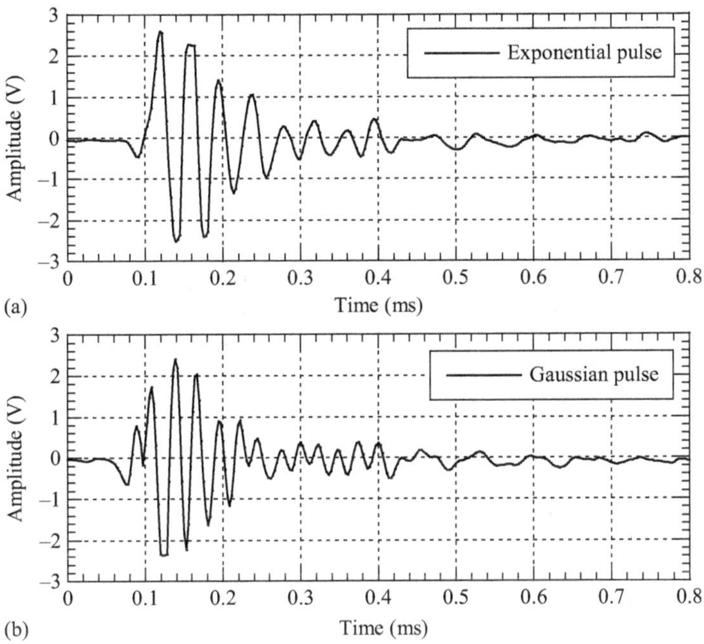

Figure 9.6 (a) Exponential and (b) Gaussian pulses extracted from data

reverberant environment as the tanks of the aquarium the close proximity of acoustic reverberant walls tends to discourage the animals from emitting high intensity biosonars because too much energy would be reflected back to the dolphins [2].

In the following, we describe methods for the analysis of recorded echolocation pulses and feature extraction. The extracted information can be used by biologists to understand the ability of dolphins to perceive their environment and to perform difficult recognition and discrimination tasks, and to relate the kind of emitted sounds to the behaviour of these fascinating mammals. Dolphins can even discriminate targets buried in ocean sediment within a range of 100 m. From the sonar/radar engineers standpoint, understanding the dolphin biosonar could help the design of more efficient systems.

The main focus of this chapter is the echolocation pulses recorded with the dolphins aligned to the hydrophone, that is, when the hydrophone is on the main axis of the dolphins. The study of measured data has been organized in four phases: classification, extraction, characterization, and estimation. In the first phase, all the recorded files have been classified by visual inspection. The time history and the time-varying spectrum of recorded data have been calculated to find the echolocation pulses. Subsequently, the interesting signals have been extracted from the files. In both audio and ultrasonic bands we found visually mainly two kinds of pulses as shown in Figure 9.6(a) and (b).

The first pulse exhibits an exponential envelope, the second pulse a Gaussian envelope. For this study, we extracted 300 echolocation pulses from audio band data and more than 400 pulses in ultrasonic band. The analysis performed on the data for the sonar clicks is similar for both bands, and then we detail it for the ultrasonic band and resume the results for both frequency ranges.

9.4 Signal estimation

9.4.1 Exponential pulse

For the sonar click of first kind we adopted a dumped exponential multicomponent signal model, that is, we model the extracted signal $x(n)$ as

$$x(n) = A_0 + \sum_{k=1}^{K} A_k e^{-\alpha_k n} \cos(2\pi f_k n + \vartheta_k) \tag{9.1}$$

where A_0 is the mean value, A_k, f_k and ϑ_k are amplitude, frequency, and initial phase of the kth component, respectively, and α_k is the decay parameter of the exponential envelope. The signal (9.1) can be expressed in the more general form:

$$x(n) = A_0 + \sum_{k=1}^{2K} \beta_k e^{-\alpha_k n} e^{j2\pi f_k n} \tag{9.2}$$

where $f_k = -f_{k+K}$, $\beta_k = \beta^*_{k+K} = A_k e^{j\vartheta_k}/2$ and $\alpha_k = \alpha_{k+K}$.

To validate our model we estimated the characteristic parameters using the Least Squares (LS) method. First of all the mean value is estimated from the data as

$$\hat{A}_0 = \frac{1}{N}\sum_{n=0}^{N-1} z(n) \tag{9.3}$$

and subtracted to the data vector $z(n) = x(n) + w(n)$, where $w(n)$ is the additive noise, so obtaining the new data $y(n) = z(n) - \hat{A}_0$. Then, the unknown parameter vector is $\boldsymbol{\theta} = [\beta_1, \ldots, \beta_{2K}, \alpha_1, \ldots, \alpha_{2K}, f_1, \ldots, f_{2K}] = [\boldsymbol{\beta}, \boldsymbol{\alpha}, \mathbf{f}]$.
Now define the cost function:

$$C(\mathbf{y}; \boldsymbol{\theta}) = \|\mathbf{y} - \mathbf{x}(\boldsymbol{\theta})\|^2 = \frac{1}{N}\sum_{n=0}^{N-1} \left| y(n) - \sum_{k=1}^{2K} \beta_k e^{-\alpha_k n} e^{j2\pi f_k n} \right|^2 \tag{9.4}$$

where N is the number of samples describing a pulse and $\mathbf{y}$ is the data vector of length N. In audio band generally $N \simeq 100$, in ultrasonic band $N > 400$. The non-linear least square (NLLS) estimator of θ is:

$$\hat{\boldsymbol{\theta}} = \arg\min_{\boldsymbol{\theta}} C(\mathbf{y}; \boldsymbol{\theta}) \tag{9.5}$$

The estimators have the following expressions:

$$(\hat{\mathbf{f}}, \hat{\boldsymbol{\alpha}}) = \arg\max_{\mathbf{f}, \boldsymbol{\alpha}} \mathbf{y}^H \mathbf{A}\left(\mathbf{A}^H \mathbf{A}\right)^{-1} \mathbf{A}^H \mathbf{y} \tag{9.6}$$

and

$$\hat{\boldsymbol{\beta}} = \left(\mathbf{A}^H \mathbf{A}\right)^{-1} \mathbf{A}^H \mathbf{y} \tag{9.7}$$

where $\mathbf{A} = [\mathbf{g}(\alpha_1) \odot \mathbf{p}(f_1) \cdots \mathbf{g}(\alpha_{2K}) \odot \mathbf{p}(f_{2K})]$, $\mathbf{a}(\alpha_k, f_k) = \mathbf{g}(\alpha_k) \odot \mathbf{p}(f_k)$, $[\mathbf{p}(f)]_n = e^{j2\pi f_k n}$, $[\mathbf{g}(\alpha_k)]_n = g(n, \alpha_k) = e^{-\alpha_k n}$, and $\odot$ represents the element-by-element Hadamard product [16]. To reduce the computational complexity of the maximization in (9.6) we use a computationally efficient algorithm based on the RELAXation method [17,18]. It allows us to decouple the problem of jointly estimating the parameters of the signal components into a sequence of simpler problems, in which we estimate separately and iteratively the parameters of each component. RELAX first roughly estimates the parameters of the strongest component. It obtains the estimate $\hat{f}_1$ from the location of the highest peak of the Periodogram [17] of the data $\mathbf{y}$, and then estimates the complex amplitude β_1 and the parameter α_1 of the strongest component using the NLLS estimators for single component [8]. Subsequently, the contribution of the strongest component is subtracted from the data, and the parameters of the new strongest second component are estimated. The procedure is iteratively repeated until "practical convergence" is achieved. This convergence is measured on the cost function:

$$CF\left(\left\{\hat{f}_k, \hat{\alpha}_k, \hat{\beta}_k\right\}_{k=1}^{P}\right) = \sum_{n=0}^{N-1} \left| y(n) - \sum_{k=1}^{P} \hat{\beta}_k e^{-\hat{\alpha}_k n} e^{j2\pi \hat{f}_k n} \right|^2 \tag{9.8}$$

where $P = 2$. Convergence is determined by checking the relative change of the cost function $CF(\cdot)$ between the jth and $(j+1)$th iterations. In our numerical simulations iterations were terminated when the relative change was lower than $\varepsilon = 10^{-4}$, as in [17]. When the convergence is achieved, the first two components are subtracted from the data and the parameters of the third one are estimated. The procedure is again iteratively repeated until convergence is achieved with the same cost function where now $P = 3$. The overall algorithm is repeated until the convergence for $P = 2K$ is achieved. Details on the RELAX are in [8,17,18].

9.4.2 Gaussian pulse

For the sonar click of second kind, we adopted a dumped Gaussian multicomponent signal model, that is, we model the extracted signal $x(n)$ as

$$x(n) = A_0 + \sum_{k=1}^{K} A_k e^{-\alpha_k (n - n_{0k})^2} \cos(2\pi f_k n + \vartheta_k) \tag{9.9}$$

where A_0 is the mean value, A_k, f_k, and ϑ_k are amplitude, frequency and initial phase of the kth component, respectively. The model (9.9) is very similar to that proposed by Kamminga and Stuart in [19], where the authors use the Gabor functions. In that work the number of components was fixed to two, the principal component and the reverberation; here K can be greater than two to fit better the observed data.

Again, the signal (9.9) can be expressed in the more general form:

$$x(n) = A_0 + \sum_{k=1}^{2K} \beta_k e^{-\alpha_k (n-n_{0k})^2} e^{j2\pi f_k n} \tag{9.10}$$

where $f_k = -f_{k+K}$, $\beta_k = \beta^*_{k+K} = A_k e^{j\vartheta_k}/2$, $\alpha_k = \alpha_{k+K}$ and $n_{0k} = n_{0k+K}$.

The difference between models (9.9) and (9.1) is the function characterizing the pulse envelope. In the model (9.1) it is an exponential function while in model (9.9) is a Gaussian function, that is $[\mathbf{g}(\alpha_k, n_{0_k})]_n = g(n, \alpha_k, n_{0_k}) = e^{-\alpha_k (n-n_{0k})^2}$. The exponential is characterized only by one parameter, the decay α, the Gaussian function by two parameters, the scale parameter α and the mean value n_0. Therefore for the Gaussian model there is one more parameter to estimate. In this case as well we applied the NLLS estimation method and we implemented the RELAX algorithm to simplify the search for the maximum. The algorithm is very similar to that applied for the exponential shaped pulse.

The periodograms of an exponential pulse and a Gaussian pulse are plotted in Figure 9.7(a) and (b). For the analysed exponential pulse, the main component is located around 25 kHz, for the Gaussian pulse around 38 kHz.

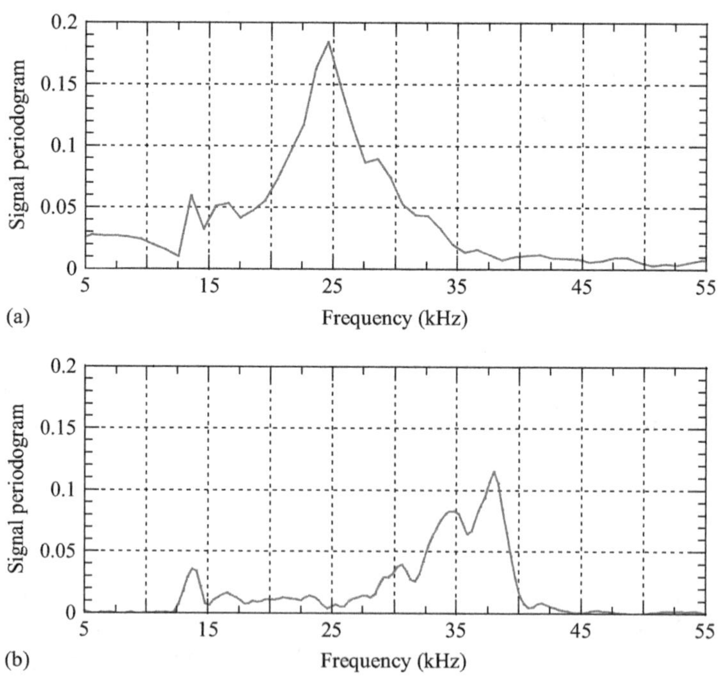

Figure 9.7 Signal periodogram for the (a) exponential and (b) Gaussian pulse in Figure 9.5(a) and (b), respectively

9.5 Estimation results

9.5.1 Exponential pulse

In our analysis we set $K=2$, 3, and 4. We obtained a good fitting already for $K=2$. Here we show the results for $K=4$. In Figure 9.8 we show the scatterplot for the first two frequencies and exponential decays. It is evident that the first component (circles) is centred around 20–25 kHz and spans almost the whole considered interval for the value of the exponential decay α_1. The frequency of the second component is spread-out on the interval 10–35 kHz. These results are confirmed by the histograms of frequencies and decays plotted in Figures 9.9 and 9.10. The first frequency (Figure 9.9(a)) has a Gaussian-like histogram with a mean value $\eta_{f_1} = 23.59$ kHz and a standard deviation $\text{std}\{f_1\} = 5.88$ kHz. Conversely the second frequency (Figure 9.9(b)) is almost uniformly distributed in the range [16–32 kHz] with a mean value $\eta_{f_2} = 24.28$ kHz and a standard deviation $\text{std}\{f_2\} = 8.32$ kHz. The exponential decays exhibit Gaussian-like histograms with parameters $\eta_{\alpha_1} = 0.0177$, normalized variance $\text{std}\{\alpha_1\} = 0.0066$, $\eta_{\alpha_2} = 0.0227$, and normalized variance $\text{std}\{\alpha_2\} = 0.010$ (Figure 9.10). The third and fourth frequency components are almost uniformly distributed as well.

The mean and the standard deviation of each parameter have been calculated as:

$$\hat{\eta}_\theta = \frac{1}{N_e}\sum_{i=0}^{N_e-1}\theta_i \tag{9.11}$$

and

$$\text{std}\{\theta\} = \sqrt{\frac{1}{N_e}\sum_{i=0}^{N_e-1}(\theta_i - \hat{\eta}_\theta)^2} \tag{9.12}$$

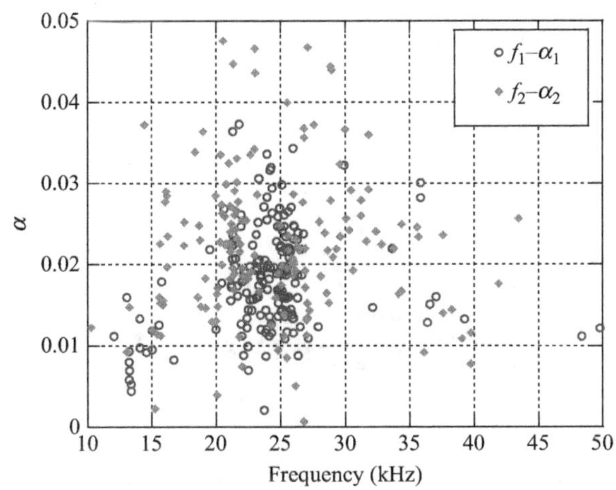

Figure 9.8 Scatterplot of frequency and exponential decay of first and second component, exponential model, K = 4

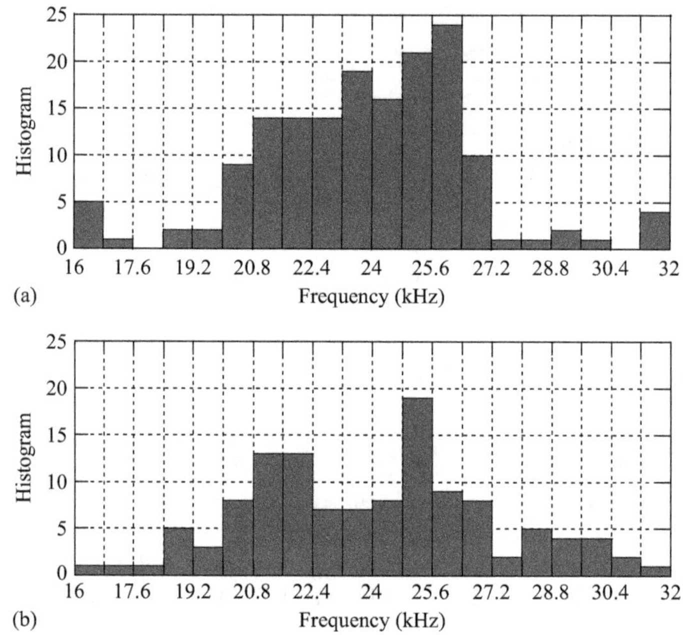

Figure 9.9 *(a) and (b) Histograms of the frequency of first and second components, exponential model,* K = *4*

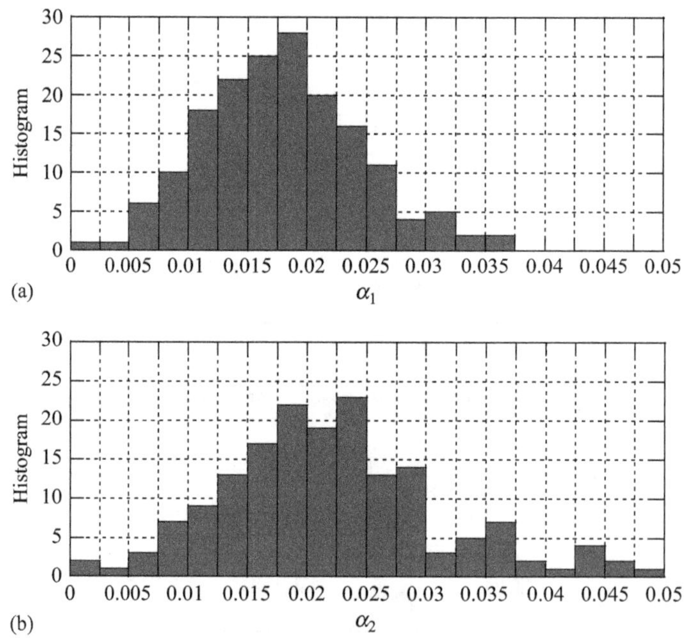

Figure 9.10 *(a) and (b) Histograms of the exponential decay parameter of first and second components, exponential model,* K = *4*

where N_e is the number of estimates and θ_i the ith estimate of the parameter of interest.

In Figure 9.11 we report the scatterplot of frequencies and amplitudes for the first two components. The amplitude is maximum when the frequency is between 20 and 25 kHz.

From the results in Figures 9.9–9.11, we can observe that the component characterizing the exponential sonar clicks is the first one; the other components simply contribute to improving the fitting. This means that, due to the almost uniform distribution of the frequency of the second component, knowledge of this frequency does not provide useful information for recognizing the sonar pulse of one dolphin specie from another.

The mean values of the frequencies of all the four components are beyond the audio band. In Figure 9.12 the observed and estimated signals are plotted for a sonar click for $K=4$. The fitting of the exponential model is good.

9.5.2 Gaussian pulse

Similar analysis has been carried out on the clicks of the second kind and the results are reported in Figures 9.13–9.17 for $K=4$. The frequency of the first component is concentrated in the interval [21–27 kHz] with mean value $\eta_{f_1} = 25.83$ kHz and normalized variance $\mathrm{var}\{f_1\} = 0.186$, the frequency of the second component is almost uniformly distributed in [14–40 kHz] with mean value $\eta_{f_1} = 27.21$ kHz and normalized variance $\mathrm{var}\{f_1\} = 0.2723$ (Figures 9.13 and 9.15). Both the scale factors exhibit a histogram with an exponential-like behaviour in the range [0–0.02] as shown in Figures 9.14 and 9.15. Even the distributions of the time delays n_{0_1} and n_{0_2} of first and second components have a very similar Gaussian shape, but the

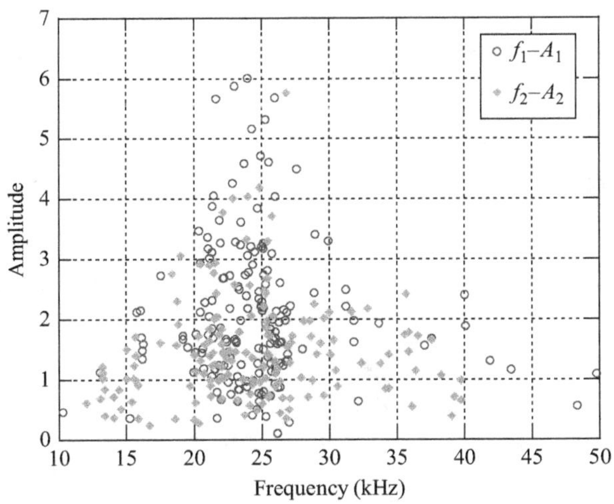

Figure 9.11 Scatterplot of frequency and amplitude of first and second component, exponential model, K = 4

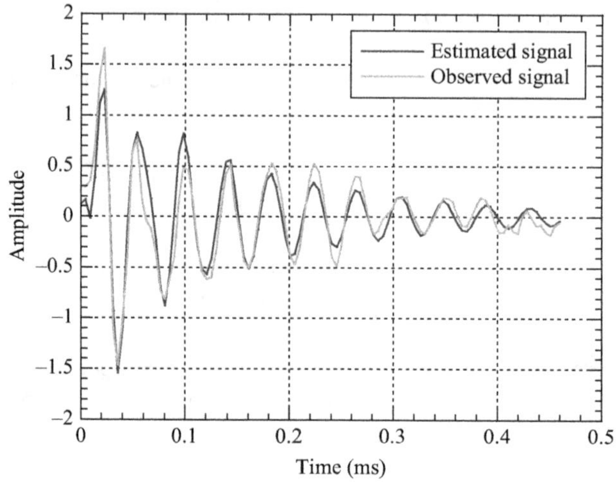

Figure 9.12 Fitting of an exponential pulse with the model (9.6) and K = 4

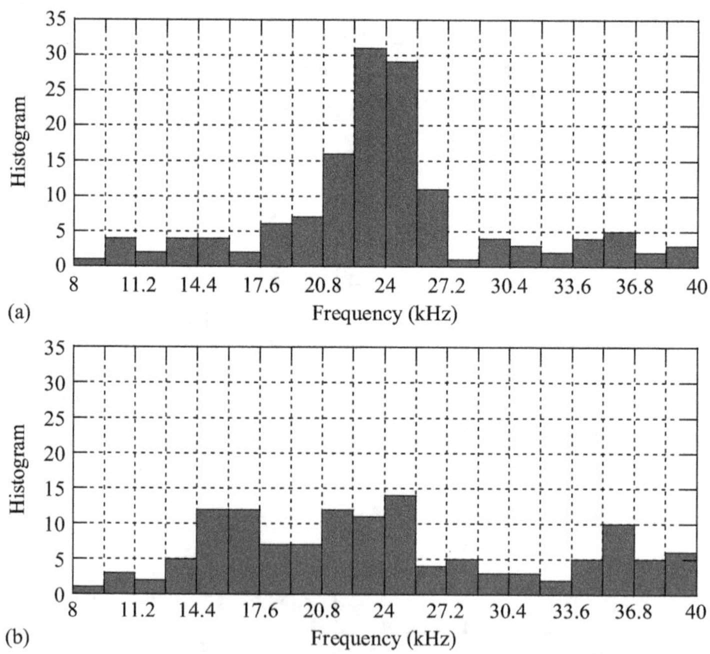

Figure 9.13 (a) and (b) Histograms of the frequency of first and second component, Gaussian model, K = 4

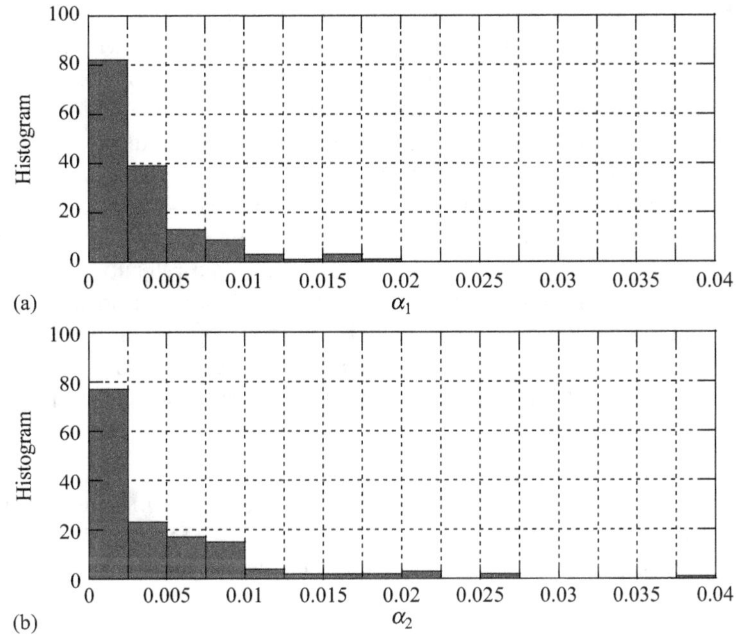

Figure 9.14 (a) and (b) Histograms of the scale parameter of first and second component, Gaussian model, K = *4*

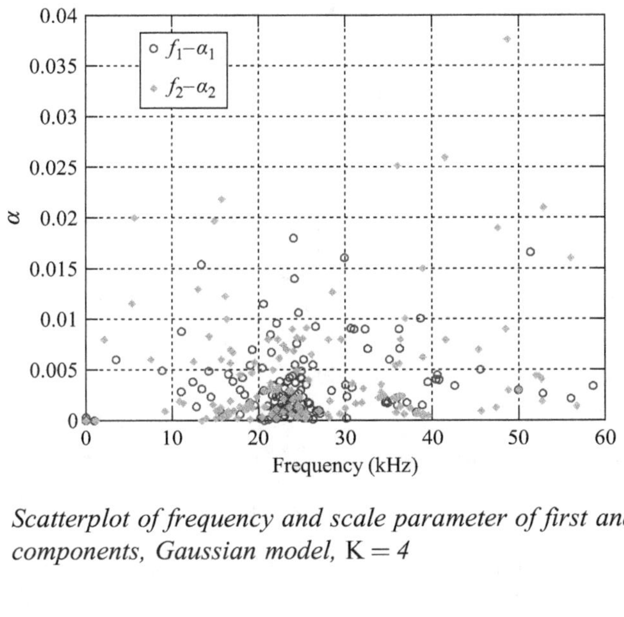

Figure 9.15 Scatterplot of frequency and scale parameter of first and second components, Gaussian model, K = *4*

mean value of the second component is greater than the first one; that is, the second Gaussian envelope is delayed with respect to the first one as shown in Figure 9.16; as a matter of fact $E\{n_{0_1}\} = 0.16$ ms and $E\{n_{0_2}\} = 0.17$ ms. The maximum amplitude corresponds to the components around 24 kHz as shown in the scatter-plot in Figure 9.17. Again, the dominant component is in the ultrasonic band.

We did not observe very high frequency peaks in the sonar clicks emitted by the analysed Mediterranean Bottlenose dolphins as reported in the literature for Oceanic Bottlenose dolphins [2]. This phenomenon could be due mainly to the difference in the environment. It is necessary to observe that those data refer to specimen living in the ocean and so in deep water and they use to move on long distances. To orientate they use high frequency and high power clicks. In fact, the dolphins cannot emit high power signals at low frequency [2]. The cetaceans we are studying live in shallow waters, therefore they can use low power signals and consequently low frequency.

9.5.3 Audio band

In analysing the data recorded in the frequency range [0–180 kHz] we did not find significant signal content at very low frequencies. This fact can be easily understood by observing that usually in the dolphin emissions higher frequency signals are characterized by higher power. The gain of the amplifier was manually changed during the recording in order to guarantee a good amplification and the absence of

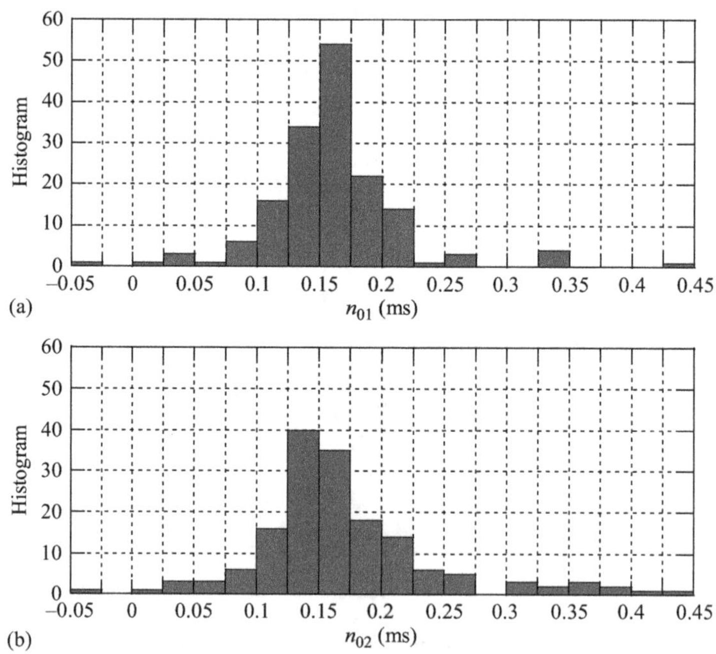

Figure 9.16 (a) and (b) Histograms of time delay of first and second components, Gaussian model, K = 4

clipping even in the presence of strong emissions. By doing so, in the wide frequency range data, the low power-low frequency pulses are covered by the electrical noise of the recording device.

Using the digital card of the laptop for audio signals, we recorded some files only in the audio band [0–16 kHz]. From these files, we extracted several exponential shaped sonar clicks. We analysed these sonar click trains as in the ultrasonic band for $K = 2$. The results are summarized in Figure 9.18 where the scatterplot of

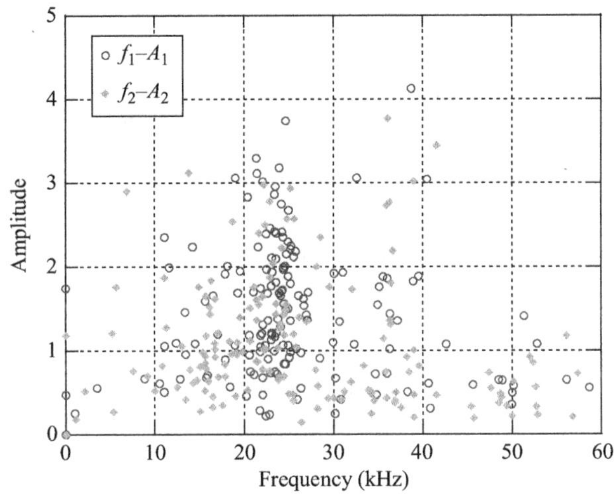

Figure 9.17 Scatterplot of frequency and amplitude of first and second components, Gaussian model, K = *4*

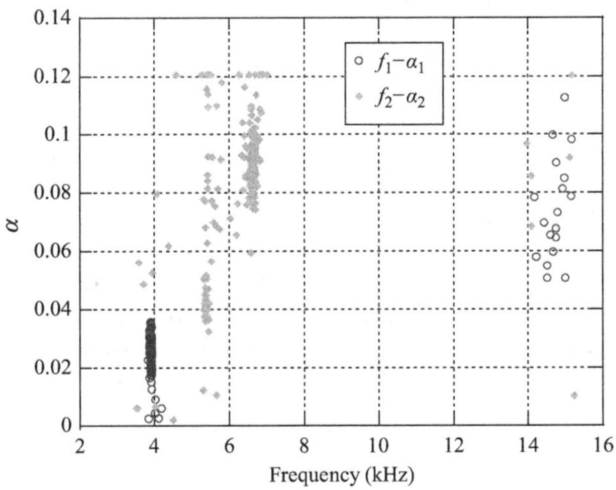

Figure 9.18 Scatterplot of frequency and exponential decay of first and second component, exponential model, K = *2, audio band*

the estimated parameters (α_1, f_1) and (α_2, f_2) is reported. From this figure it is evident that the frequency of the first peak is almost constant around 3.8 kHz for each pulse while its exponential decay (α_1) varies (vertical dark line of circles) in the range [0, 0.038]. The frequency of the second peak seems to have two more frequent values around 5.3 and 6.5 kHz. The decay parameter varies sensibly in the range [0, 0.12] (light gray line of diamonds). On the graph, there are some isolated points up to 14 kHz due to a minority of very short pulses.

9.6 Conclusions

After 35 million years of trials, training and errors, the dolphins have developed a very efficient biosonar for detecting, classifying and localizing targets the size of a sardine, in cluttered background, over ranges from 0 m to about 150 m, in all the sea states, and all the maritime environments, from the open ocean to rivers and estuaries.

In this chapter, we described the analysis carried out on the sonar clicks emitted by Mediterranean Bottlenose dolphins in both audio and ultrasonic bands. We found that most of the sonar clicks emitted when the dolphin is in front of the hydrophone can be modelled by exponential or by Gaussian broadband multi-component signals. The parameters of these two models have been estimated. The components characterizing each pulse are generally the first or the first two most powerful ones and the fitting with the data seems to be very good in both audio and ultrasonic band. Actually, the meaning of the sonar clicks in the audio band signals is not clear. Maybe, as reported by Zimmer [20], they can be "machinery noise", that is noise produced by dolphins in emitting the ultrasonic pulses used for the echolocation. In ultrasonic band the most powerful frequency component is located around 24 kHz, almost four octaves under the frequency peak measured for the Oceanic Bottlenose dolphins. This phenomenon can be mainly due to the differences in the Oceanic and Mediterranean environments. However, the efficiency of the dolphin sonar is not only due to the broadband characteristic of the sonar click signals, then to the very high range resolution. Dolphins are able to use as well a multi-perspective, multi-waveforms approach to sense the targets, moving around their preys and changing their trajectories [21], the power and the Pulse Repetition Frequency (PRF) based upon what they have learned from the previous sonar clicks [11,22]. Finally, they use their trained brain to control the whole biosonar process allowing for versatility and continuous learning [4]. It is, of course, not trivial to build a technological sonar similar to that of the marine mammals but the idea of cognitive radars (and so sonars) has been already proposed in 2006 by Haykin in [23] and some progress has been made since then along that path.

References

[1] W.W.L. Au, "Dolphin sonar detection and discrimination capabilities", *Journal of the Acoustical Society of America*, vol. 115, no. 5, Apr. 2004, p. 2614.

[2] H.L. Roitblat, W.W.L. Au, P.E. Nachtigall, R. Shizumura, and G. Moons, "Sonar recognition of targets embedded in sediment", *Neural Networks*, vol. 8, nos. 7–8, 1995, pp. 1263–1273.

[3] T.W. Cranford, W.R. Elsberry, W.G. Van Bonn, *et al.*, "Observation and analysis of sonar signal generation in the Bottlenose dolphin (*Tursiops truncatus*): evidence for two sonar sources", *Journal of Experimental Marine Biology and Ecology*, vol. 407, no. 1, 2011, pp. 81–96.

[4] W.W.L. Au and S.W. Martin, "Why dolphin biosonar performs so well in spite of mediocre equipment", *IET Radar, Sonar & Navigation*, vol. 6, no. 6, 2012, pp. 566–575.

[5] W.W.L. Au, *The Sonar of Dolphins*, Springer-Verlag, New York, 1993.

[6] W.W.L. Au, B.K. Blanstetter, K.J. Benoit-Bird, and R.A. Kastelein, "Acoustic basis for fish prey discrimination by echolocating dolphins and porpoises", *Journal of the Acoustical Society of America*, vol. 126, no. 1, 2009, pp. 460–467.

[7] P. Dobbins, "Dolphin sonar-modeling a new receiver concept", *Bioinspiration & Biomimetics*, vol. 2, no. 1, 2007, pp. 19–29.

[8] M. Greco and F. Gini, "Analysis and modeling of echolocation signals emitted by Mediterranean Bottlenose dolphins", *Journal of Applied Social Psychology*, vol. 2006, no. 1, 2006, pp. 1–10.

[9] K.S. Norris, J.H. Prescott, P.V. Asa-Dorian, and P. Perkins, "An experimental demonstration of echolocation behavior in the porpoise, *Tursiop truncatus*",*Biological Bulletin*, vol. 120, no. 2, 1961, pp. 163–176.

[10] W.W.L. Au, D.A. Carder, R.H. Penner, and B.L. Scronce, "Demonstration of adaptation in beluga whale echolocation signals", *Journal of the Acoustical Society of America*, vol. 77, no. 2, 1985, pp. 726–730.

[11] J.E. Sigurdson, "Biosonal dynamics of the Bottlenose dolphin in VSW search and detection tasks", *Journal of the Acoustical Society of America*, vol. 102, no. 5, Jun. 1998.

[12] E.L. Ferguson, "Precise sound source localisation of dolphin biosonar pulse transmissions", *IEEE OCEANS 2010*, Sydney, pp. 1–8, 2010.

[13] M.P. Olivieri, "Bio-inspired broadband SONAR technology for small UUVs", *MTS/IEEE OCEANS 2002*, vol. 4, pp. 2135–2144, 2002.

[14] D.L. Herzing, "Clicks, whistles and pulses: passive and active signal use in dolphin communication", *Acta Astronautica*, vol. 105, no. 2, Dec. 2014, pp. 534–537.

[15] W.M. Carey, "Sound sources and levels in ocean", *IEEE Journal of Oceanic Engineering*, vol. 31, no. 1, 2006, pp. 61–75.

[16] P. Stoica and R. Moses, *Introduction to Spectral Analysis*, Prentice-Hall, Ed., Upper Saddle River, NJ, 1997.

[17] J. Li and P. Stoica, "Efficient mixed-spectrum estimation with application to target feature extraction", *IEEE Transactions on Aerospace and Electronic Systems*, vol. 44, no. 2, Feb. 1996, pp. 281–295.

[18] F. Gini, M. Greco, and A. Farina, "Multiple target detection and estimation by exploiting the amplitude modulation induced by antenna scanning.

Part I: parameter estimation", *IEEE Transaction on Aerospace and Electronic Systems*, vol. 39, no. 4, Oct. 2003, p. 1316.

[19] C. Kamminga and C.A.B. Stuart, "Wave shape estimation of delphinid sonar signals, a parametric model approach", *Acoustics Research Letters*, vol. 19, no. 4, 1995, pp. 70–76.

[20] W. Zimmer, "Private communication", February 2004.

[21] Y. Pailhas, C. Capus, and K. Brown, "Dolphin-inspired sonar system and its performance", *IET Radar, Sonar & Navigation*, vol. 6, no. 8, 2012, pp. 753–763.

[22] L.N Kloepper, A.B Smith, P.E Nachtigall, J.R Buck, J.A. Simmons, and A.F. Pacini, "Cognitive adaptation of sonar gain control in the Bottlenose dolphin", *PLoS ONE*, vol. 9, no. 8, 2014, p. e105938. doi: 10.1371/journal.pone.0105938.

[23] S. Haykin, "Cognitive radar: a way of the future", *IEEE Signal Processing Magazine*, vol. 23, no. 1, 2006, pp. 30–40.

Chapter 10

Human echolocation – spatial resolution and signal properties

Liam Norman[1] *and Lore Thaler*[1]

10.1 Introduction and background

Perception is the process by which an organism uses sensory input to understand the world around them. According to some theories on perception, such as Gibson's [1], the perceiver plays an active role in achieving this – e.g., by choosing to orient their head towards a sound source or to turn round an object in their hands. The senses that are most important for human perception of the distal environment, however, rely on some external source of energy, in the form of light or sound, to excite the environment in order to be useful. Some humans, however, have developed a mode of perception – echolocation – that is unique in that it does not require an external source of energy to reveal something about the environment. Echolocation is the perception of objects through the sound waves reflected from their surfaces, and typically an echolocator may choose to create their own sound emissions that induce these reflected sound waves. Thus, echolocation affords an individual with the unique ability to produce the energy that "excites" the world around them [2]. Given this unique characteristic of echolocation it is unsurprising that it is a technique that is used by some non-human species that have evolved to live in environments with low light levels (e.g., bats, toothed whales) [3,4], but it is quite remarkable that some blind humans have also developed this ability, often by independently training themselves. This is especially impressive given that the human auditory system is predisposed to actively suppress the perception of echoes – a quality often termed echo suppression or the precedence effect [5,6]. Nonetheless, some humans have become proficient at using echoes to perceive their environment and, given the unique nature of echolocation, the degree to which it can be successfully used is determined both by the acoustic properties of the sonar emission as well as the proficiency with which the returning echoes can be interpreted. This chapter will focus on the echolocation abilities of humans, describing the acoustic properties of their emissions as well as the acuity with which they are able to discriminate certain object properties – distance,

[1]Department of Psychology, Durham University, Durham DH1 3LE, UK

angle (horizontal and vertical), size, shape and material – and the various cues that they might use to do this.

Historical accounts of blind individuals who were seemingly able to avoid obstacles and judge their distances were initially not explained by an ability to perceive echoes [7], and instead an explanation involving "facial sense" (or "facial vision") was put forward. This explanation involved the supposition that close proximity to an object increased the pressure exerted on the individual's face. This explanation, however, was rejected following experiments by Supa *et al.* [8], in which humans who allegedly used a "facial sense" were required to walk towards an obstacle and stop before colliding with it. When their ears were occluded, their ability to do this was significantly impaired, but not when their facial skin was covered, and other similar experiments also concluded that this recognised ability of a "facial sense" was in fact an auditory-based ability [9–11]. These findings on humans mirrored those on the navigational abilities of bats by Donald Griffin, in which it was shown that obstructing either the bat's mouth or ears, but not their eyes, impaired their ability to navigate in flight [12]. The term "echolocation", coined by Griffin [13], was thus adopted to describe navigation and obstacle avoidance by humans and nonhuman animals.

10.2 Acoustic properties of human sonar emissions

Echolocation is classically thought of as an active mode of perception, in that an echolocator typically emits a signal that excites the environment. It is also possible, however, for humans to use echolocation passively, relying on cues such as the echoes that result from external sound sources as well as the build-up of sound pressure that results from one being in close proximity to a large object [14,15]. Nonetheless, a very active form of echolocation is used by many blind people through the production of sound emissions followed by the interpretation of their returning echoes. The canonical form of emission generated by human echolocators is an oral click, which is usually produced by forming and releasing a vacuum between the tongue and palate by first pressing the tongue against the palate and then sharply moving the tongue down, away from the palate [16,17], although other types of emissions have been reported to be used (e.g., finger-snapping, hand-clapping, cane-tapping and other vocalisations). The waveform of a typical oral click can be described as a sinusoidal function modulated by a decaying exponential, although the exact nature of the oral click does vary between individuals and the procedure that they use to generate it. A click (including the exponentially decaying tail of the waveform) can last up to 15 ms [16] or be as short as 3 or 5 ms [18], with the number of clicks made varying between 1 and 5 per second and the sound levels between 60 and 108 dB [2]. In terms of the spectral content of these oral clicks, there is energy at multiple parts of the audible range, with peak frequencies typically varying between 2 and 8 kHz [2,19], which is low compared to echolocating bats, who can produce emissions in the ultrasonic range (i.e., frequencies higher than 20 kHz) [15].

The oral click might be an effective form of emission for echolocation compared to other forms of natural sound emissions that are available to humans [16,17]. In comparing the use of an oral click to that of a finger-snap, for example, Tonelli *et al.* [20] showed that novice echolocators learned just as well with either mode of emission, but at close distances to the object (30 cm) the oral click seemed to be the most useful. This oral click advantage may arise at such short ranges because the sound source of the oral click, compared to that of the finger-snap, ensures that an object directly in front of the echolocator at head-height can be effectively stimulated. Although a brief transient (i.e., click, finger snap) might be satisfactory in most natural settings, there is some evidence that acoustic emissions of a longer duration (i.e., 500 ms or longer) [21] might also be effective [21–23]. Schenkman *et al.* [23], for example, tested both blind and sighted participants' ability to detect when a reflecting surface was present as the number of short noise bursts made by a loudspeaker was varied between 1 and 64 per 500 ms or was a continuous noise emission. They found that at an object distance of 100 cm, the blind participants' performance increased steadily with the number of clicks, with the best performance obtained with 64 bursts or with continuous noise, whilst for sighted participants there was a decline in performance as the number of bursts exceeded 32. Interestingly, with an increase in target distance, the blind participants' performance also declined at 32 bursts. This decline was believed to have resulted from the temporal overlap between emission and echo at the ear of the perceiver. Nonetheless, a brief oral click is accepted as the canonical form of echolocation emission, partly also because its production does not interfere with the user's breathing or movement – a rate of two oral clicks per second, for example, can be sustained comfortably for an indefinite period [16]. There is the possibility that a single type of emission is not sufficient for all instances of echolocation, and echolocators may need to adapt the type of emission they use in accordance with changing task demands, similarly to bats [3] and porpoises [24]. There is certainly strong evidence that humans engage in head movements in order to perceive attributes such as shape or distance more accurately through echolocation [25–28].

The sound emissions that are typically used by echolocators, such as oral clicks or finger snaps (and even external sounds), will reach the echolocator's ear not only through airborne vibrations, but also through bone conduction [29]. It is possible that this conduction will persist beyond the time it takes for the returning echo to reach the echolocator's ear [2,30], thus adding to the overall level of interference when perceiving the signal. The signal arriving at the ear through bone conduction, however, is predominantly low in frequency, and the interaction between this and the returning echo at different temporal delays will result in a change in the fundamental frequency of the perceived sound that Schörnich *et al.* [2] theorised could be used as a cue by participants to judge distance to a reflecting object. The predominantly low frequency components of the emission present at the ear through bone conduction are one reason why oral clicks that contain a high level of energy in higher frequencies may be favoured over those that are lower in frequency [2], as the presence of high frequency sound at the ear would be a better indicator of the presence of an echo [31].

10.3 Environmental factors

In addition to the acoustics of the emission, the acoustics of the environment have been shown to affect the ability of humans to echolocate – specifically in the form of natural reverberations and additional reflectors. Although it had been thought by some that such additional factors would impair the ability to detect and discriminate objects, in a manner similar to the effect of background noise on sound localisation [15], Schenkman and Nilsson [21] in fact showed that a reflecting object could be detected up to a distance of 100 cm in an anechoic chamber, but up to 200 cm in a conference room. Furthermore, in a study by Tonelli *et al.* [20], in which the reverberation time was as long as 1.4 s [20], participants' ability to echolocate (both in the precision and accuracy of depth judgement) was still better in a reverberating room compared to an anechoic one. One theory for the cause of better performance in a more natural reverberating room is that "late echoes" – those echoes that reflect from walls and surrounding objects – may provide additional cues that aid in solving the task [20]. Gibson [1] posits that a surplus of perceptual information, that approximates ecological conditions, can make perception more reliable and veridical, and this principle may apply here. Nonetheless, to date these findings have only been obtained in people who were relatively new to echolocation, so it remains unclear as to whether these findings also apply to those who are more proficient.

10.4 Localising objects in space

10.4.1 Localising objects in depth (ranging): signal properties

The returning echo from a sonar emission such as an oral click can convey several pieces of information about the spatial content of the environment. The typical pattern of activity at the ear of an echolocator consists of the following successive sounds: the emission only, the superimposition of the emission and echo and, finally, the echo only [15]. For echoes returning from distant objects the superimposition of emission and echo may be absent, and there may be a period of silence instead. One application of active echolocation is in the detection of an object in front of the echolocator and inferring its distance from the temporal delay between the emission and echo. Figure 10.1(b) and (d), for example, show examples of waveforms and spectrograms from binaural recordings made at the ears of an echolocator (upper panel left ear, lower panel right) when making oral clicks at a target present at a distance of 150 cm (b) or 85 cm (d). The echo shown in (b) has a lower amplitude to that in (d) and is delayed by an additional 3–4 ms. The echo returning from a reflecting object at a longer distance might be perceivable as a signal separate from the emission – and distance might be inferred from this delay – but with shorter distances the echo and emission might fuse either acoustically or perceptually. Specifically, in the case of acoustic fusion, constructive and deconstructive interference between the emission and echo take place when they are temporally superimposed, resulting in physical changes in frequency and intensity of the composite sound. In the case of perceptual fusion, illusory changes in the perception of pitch arise from the auditory system's inability to

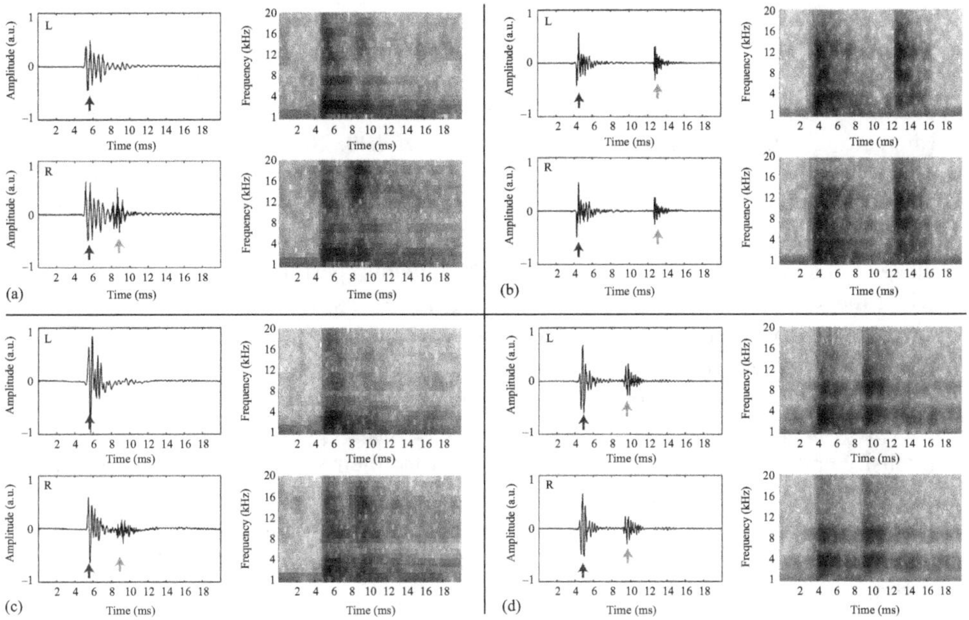

Figure 10.1 *Waveforms, plotting amplitude (a.u. = arbitrary units) against time (ms) and spectrograms denoting frequency (kHz) content as function of time (ms). In spectrograms darker shades indicate more energy in that frequency band at that moment in time. All figures are based on binaural recordings of clicks and echoes for four different echolocators (a–d). Recordings were made either at the entrance of the echolocators' ear canals (a, c, and d) or next to their ears, that is, on each side of the head, but placed outside the pinna, (b) while they made clicks and listened to echoes. Black arrows in waveform plots highlight clicks, and grey arrows highlight echoes. The recording sample frequency was 96 kHz for data shown on the right (b and d), and 44.1 kHz for data on the left (a and c). Spectrograms were calculated using a 1-ms window with 0.8-ms overlap in steps of 1 kHz. For (a) and (c), a sound-reflecting surface was located 60° to their right at a distance of 50 cm. For (b), a sound-reflecting surface was located straight ahead at a distance of 150 cm. For (d), a sound-reflecting surface was located straight ahead at a distance of 85 cm. Reproduced with permission from [18]*

resolve small delays between the emission and the echo (a phenomenon also referred to as repetition pitch) [32]. The presence of an echo at close temporal proximity to the emission, therefore, may result in an acoustic and/or perceptual composite sound that has certain perceptual qualities that could still be used to infer distance to a reflector. Importantly, however, only the emission-echo delay has an unambiguous physical relationship to reflector distance, whereas the frequency and intensity of the resultant sound will also depend on size, shape and material of the reflector.

Schenkman and Nilsson [33] played a set of pre-recorded sounds to both blind and sighted participants with the aim of determining which of the two cues (frequency or intensity) was more important for echolocation. The emissions in the sounds were not oral clicks, but 500 ms white noise bursts with a target object (diameter of 50 cm) placed at distances of either 100, 200 or 300 cm, with participants being required to detect the presence of the object in a 2 alternative forced choice (2AFC) task. With unprocessed sounds, blind participants detected the object almost perfectly in the 100 and 200 cm, but their performance dropped very close to chance in the 300 cm condition. Participants' performance dropped by 5% when the sounds were processed to remove the intensity component, but participants' performance dropped by almost 20% when the frequency component was removed. This was true for both blind and sighted participants, with blind participants outperforming the sighted participants only when the frequency component was present. It is difficult to know whether these results generalise, however, as the intensity and frequency cues were not equated for their salience. Arias and Ramos [34] suggested that the level of the reflected signal was not important for discriminating the distance of the reflector but, as shown through their pitch-matching experiments of both clicks and white noise, sighted participants and one blind participant did indeed perceive echoes from closer objects as having a higher pitch than those from distant ones, and an emission-echo delay of 5 ms was optimal for evoking an illusory perception of a change in pitch [34,35]. Thus, pitch might be a useful component to infer the presence of a reflector when the delay between emission and echo is short.

10.4.2 Localising objects in depth (ranging): spatial resolution of human echolocators

Schörnich *et al.* [2] used a psychophysical design to measure sensitivity to depth through echolocation in sighted participants, but who were extensively trained in echolocation to a level of performance that stabilised across sessions. They used a virtual echo-acoustic space paradigm in which each participant's click was recorded by a microphone, convolved with their individual head-related transfer function and played back to them with artificially induced time delay and intensity attenuation. Head related transfer functions (HRTFs) had been measured individually for each participant using calibrated sound sources placed at various locations in space and binaural microphones placed in a participant's ear canals. The different time delays and attenuation factors simulated sounds reflected by objects presented at different distances. They measured participants' sensitivity to changes in distance at three reference ranges: 170, 340 and 680 cm. The just-noticeable difference (JND – the minimum change in a stimulus attribute, in this case distance, required to elicit a change in perception) values at 340 cm were very consistent at approximately 80 cm across the five tested participants. The JNDs at 170 and 680 cm were more variable, but on average these were 40 and 125 cm, respectively. Two participants, however, had JNDs as high as 200 cm in the 680 cm range condition. These results can be compared to those acquired earlier by Kellogg [36], who showed using a real setup

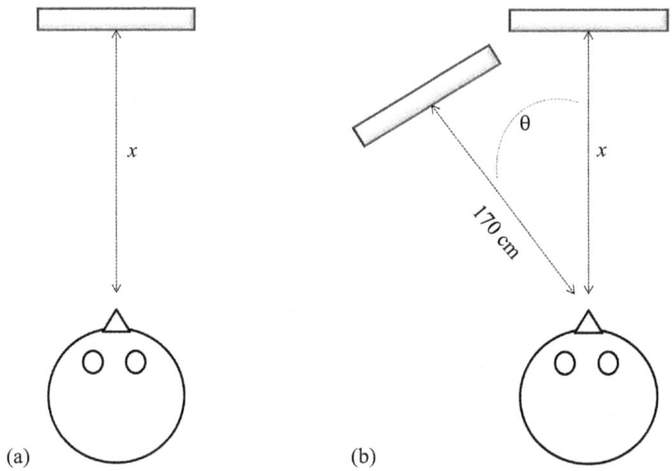

Figure 10.2 Virtual stimulus setup used in Schörnich et al.'s [2] experiment to measure localising objects in depth (ranging). Note that reflectors used in the experiment were simulated as 'perfect acoustic mirrors', i.e., they were simulated with an artificially induced time delay and intensity attenuation. Thus, physical reflectors drawn here are for illustration only. Participants' mouth-clicks were recorded and played back to them with an artificially induced delay and intensity attenuation to simulate the presence of a reflector at a variable distance (x). Participants' sensitivity to changes in the reflector's depth was measured at three levels of depth (x = 170, 140 or 680 cm) either without a secondary lateral reflector (a) or with a secondary lateral reflector (b). The lateral reflector, when present, was positioned at a fixed depth of 170 cm from the participant and offset by a horizontal angle (θ) of either 15°, 30° or 45°. Participants' sensitivity to localising the primary reflector improved when the lateral reflector was present at an angle of 45°, but not 15° or 30°

as opposed to a virtual one, that a blind practiced echolocator was sensitive to changes in the depth of an object by 10 cm at a distance of 60 cm. Interestingly in Schörnich *et al.*'s [2] experiment, they found that the presence of a second virtual reflector at a fixed distance of 170 cm, but offset laterally at a horizontal angle (azimuth) of 45° (see Figure 10.2), improved the participants' sensitivity to changes in the target reflector's distance (e.g., one participant improved from a JND of 27 cm to one of 13 cm at a reference range of 170 cm).

This improvement was greater with increasing azimuth angle of the second reflector. It was theorised that this second reflector served as a calibrating tool for the participants – a point of temporal reference against which to judge the echo from the target reflector. The second reflector did not improve performance if it was at an azimuth of 15° or 30°.

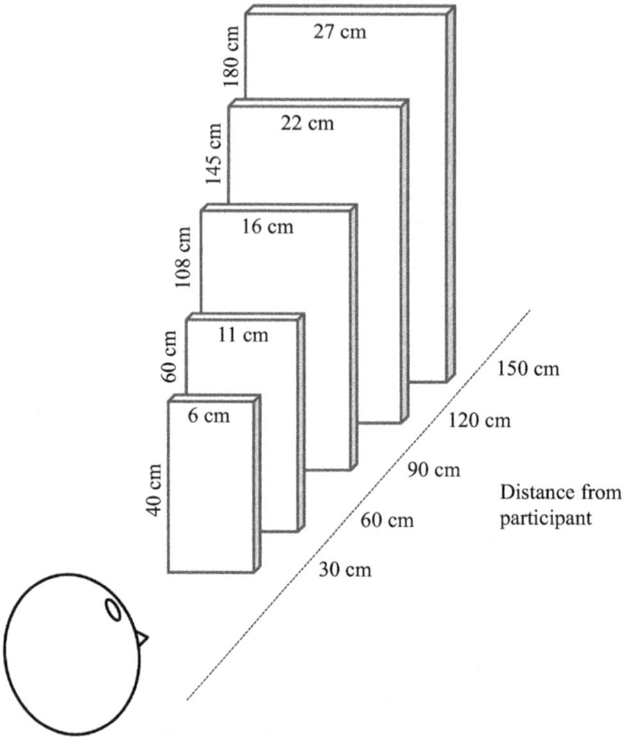

Figure 10.3 Illustration of the stimuli and experimental setup used by Tonelli et al. [20] to measure the resolution of localising objects in depth. Participants stood in front of a bar positioned at one of five locations in depth (30, 60, 90, 120 or 150 cm) and had to judge the depth that the bar was positioned at using echolocation with either a mouth-click or finger snap. The acoustic size of the bar (the acoustic angle it subtended) was kept constant (10° azimuth and 62° elevation) by using bars with greater height and width at greater positions in depth

Tonelli *et al.* [20] also trained sighted individuals without any prior experience echolocating in a depth discrimination task. Unlike the design implemented by Schörnich *et al.* [2], that used by Tonelli *et al.* [20] involved depth discrimination of real (i.e., non-virtual) objects in front of the echolocators. The reflector used was a sheet of poly-methyl methacrylate that was placed at a distance of either 30, 60, 90, 120 or 150 cm, with the size of the object being adjusted to keep its acoustic size constant at 10° in azimuth and 62° in elevation (see Figure 10.3). The acoustic size of an object refers to the acoustic angle subtended by that object, and this will change for an object of fixed size as a function of its distance from the perceiver. Participants in Tonelli *et al.*'s experiment [20] had to indicate which of these five positions they believed the object to be at and required only one hour of training for their performance to improve. Additionally, over two training sessions their error in precision in estimating distance decreased from approximately 25 cm to approximately 15 cm when tested in a reverberant room (performance was slightly worse in an anechoic room).

10.4.3 Localising objects in the horizontal plane (azimuth): signal properties

Echolocation allows the user the ability to detect the position of objects in the horizontal plane, known as azimuth. Figures 10.1(a) and (c) illustrate how the relative amplitudes of the echoes in the right and left ear convey information about the lateral position of the reflecting object relative to the orientation of the echolocators' head. In both cases the object was positioned at a horizontal angle (or azimuth in spherical coordinates) of 60° clockwise at a distance of 50 cm, and it is clear that the returning echo is much stronger in the right ear compared to the left. Another physical cue to lateral position is the interaural timing difference – that is, an object located to the right of the echolocator will reflect echoes to their right ear at an earlier time compared to those that reach the left ear.

Rowan *et al.* [22,37] measured sighted participants' ability to identify whether a 55 × 55 cm board was presented 17° to the left or right of centre in virtual acoustic space at various distances between 60 and 180 cm. By using different bands of noise as sound emissions, they showed that the binaural cue that was used to discriminate between azimuthal differences in their study was within the frequency range of 2–12 kHz. This would highlight the importance of interaural level differences for localisation. Additionally, by selectively removing binaural cues and measuring an associated drop in performance, they concluded that participants do indeed use a binaural cue that likely corresponds to the interaural level difference to resolve azimuthal differences. Rowan *et al.* [37] did, however, state that one participant in their experiment was able to use a monaural low-frequency cue to discriminate between the azimuthal positions. The participant anecdotally reported hearing a change in pitch between the left and right positions, but it was not determined experimentally whether this cue was indeed being used by this participant. It is unclear if these findings will generalise to click-emissions.

10.4.4 Localising objects in the horizontal plane: spatial resolution of human echolocators

Thaler *et al.* [38] showed using a 2AFC task that an early-blind blind echolocator could detect a change in azimuth of 4° of a pole (width 6 cm × height 180 cm) placed at a distance of 150 cm. Teng *et al.* [39] measured the sensitivity of blind echolocators to the horizontal offset of two flat circular disks (20.3 cm in diameter) at a distance of either 50 or 100 cm (see Figure 10.4). Participants had to indicate whether the top disk was positioned to the left or right of the bottom disk. At the 50 cm distance the top disk could be horizontally offset from the bottom disk between 1.1° and 13.2° to the left or right. At the 100 cm distance it could be offset between 0.57° and 3.4° to the left or right. The three best echolocators could discriminate between angular offsets of less than 2° with 75% success, but there was

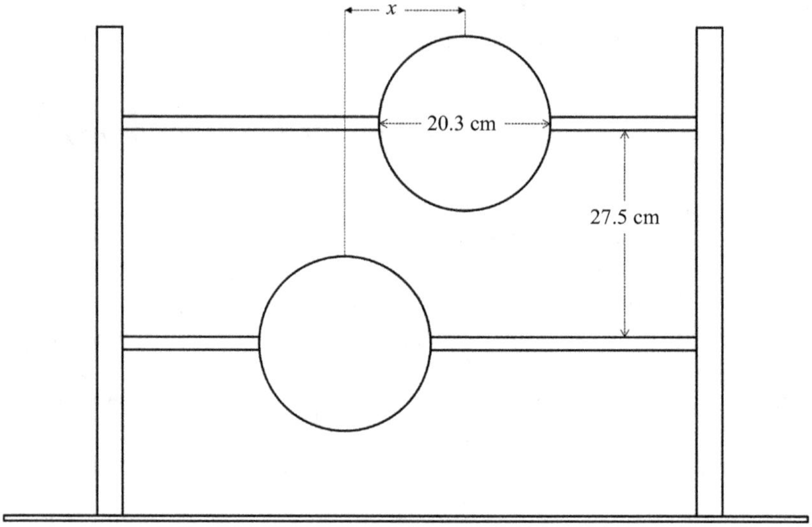

Figure 10.4 Illustration of the stimulus apparatus used by Teng et al. [39] to measure the spatial resolution of localising objects in the horizontal plane. Participants had to judge whether the top disk was offset to the left or right of the bottom disk. The horizontal displacement between the two disks (x) was varied such that the auditory angle subtended between the two disks varied between 1.1° and 13.2° at a distance of 50 cm from the participant or between 0.57° and 3.4° at a distance of 100 cm. The greater distance of 100 cm was used to avoid ceiling effects in performance of the three participants with the highest levels of performance

considerable variability between participants. This variability was partially predicted by the age of blindness-onset – with those who lost their vision at an earlier age having better performance in this task. This level of acuity has been compared to that of sound source localisation in the frontomedial plane [15] and corresponds approximately to monocular visual acuity in the same task at a retinal eccentricity of 35° [39]. In comparison, Wallmeier *et al.* [40] tested sighted participants' ability to discriminate the azimuth of an object in virtual acoustic space. Using a 2AFC task and an adaptive psychophysical method, they found that on average participants could discriminate between azimuth changes of 6.7° (with a range between participants' ability of 4.8°–9.2°) at a distance of 200 cm.

10.4.5 Localising objects in the vertical plane (elevation)

As for the cues used for possibly resolving object positions in the vertical plane (elevation in spherical coordinates), in general the incoming soundwaves are diffracted and reflected by the individual shape of one's pinna [41] and, as a result, contain complex spectral cues that denote elevation as well as allow front and rear disambiguation [42,43]. It is unclear to date to what degree this might apply within the context of echolocation, but future research should look at this issue.

10.5 Discriminating object size, shape and material

It is clear that human echolocators are able to resolve small changes in an object's position in space, both in terms of its depth and lateral position. It is also important, however, for human echolocators to be able to infer certain properties of an object in addition to its spatial position – such as size, shape and material – in order to appropriately interact with those objects. Teng and Whitney [44] measured size discrimination abilities of trained sighted echolocators and compared their performance to that of an expert blind echolocator. Participants judged whether the larger of two discs was presented above or below the other, with one having a fixed diameter of 25.4 cm and the other varying between 5.1 and 22.9 cm (see Figure 10.5). At a distance of 33 cm, the difference in the acoustic angle subtended by the two discs thus varied between 4.4° and 31.7°. Initially, participants had great difficulty discriminating even the largest differences in size, but performance improved markedly after a single training session of 100 trials. On average, performance reached an asymptote of approximately 80% for the angular size difference of 31.7°. To compare performance of the trained sighted participants to that of the expert blind participant, the highest skilled sighted participant reached a 75% performance threshold of 14.5°, whereas the same score for the expert blind participant was 8.0°. Thaler *et al.* [45] found similar levels of performance in sighted participants with the same task, and Rice and Feinstein [46] showed that blind participants could use echolocation to discriminate between object sizes at a surface area ratio as low as 1:1.07. Stoffregen and Pittenger [47] state that the size of an object can be inferred through echolocation by any differences in the level or spectrum of the reflected sound that are not accounted for by the object's distance and material. Remarkably, Milne *et al.* [48] have shown that a blind expert echolocator is able to infer the physical size of an object independently of its acoustic size. The expert echolocator in Milne *et al.* [48] study, for example, was able to correctly identify the size of a large object presented at a distance that equated its acoustic angle to that of a smaller object presented at a closer distance (see Figure 10.6). This is known as size constancy, and is a normal property of visual perception [49].

The shape of an object can also be inferred using echolocation. Milne *et al.* [26], for example, showed that expert echolocators were able to identify the shape of an object (square, triangle or two differently oriented rectangles, each with a surface area of approximately 1,600 cm^2 presented at a distance of either 40 cm or 80 cm; see Figure 10.7) at an accuracy of 75% correct identification. Importantly, this was only possible when they were able to move their head during echolocation. Non-echolocators, however, both blind and sighted, were unable to make

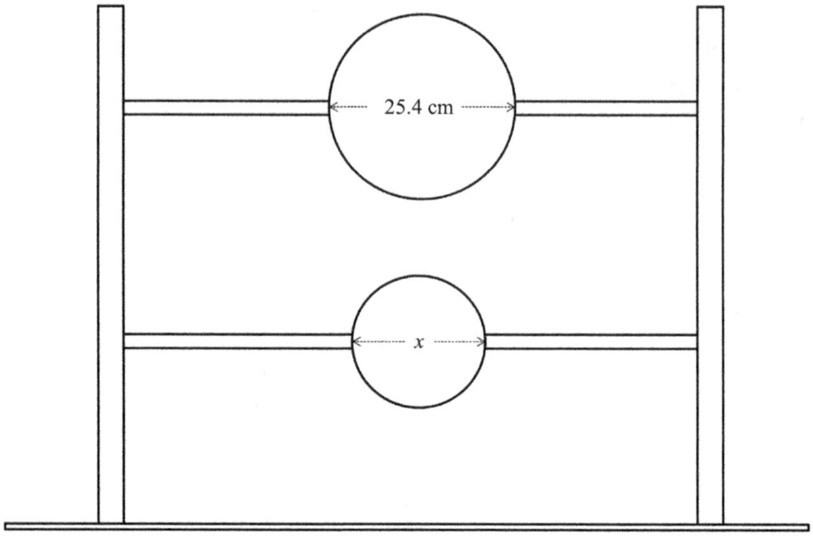

Figure 10.5 Illustration of the stimulus apparatus used by Teng and Whitney [44] to study size discrimination. The larger disk had a fixed size of 25.4 cm and would be positioned either on the upper of lower part of the apparatus, whereas the smaller disk had a variable diameter (x) between 5.1 and 22.9 cm. The auditory angle subtended by the difference in size between the two disks thus varied from 4.4° to 31.7° at a distance of 33 cm. Participants had to judge whether the larger disk was positioned on the upper or lower part of the apparatus after producing mouth-clicks

these discriminations at any level above chance. In contrast, Hausfeld *et al.* [50] study found that sighted participants could identify significantly above chance which of three simple geometric shapes (circle, triangle or square, with surface areas of approximately 177 cm^2) was presented at a distance of 25 cm from their forehead. The participants in that study were not practised echolocators and they performed less accurately in comparison to a blind expert echolocator, but it remains somewhat unclear as to the specific cues that are being used to solve such a task. The discrepancy between the results of Milne *et al.* [26] and Hausfeld *et al.* [50] might be due to the fact that performance feedback was given to those participants in the latter and/or that objects were presented at a closer distance.

In addition to object shape, there is also evidence that an object's material can be identified through echolocation [50–52], as different materials such as carpet or wood reflect different sound frequencies in different quantities. Milne *et al.* [51] showed that both echolocating experts as well as non-experts were able to reliably identify echoes from synthetic foliage, fleece or a whiteboard, and previous research has shown that participants report using pitch and timbre to identify materials such as wood and carpet [50,52].

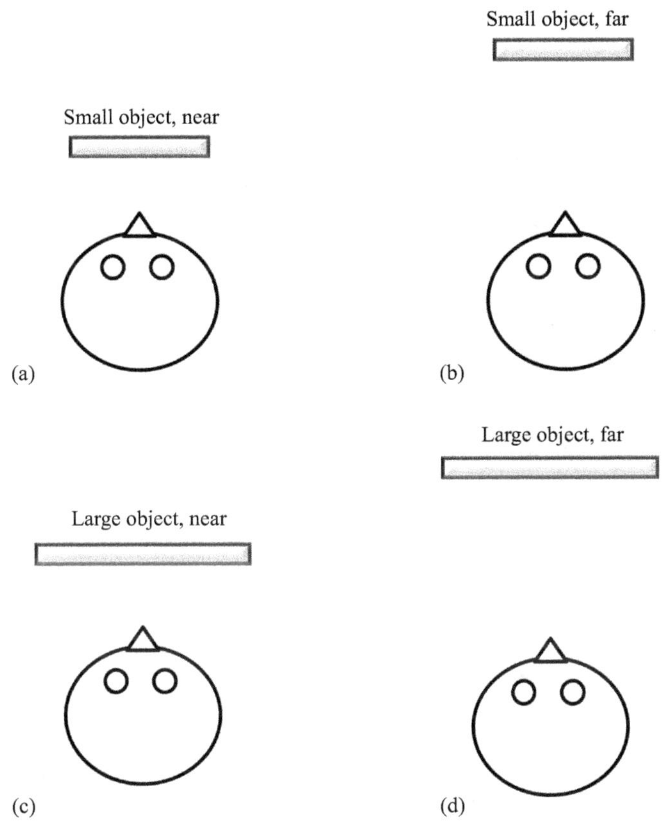

Figure 10.6 Illustration of the stimuli and experimental setup used by Milne et al. [48] to measure size constancy in echolocation. On a given trial, either a small or large object was positioned in front of the participant at either a near or far distance. The dimensions of the large object were such that, when positioned at the far distance (d), the object's acoustic size was identical to that of the smaller object when positioned at the near distance (a). Participants had to identify, after producing mouth-clicks, whether the small or large object was present in front of them, irrespective of its distance. Although blind and sighted novice echolocators were not able to do this, an expert echolocator was able to integrate information about acoustic size and distance in order to correctly identify the true physical size of the object, thus showing size constancy. Both rectangles and circles were used as stimuli in the experiment

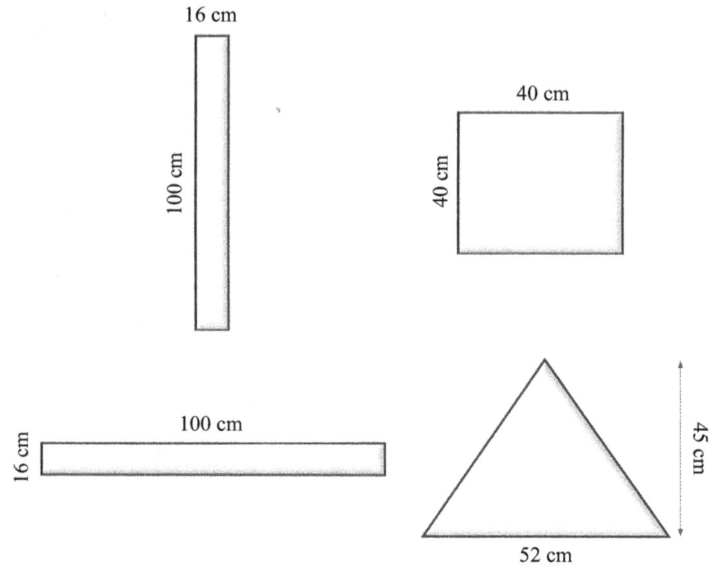

Figure 10.7 Illustration of the stimuli used by Milne et al. [26] to study shape identification by expert echolocators and novices. The shapes were two-dimensional and made of foam board with a surface covering of aluminium foil. The shapes were positioned on a pole with a 0.6 cm diameter, which was not detectable through echolocation, at a distance of 40 cm from the participant. Participants were presented with one of the four shapes on each trial and had to identify which shape it was. Although blind and sighted novice echolocators were not able to do this, expert echolocators were able to discriminate the shapes at a performance level of 75% correct when they were able to move their head during echolocation

10.6 Concluding comments and future directions

To conclude, echolocation offers some humans the ability to perceive their environments without the need for an external source of energy. Humans achieve this by producing emissions, usually in the form of an oral click, and interpreting the returning echoes that are reflected from objects in the environment. Many properties of these objects can be inferred using echolocation, including distance, angular position, size, shape and material, using cues such as frequency, level, delay between emission and echo, and interaural differences in time and level of the returning echoes, and potentially other factors such as sensory-motor contingencies that are not yet fully understood. Important factors that modulate the ability to echolocate include the reverberation conditions of the surrounding environment and the presence of additional reflecting surfaces, whether the echolocator is able to move freely whilst making oral clicks and also the acoustic properties of the emissions (e.g., spectral composition, temporal duration). It is quite difficult to summarise the acuity that echolocation affords, however, given the large variability in participants' abilities, and the differences in the types of task and methods used to measure these abilities. What is clear, however, is that the acuity of some expert echolocators in discriminating certain properties such as object distance or angular position can be extraordinarily fine, and often non-expert sighted individuals can be trained to an impressive level of accuracy as well. Experiments in the future are likely to establish more common ground and likely give us a more accurate estimation of the perceptual acuity afforded by echolocation in humans and greater understanding over the factors that modulate how well echolocation can be applied.

The first scientific investigations into human echolocation were conducted in the 1940s, i.e., at about the same time when research into bat echolocation commenced. In fact, as stated in the introduction, Griffin [13] coined the term echolocation with respect to both humans and nonhuman animals. Nonetheless, research into human echolocation has progressed at a slower pace than research into bat echolocation, for example. One possible reason for this is that there are few humans who use echolocation regularly and/or at high skill levels, making nonhuman echolocating species possibly more attractive and more available systems to investigate. In the last 5–10 years, however, research into human echolocation has experienced a resurgence. One possible reason is that humans can communicate verbally and thus provide opportunities for experimentation not available in nonhuman echolocating species. Furthermore, research has shown that some humans can achieve levels of performance that are unexpected considering that they are working in the audible (sonic) sound spectrum [39], thus highlighting opportunities for research not previously considered.

Based on the results of the experiments summarised in this chapter, it is evident that echolocation can provide humans with a range of information about the distal environment that is not limited to spatially localising an object. Specifically, the same echolocation process is used to reveal information about size, shape and material of objects as well as their spatial location. Furthermore, even though there have been only few studies to date investigating contextual factors, e.g., echolocation in anechoic environments vs. 'regular' environments, it seems that echolocation performs well across acoustic contexts (see Section 10.3). Developers of artificial systems might therefore benefit from studies on natural human echolocation in the development of 'multi-purpose' systems that use echolocation to provide multi-faceted information about the distal environment in various conditions.

It can also be learned from human echolocation that a relatively high level of spatial resolution in object localisation can be achieved with only a single sound emitter (usually the mouth) and two receivers (the ears), all working in the audible sound spectrum. Achieving good performance through echolocation in humans, therefore, is likely to be primarily limited by the neural systems processing the signals. This poses an exciting new approach to artificial sonar systems, as the focus is shifted away from sophistication of ‘hardware’ in creating and receiving signals towards ‘software’ processing and interpreting signals.

As stated before, natural human echolocation is fundamentally limited to sound information in the sonic range (<20 kHz), whereas an artificial system could go beyond these limitations to potentially achieve higher spatial acuity and directional specificity. In this context, Sohl-Dickstein *et al.* [53] showed that an artificial system could be used in combination with a human ‘processor’, in that the system generates and records ultrasonic emissions and echoes and plays resultant sounds back to a human participant with sounds having been slowed down to bring them into the range of human hearing. Using such a system participants were able to successfully, and intuitively, interpret echo-acoustic cues to make judgements about an object’s spatial location, suggesting that there can be a synergistic relationship between natural human echolocation and an artificial echolocation system. This presents further opportunities for research into the processes driving human echolocation and applications.

References

[1] Gibson, J. (1979). *The Ecological Approach to Visual Perception*. Boston, MA: Houghton Mifflin.

[2] Schörnich, S., Nagy, A. and Wiegrebe, L. (2012). “Discovering your inner bat: echo-acoustic target ranging in humans”. *Journal of the Association for Research in Otolaryngology*, 13, 673–682.

[3] Griffin, D. R. (1958). *Listening in the Dark: The Acoustic Orientation of Bats and Men*. New Haven, CT: Yale University Press.

[4] Jones, G. (2005). “Echolocation”. *Current Biology*, 15, 484–488.

[5] Wallach, H. Newman, E. B. and Rosenzweig, M. R. (1949). “The precedence effect in sound localization”. *American Journal of Psychology*. 62, 315–336.

[6] Litovsky, R. Y., Colburn, H. S., Yost, W. A. and Guzman, S. J. (1999). “The precedence effect”. *The Journal of the Acoustical Society of America*, 106, 1633–1654.

[7] Jourdain, M. (1916). *Diderot’s Early Philosophical Works*. Chicago, IL: Open Court.

[8] Supa, M., Cotzin, M. and Dallenbach, K. M. (1944). “Facial vision: the perception of obstacles by the blind”. *American Journal of Psychology*, 57, 133–183.

[9] Worchel, P. and Dallenbach, K. M. (1947). “Facial vision: perception of obstacles by the deaf-blind”. *American Journal of Psychology*, 60, 502–553.

[10] Cotzin, M. and Dallenbach, K. (1950). “Facial vision: the role of pitch and loudness in the perception of obstacles by the blind”. *American Journal of Psychology*, 63, 485–515.

[11] Köhler, I. (1964). “Orientation by aural clues”. *American Foundation of the Blind Research Bulletin*, 4, 14–53.

[12] Griffin, D. and Galambos, R. (1941). "The sensory basis of obstacle avoidance by flying bats". *Journal of Experimental Zoology*, 86, 481–506.

[13] Griffin, D. R. (1944). "Echolocation by blind men, bats and radar". *Science*, 100, 589–590.

[14] Ashmead, D. H. and Wall, R. S. (1999). "Auditory perception of walls via spectral variations in the ambient sound field". *Journal of Rehabilitation Research and Development*, 36, 313–322.

[15] Kolarik, A. J., Cirstea, S., Pardhan, S. and Moore, B. C. (2014). "A summary of research investigating echolocation abilities of blind and sighted humans". *Hearing Research*, 310C, 60–68.

[16] Rojas, J. A. M., Hermosilla, J. A., Montero, R. S. and Espi, P. L. L. (2009). "Physical analysis of several organic signals for human echolocation: oral vacuum pulses". *Acta Acustica united with Acustica*, 95, 325–330.

[17] Rojas, J. A. M., Hermosilla, J. A., Montero, R. S. and Espi, P. L. L. (2010). "Physical analysis of several organic signals for human echolocation: hand and finger produced pulses". *Acta Acustica united with Acustica*, 96, 1069–1077.

[18] Thaler, L. and Goodale, M. A. (2016). "Echolocation in humans: an overview". *Wiley Interdisciplinary Reviews: Cognitive Science*, 7, 382–393.

[19] Thaler, L. and Castillo-Serrano, J. G. (2016). "People's ability to detect objects using click-based echolocation: a direct comparison between mouth-clicks and clicks made by a loudspeaker". *PLoS ONE*, 11, e0154868.

[20] Tonelli, A., Brayda, L. and Gori, M. (2016). "Depth echolocation learnt by novice sighted people". *PLoS ONE*, 11, e0156654.

[21] Schenkman, B. N. and Nilsson, M. E. (2010). "Human echolocation: blind and sighted persons' ability to detect sounds recorded in the presence of a reflecting object". *Perception* 39, 483–501.

[22] Rowan, D., Papadopoulos, T., Edwards, D., *et al.* (2013). "Identification of the lateral position of a virtual object based on echoes by humans". *Hearing Research*, 300, 56–65.

[23] Schenkman, B. N., Nilsson, M. E. and Grbic, N. (2016). "Human echolocation: acoustic gaze for burst trains and continuous noise". *Applied Acoustics*, 106, 77–86.

[24] Tyack, P. (2015). "Echolocation: clicking for supper". *Elife*, 4, e07690.

[25] Rosenblum, L. D., Gordon, M. S. and Jarquin, L. (2000). "Echolocating distance by moving and stationary listeners". *Ecological Psychology*, 12, 181–206.

[26] Milne, J. L., Goodale, M. A. and Thaler, L. (2014). "The role of head movements in the discrimination of 2-D shape by blind echolocation experts". *Attention, Perception & Psychophysics*, 76, 1828–1837.

[27] Wallmeier, L. and Wiegrebe, L. (2014a). "Ranging in human sonar: effects of additional early reflections and exploratory head movements". *PLoS ONE*, 9, e115363.

[28] Wallmeier, L. and Wiegrebe, L. (2014b). "Self-motion facilitates echo-acoustic orientation in humans". *Royal Society Open Science*, 1, 140185.

[29] Stenfelt, S. and Goode, R. L. (2005). "Transmission properties of bone conducted sound: measurements in cadaver heads". *Journal of the Acoustical Society of America*, 118, 2373–2391.

[30] Patterson, R. D. (1976). "Auditory filter shapes derived with noise stimuli". *Journal of the Acoustical Society of America*, 59, 640–654.

[31] Kuc, R. and Kuc, V. (2016). "Modelling human echolocation of near-range targets with an audible sonar". *Journal of the Acoustical Society of America*, 139, 581–587.

[32] Bilsen, F. (1966). "Repetition pitch: monaural interaction of a sound with the repetition of the same, but phase shifted sound". *Acustica*, 17, 295–300.

[33] Schenkman, B. N. and Nilsson, M. E. (2011). "Human echolocation: pitch versus loudness information". *Perception*, 40, 840–852.

[34] Arias, C. and Ramos, O. A. (1997). "Psychoacoustic tests for the study of human echolocation ability". *Applied Acoustics*, 51, 399–419.

[35] Bilsen, F. and Ritsma, R. (1969/1970). "Repetition pitch and the implication for hearing theory". *Acoustica*, 22, 63–73.

[36] Kellogg, W. N. (1962). "Sonar system of the blind". *Science*, 137, 399–404.

[37] Rowan, D., Papadopoulos, T., Edwards, D. and Allen, R. (2015). "Use of binaural and monaural cues to identify the lateral position of a virtual object using echoes". *Hearing Research*, 323, 32–39.

[38] Thaler, L., Arnott, S. R. and Goodale, M. A. (2011). "Neural correlates of natural human echolocation in early and late blind echolocation experts". *PLoS ONE*, 6, e20162.

[39] Teng, S., Puri, A. and Whitney, D. (2012). "Ultrafine spatial acuity of blind expert human echolocators". *Experimental Brain Research*, 216, 483–888.

[40] Wallmeier, L., Geßele, N. and Wiegrebe, L. (2013). "Echolocation versus suppression in humans". *Proceedings of the Royal Society B*, 280, 1769.

[41] Shaw, E. A. G. (1980). The acoustics of the external ear. In: Studebaker, G. A. and Hochberg, I. (eds.). *Acoustical Factors Affecting Hearing Aid Performance and Measurement (109–125)*. Baltimore, MD: University Park Press.

[42] Musicant, A. D. and Butler, R. A. (1984). "The influence of pinnae-based spectral cues on sound localization". *Journal of the Acoustical Society of America*, 75, 1195–1200.

[43] Middlebrooks, J. and Green, D. M. (1991). "Sound localization by human listeners". *Annual Review of Psychology*, 42, 135–159.

[44] Teng, S. and Whitney, D. (2011). "The acuity of echolocation: spatial resolution in sighted persons compared to the performance of an expert who is blind". *Journal of Visual Impairment and Blindness*, 105, 20–32.

[45] Thaler, L., Wilson, R. C. and Gee, B. K. (2014). "Correlation between vividness of visual imagery and echolocation ability in sighted, echo-naive people". *Experimental Brain Research*, 232, 1915–1925.

[46] Rice, C. E. and Feinstein, S. H. (1965). "Sonar system of the blind: size discrimination". *Science*, 148, 1107–1108.

[47] Stoffregen, T. A. and Pittenger, J. B. (1995). "Human echolocation as a basic form of perception and action". *Ecological Psychology*, 7, 181–216.

[48] Milne, J. L., Anello, M., Goodale, M. A. and Thaler, L. (2015a). "A blind human expert echolocator shows size constancy for objects perceived by echoes". *Neurocase*, 21, 465–470.

[49] Boring, E. G. (1964). "Size constancy in a picture". *American Journal of Psychology*, 77, 494–498.

[50] Hausfeld, S., Power, R. P., Gorta, A. and Harris, P. (1982). "Echo perception of shape and texture by sighted subjects". *Perceptual and Motor Skills*, 55, 623–632.

[51] Milne, J. L., Arnott, S. R., Kish, D., Goodale, M. and Thaler, L. (2015b). "Parahippocampal cortex is involved in material processing via echoes in blind echolocation experts". *Vision Research*, 109(Part B), 139–148.

[52] DeLong, C. M., Au, W. W. and Stamper, S. A. (2007). "Echo features used by human listeners to discriminate among objects that vary in material or wall thickness: implications for echolocating dolphins". *Journal of the Acoustical Society of America*, 121, 605–617.

[53] Sohl-Dickstein, J., Teng, T., Gaub, B., *et al.* (2015). "A device for human ultrasonic echolocation". *IEEE Transactions on Biomedical Engineering*, 62, 1526–1534.

Chapter 11

Polarization tensors and object recognition in weakly electric fish

William R.B. Lionheart[1]
and Taufiq K. Ahmad Khairuddin[2]

11.1 Electrosensing in fish and electrical imaging by humans

Weakly electric fish have a single electric discharge organ and hundreds of voltage sensing receptors on their body to perform active electrosensing. Living in shaded parts of tropical rivers, in often turbid water, they use this additional sense for navigation as well as to characterize objects and locate prey [1–3]. While many salt and fresh water fishes use passive electroreception to pick up electrical signals from other animals, active electrosensing has been observed to have evolved only in knifefishes (*Gymnotiformes*) in South America and in elephantnose fishes (*Notopteroidei*) in Africa (Figure 11.1). Passive electroreception has also been observed in monotremes including the platypus in water, and bees have been reported to sense static electric fields in air [4].

Some species including Peters' elephantnose fish *Gnathonemus petersii* generate a broad spectrum pulse signal while others, such as the black ghost knifefish *Apteronotus albifrons*, use a signal close to a sine wave [5,6]). Typically fish using narrow band signals live in close proximity to others of the same species and operate on distinct bands avoiding interference. See Moller [7] for a comprehensive treatment of the behaviour of electric fishes.

Detection of objects with a contrast in conductivity and permittivity using low-frequency electrical measurement is also used by human technology. The oldest examples come from geophysics. Early work on detecting subsurface resistivity changes began in the 1900s; a technique now known as surface resistivity mapping. Current (typically a square wave pulse) is applied via pairs of electrodes on the surface of the earth while other electrodes are used to measure voltage and anomalies, such as where the ground is more conductive due to aqueous solutions,

[1]School of Mathematics, The University of Manchester, Manchester M13 9PL, UK
[2]UTM Centre for Industrial and Applied Mathematics, Department of Mathematical Sciences, Universiti Teknologi Malaysia, UTM Johor Bahru, 81310 Johor, Malaysia

Figure 11.1 An electrosensing elephantnose fish Gnathonemus phantasticus. *Photo courtesy of Chester Zoo*

show up on a contour plot of equipotentials. With the availability of fast computers, geophysicists began, in the 1980s, to solve the inverse problem of reconstructing a depth-dependent conductivity from surface measurement; a technique called Electrical Resistivity Tomography (ERT). Polarizable minerals are also of interest, corresponding to a contrast in dielectric permittivity, and can be found from the transient response to pulses or from the magnitude and phase measurement of voltages resulting from sinusoidal currents, a technique called Induced Polarization Tomography. Very similar techniques are used in medicine. An array of electrodes is attached to the skin and sinusoidal current is driven between pairs of these electrodes while voltages are measured on the others. This technique is known as Electrical Impedance Tomography (EIT) and is used especially for monitoring lungs in intensive care [9]. Similar techniques are used in industrial process monitoring to monitor multiphase flows in pipes and mixing vessels. One example of this is Electrical Capacitance Tomography [10]. Typically the process of image reconstruction involves solving the electric field equations using a computationally involved finite element model and the resistivity or conductivity is iteratively updated by solving a linearized system of equations.

When a weakly electric fish moves through the water approaching an object (see Figure 11.2), its single but moving electric discharge acts in a similar way to switching between driven electrodes in an ERT/EIT system. The main difference is that moving and changing the surface of the fish are equivalent to a changing boundary shape and electrode positions in EIT. Ammari *et al.* [11] have recently studied electrosensing fish in a two-dimensional approximation where the same model as EIT has been used. Details of how the fish perform the location and characterization are currently unknown. According to von der Emde and Fetz [2],

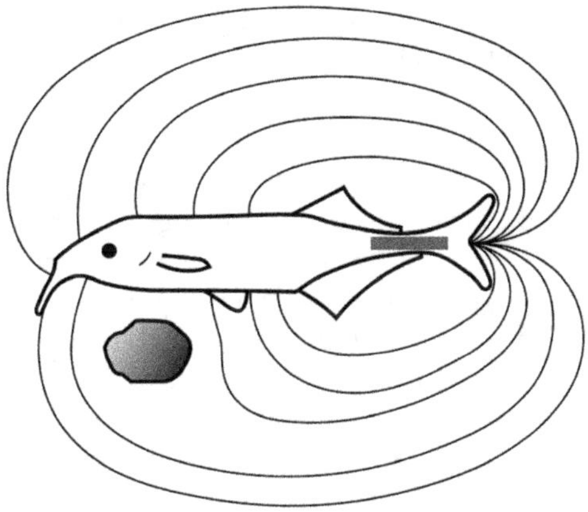

Figure 11.2 Diagrammatic representation of Peters' elephantnose fish showing current stream lines from the electric discharge organ distorted by a low conductivity target. Image adapted from [8] Copyright CC-BY-4.0

weakly electric fish can be trained to recognize and discriminate between conducting and insulating objects of different shapes. Previous studies [3] have also shown that fish can distinguish objects in a variety of orientations.

It is unlikely that a fish brain implements the equivalent of the computationally intensive algorithms used for ERT/EIT, and it is hypothesized that simpler algorithms with fewer degrees of freedom in the model are more likely. Indeed for the location of a single object, possibly their prey, it would be enough to locate the object at a fairly long range; then when closer classify the object using features such as its conductivity contrast to the water, its approximate size, and its approximate shape. If it is not spherical it would be desirable to determine which way it is oriented. While we might hope in the future that more detailed studies will be undertaken of the neuroscience of electrosensing fishes that shed light on how they process their sensory data, our current focus is on devising algorithms that would do the same job at a low cost. We can at least then hope that further behavioural experiments on the capabilities of the fish might rule out some of these ideas. In the mean-time as far as biomimetics is concerned, we have the tantalizing example that such tasks can be performed with a relatively small, low power consumption 'computer' and this can inspire us to find ways to do that ourselves.

11.2 Algorithms for electrical imaging

In general, for problems of low frequency electrical imaging in geophysics, medicine and industry, it is assumed that the magnetic field is negligible, so the electric field can be represented as the gradient of a scalar voltage potential u. For a time harmonic problem at a fixed angular frequency ω, and complex conductivity (or *admittivity*) $\gamma = \sigma + i\omega\varepsilon$, we have

$$\nabla \cdot \gamma \nabla u = 0 \tag{11.1}$$

where there are no current sources. Here σ is the conductivity and ε the permittivity.

Generally one assumes that the conductive body Ω is bounded. The current is applied and then, the voltage is then measured via a system of electrodes on the boundary $\partial\Omega$. Of course there is some coupling (electrostatic and inductive) externally but this is generally assumed to be negligible. The data therefore in electrical imaging is a knowledge of the Neumann boundary data, the current density $j = -\gamma \partial u/\partial \mathbf{n}$, where $\mathbf{n}$ is the outward normal, and $u|_{\partial\Omega}$ the voltage restricted to the boundary. These data are known for a number of applied currents j typically by exciting pairs of electrodes in turn. From this one attempts to deduce a γ consistent with (11.1). For an introduction to the algorithms used for this in the general case see [12,13]. One key ingredient in most reconstruction methods is the linearization of the forward problem, that is, for a perturbation $\gamma + \delta\gamma$ that results in a voltage $u + \delta u$, we have to find the best linear approximation to δu on $\partial\Omega$ in terms of γ. A simple form of this is when the original conductivity is uniform, say $\gamma = 1$, then we have $\nabla^2 u = 0$ and

$$\nabla^2 \delta u = -\nabla u \cdot \nabla \delta\gamma + O(||\delta\gamma||^2), \tag{11.2}$$

with the boundary condition $\partial \delta u/\partial n = 0$. This can be seen to be Poisson's equation where the source term depends on the original voltage u and the perturbation in γ. An isolated spherical object at a point z can be approximated by a delta function $\delta\gamma = \delta(x - z)$ and in this case, the right-hand side of (11.2) is a dipole source.

In the laboratory, a typical experiment performed by EIT researchers is to make a shallow horizontal circular tank filled with a salt solution with the inside circular portion covered by metal electrodes forming vertical strips. The excitation pattern drives pair of electrodes with a current source. If one plots the voltage difference as a small object (such as a conducting or insulating rod) moved around the tank with some experience, one can find the location of the rod simply by imagining the field lines corresponding to ∇u and the field that would result from the dipole source implied by (11.2). With multiple objects, especially close together, this becomes much harder to do simply by experience. The Laplacian operator in (11.2) is highly smoothing as one goes away from the source, so the perturbations in conductivity away from the boundary (where measurements are made) are difficult to interpret. This is a simple way of stating that the EIT reconstruction problem (or the inverse conductivity problem as it is also called) is extremely ill-posed.

Sophisticated reconstruction algorithms used in geophysics, industry, and medicine involve solving the forward problem for some system of partial differential equations, which includes the mathematical modelling of the electrodes [12]. This is often done using the Finite Element Method (FEM) in which domains with a complicated shape (like the human body) can be meshed using an irregular division in to, for example tetrahedra, and the conductivity within each tetrahedron can vary.

The linearized inverse problem is typically solved with some regularization to combat the ill-conditioning and the process is repeated until the predicted and measured boundary data fits within some tolerance. Details of contemporary implementations of such algorithms are given in [12]. In some problems, one seeks to locate fairly homogeneous objects with an irregular boundary. In those cases, it may be sensible to use the Boundary Element Method (BEM) rather than FEM so that only the unknown surfaces are discretized. An economical representation for the unknown surface is provided by using level sets of smooth functions [14].

Now we consider what we believe are the requirements of an algorithm for an electrosensing fish. There is much hard experimental biology and neurophysiology required to determine exactly how a fish performs target location and characterization, and indeed what are their capabilities and limitations. In the spirit of this book however we are looking for inspiration from non-human animals and the way that such remote sensing problems have been solved by evolution and by the learnt methods of individual fish. It would be highly unlikely that a fish brain would encode a FEM and the necessary numerical linear algebra to solve the forward and inverse problem this way. Such a 'computational' approach is more suited to conventional digital computers than neural systems and it is hard to see how intermediate steps in the development of such a system (such as finite element models with very few elements) would confer a selective advantage to the organism.

By contrast, recognition of patterns is widely prevalent in biology and we notice from our own experience that our own visual processing system matches our high spatial resolution view of objects to simplified representations that can be more easily understood. This visual recognition is remarkably robust in the face of varying lighting conditions and in 'cluttered' scenes. The ability of EIT researchers to learn to locate objects from a display of boundary voltage data is a simple example of this. We might reasonably expect that a weakly electric fish might use something we would understand as a low-dimensional representation of a target object and seek such a representation for our own use. Ideally we would be able to separate the characteristics of the object, its shape, and conductivity, from its relative size and its position and orientation in space. This would allow us to recognize that we have seen a similar object before, but perhaps at a different relative position to us, swimming in a different direction and of a different size. Although from the fish's point of view, they might remember that they found it easy prey and, if they crept up behind it, a tasty meal. Within computer vision research, it is common to seek computational methods to perform the same tasks as the human visual system, without necessarily implying that these methods are implemented in the brain. See for example [15] and references therein.

One possible mechanism to recognize shapes of objects from electrical data at a distance and independent of the orientation of the object, is to use what is called as the Polarization Tensor (PT) of the object [16], which can be thought of as the 'best fitting electrical ellipsoid'. The PT gives rise to an approximation to the perturbed voltage due to an object. All information about the shape and orientation of the object is given by the PT while the location information is obtained only from the incident field and the position of the detectors.

The hypothesis is that the fish could learn to recognize eigenvalues of the PT associated with specific objects. Moreover, the principal directions (eigenvectors) of the PT would give the orientation of the object in space. Beyond the best fitting ellipsoid, the shape of the object is completely characterized by the Generalized Polarization Tensors (GPT), a series of tensors with increasing rank. Unlike the perturbation formula (11.2), the GPT does not assume that the change in the conductivity is small.

11.3 Polarization tensors

The idea of the PT appeared in the classical problems of potential theory, specifically gravitation, magnetics and electrostatics. For example, Pólya [17] used the idea in the investigation of fluid mechanics, where the PT was called the *virtual mass*. These ideas were extended to what we now call the Pólya–Szegő PT and has been introduced in the studies of hydrodynamics and theory of electricity by [18,19]. Furthermore, in low frequency electromagnetic scattering, Kleinman and Senior [20] used a slightly different PT, which they called the polarizability tensor, to formulate the induced electric and magnetic dipole moments. For electric and magnetic dipole moments of the homogeneous Maxwell's equations, some examples of the polarizability tensor has been given also by Kleinman and Senior in [21]. Additionally, in more recent applications, the PT has been used for examples, to study composites [22,23], multipole moments in electromagnetism [24], ERT/EIT [22], and acoustic scattering problems [25].

In the ERT case, by using the method of asymptotic expansion and layer potential technique, Ammari and Kang [22] have derived an asymptotic formula representing the perturbation in the electrical field due to the presence of a conducting object. The leading order term of the formula is given by the PT, and this is actually the lowest order term of what they call the GPT for the conducting object. The formula of the GPT (which includes the PT) given in [22] has the advantage that it can be computed from only a knowledge of the geometry and the conductivity of the object. Ammari and Kang [22] go on to show that the full GPT completely specifies the shape of the object.

For a two-dimensional object with a fixed conductivity, a numerical method for computing its GPT based on the formula in [22] can be found in [26]. We gave a numerical method for computing the PT of the three-dimensional object in [27] and have extended the method by using methods for discretization of boundary integral operators from BEM to improve the approximation of the PT for the three dimensional objects in [28,29]. In [28,29], the PT for the three dimensional objects has been computed by using the library *BEM++* (see [30]).

Using the lowest order or a few of the lower order terms in the GPT to characterize objects offers fewer degrees of freedom and hence potentially a lower computational effort than full object reconstruction. A few studies have used this idea as a hypothetical mechanism for electrosensing fish [11,16,31–34] and as a possible method for inductive metal object location and characterization [33–45].

In this chapter, we will further discuss the characterization of objects in electrosensing fish and metal detection by using the PT. Here, we will specifically consider the PT of the low frequency electromagnetic metal detection problem. Our own interest is actually in identifying buried landmines.

11.4 Generalized polarization tensor

Following [22,46], let B be a bounded domain in three-dimensional Euclidean space $\mathbb{R}^3$ containing the point $\mathbf{z}$. Let conductivity of B be k where $k > 0$ and $k \neq 1$ (we choose the conductivity of the background to be equal to 1 so that k is the ratio between conductivity of the object and conductivity of the background). Moreover, k can also be complex-valued in the case of a permittivity contrast. The complex conductivity is thus

$$\gamma = 1 + (k - 1)\chi(B)$$

where $\chi(B)$ is the characteristic function of B. We denote by u_0 the voltage in the absence of the object, that is $\nabla^2 u_0 = 0$ in $\mathbb{R}^3$ and let u be the solution to the following problem:

$$\begin{cases} \nabla \cdot \gamma \nabla u = 0 \text{ in } \mathbb{R}^3 \\ u(\mathbf{x}) - u_0(\mathbf{x}) = O(1/|\mathbf{x}|^2) \quad \text{as } |\mathbf{x}| \to \infty. \end{cases} \tag{11.3}$$

The GPT is then defined (for example by [22]) through the following far-field expansion of u as

$$(u - u_0)(\mathbf{x}) = \sum_{|i|,|j|=1}^{\infty} \frac{(-1)^{|i|}}{i!j!} \partial_{\mathbf{x}}^i \Gamma(\mathbf{x}) M_{ij}(k, B) \partial_{\mathbf{z}}^j u_0(\mathbf{z}) \quad \text{as } |\mathbf{x} - \mathbf{z}| \to +\infty \tag{11.4}$$

for $i = (i_1, i_2, i_3), j = (j_1, j_2, j_3)$ multi indices, $|i| = i_1 + i_2 + i_3$. Here, the sequence of tensors $M_{ij}(k, B)$ is actually the GPT and it is important to note that it depends only on k and B and not on u_0 or the position of the object $\mathbf{z}$. We also define $\Gamma(\mathbf{x}) = 1/|\mathbf{x}|$ as the fundamental solution of Poisson's equation, that is the field for a monopole source.

We can interpret the terms in the first derivatives of Γ as a dipole, the dipole moment being determined by the gradient of the background solution u_0 and the relevant coefficients in M. The series as a whole can be considered as a *multipole expansion* of the perturbation in electric potential, in multipoles at the point $\mathbf{z}$ in the object. The important thing though is that we have separated the spatial dependence to be all in the u_0 terms while the dependence on the shape and conductivity of the object is in M.

The components of the GPT in (11.4) can be computed as follows [22]. They are moments

$$M_{ij} = \int_{\partial B} y^j \phi_i(\mathbf{y}) d\mathbf{y} \tag{11.5}$$

of an auxiliary field $\phi_i(y)$, which is the solution of the integral equation

$$\phi_i(\mathbf{y}) = (\lambda I - K_B^*)^{-1}(\partial x^i / \partial \mathbf{n_x})(\mathbf{y}) \tag{11.6}$$

for $\mathbf{x}, \mathbf{y} \in \partial B$. Here, I is the identity, $\mathbf{n_x}$ is the outer unit normal vector to the boundary ∂B at $\mathbf{x}$. As i is a multi-index, x^i consists of powers of the components of $\mathbf{x}$; λ is defined by $\lambda = (k+1)/2(k-1)$ and K_B^* is a singular integral operator defined using the Cauchy principal value $P.V.$ by

$$K_B^*\phi(\mathbf{x}) = \frac{1}{4\pi} P.V. \int_{\partial B} \frac{(\mathbf{x}-\mathbf{y}) \cdot \mathbf{n}_x}{|\mathbf{x}-\mathbf{y}|^3} \phi(\mathbf{y}) d\mathbf{y} \tag{11.7}$$

for any square integrable function ϕ on ∂B.

The GPT for the EIT and electric fish problems benefit from the full asymptotic expansion (11.4). It can be shown, [22], that the GPT completely determines the EIT data and hence both k and B. In other problems, only the lower order term of such an expansion is known and we have only the rank 2 polarization tensor. A PT introduced from the time harmonic Maxwell's equations by Ammari, Vogelius, and Volkov in [36] might be useful for the radar, GPR [45], and microwave tomography. PTs also exist for time harmonic scalar Helmholtz equation which could be used for acoustic imaging techniques such as sonar. The PT for metal detection considered by Ledger and Lionheart [40] is based on the eddy current approximation to the Maxwell's equations and the formulation is somewhat more complicated but is proposed as the basis for location and classification of metal objects, as we shall see in Section 11.7.

11.5 The second rank polarization tensor

The GPT is defined by all of the terms M_{ij} where i and j are multi-indices. The lowest order term in (11.4) occurs for $|i| = |j| = 1$. We can simplify the notation for this case and write $M_{(1,0,0)(1,0,0)} = m_{11}, M_{(1,0,0)(0,1,0)} = m_{12}$ and so on. This rank 2 tensor is called simply the PT or specifically the Pólya–Szegő tensor and is denoted by m.

In a given coordinate system, m can be written as a 3×3 matrix and [22] shows that it is symmetric. By construction it is clear that it is a tensor so, when the coordinate axes are rotated, the components transform in the expected way. When a different point $\mathbf{z}$ is chosen within B, m remains the same but the other terms in the GPT might change.

Specifically, let R be an orthogonal matrix and R^T is the transpose of R such that $B' = RB$. If $m(k, B)$ and $m(k, B')$ are the m tensors associated to domains B and B', respectively, for a conductivity k, then $m(k, B') = Rm(k, B)R^T$.

This means that the eigenvalues of m (considered as a symmetric matrix) are independent of rotations and so are characteristic of the target's shape and conductivity rather than its orientation in space and its position. When the eigenvalues are distinct the eigenvectors give what might be thought of as principal directions and these determine the orientation of the target in space.

For a target that is a solid ellipsoid E represented by $x_1^2/a^2 + x_2^2/b^2 + x_3^2/c^2 \leq 1$ in Cartesian coordinates, where a, b, and c are the lengths of semi-principal axes of E, m of E at conductivity k, which we denote by $m(k, E)$, is given explicitly in [22] as

$$m(k,E) = (k-1)|E| \begin{bmatrix} \frac{1}{(1-P)+kP} & 0 & 0 \\ 0 & \frac{1}{(1-Q)+kQ} & 0 \\ 0 & 0 & \frac{1}{(1-R)+kR} \end{bmatrix}, \tag{11.8}$$

where $|E|$ is the volume of E. The *depolarization factors* P, Q, and R are constants defined by the elliptic integrals

$$P = \frac{bc}{a^2} \int_1^\infty \frac{1}{t^2 \sqrt{t^2 - 1 + (\frac{b}{a})^2} \sqrt{t^2 - 1 + (\frac{c}{d})^2}} \mathrm{d}t, \tag{11.9}$$

$$Q = \frac{bc}{a^2} \int_1^\infty \frac{1}{(t^2 - 1 + (\frac{b}{a})^2)^{3/2} \sqrt{t^2 - 1 + (\frac{c}{a})^2}} \mathrm{d}t, \tag{11.10}$$

$$R = \frac{bc}{a^2} \int_1^\infty \frac{1}{\sqrt{t^2 - 1 + (\frac{b}{a})^2} (t^2 - 1 + (\frac{c}{a})^2)^{\frac{3}{2}}} \mathrm{d}t. \tag{11.11}$$

Depolarization factors or demagnetizing factors are classical terms: the depolarization factors for general ellipsoid have been studied by [47] in 1945 and with roots going back to work of Maxwell and of Poisson.

If we are given a positive definite symmetric matrix m then, we can numerically solve the overdetermined equations (11.8) and (11.9)–(11.11) to give a, b, c, and k, see [48] [49, Chapter 4] for examples. This supports the interpretation of m as the best fitting ellipsoid for some value of k.

In Table 11.1, we see a table of the PT of a variety of geometric objects B, with a calculated $m(k, B)$ for a specific k and ellipsoid E with $m(k, B) = m(k, E)$. For details of the calculations see [49].

11.6 PT in electrosensing fish

The PT forms an attractive low-dimensional position independent characterization of the targets and the eigenvalues of the PT form an orientation independent characterization. In electrosensing fish, the formulation (11.3) neglects the electrical properties of the fish itself and indeed the river bank, which is assumed to be distant, but captures the essential features of the response of the voltage field to the target B. Moreover, since the series (11.4) is asymptotic in the distance between the measurement (such as the fish) and the target, it is more suited to characterization of objects at a distance.

Table 11.1 Objects and ellipsoids with the same PT for a fixed conductivity k. From [49]

Object, S	k	$m(k,S)$	Ellipsoid, E $\left(\frac{x}{a}\right)^2+\left(\frac{y}{b}\right)^2+\left(\frac{z}{c}\right)^2=1$
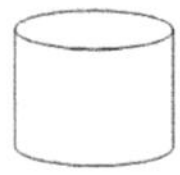 **Cylinder** $d=3, h=3$	5×10^{-5}	$\begin{bmatrix} -33.80 & 0.00 & 0.00 \\ 0.00 & -33.80 & 0.00 \\ 0.00 & 0.00 & -33.52 \end{bmatrix}$	$a=1.7425$ $b=1.7425$ $c=1.7669$
 **Hemisphere** $d=3$	1×10^{-2}	$\begin{bmatrix} -9.70 & 0.00 & 0.00 \\ 0.00 & -9.70 & 0.00 \\ 0.00 & 0.00 & -15.42 \end{bmatrix}$	$a=1.5240$ $b=1.5240$ $c=0.7699$
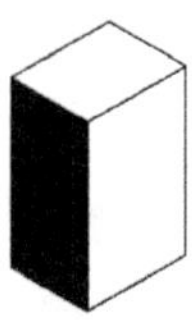 **Cuboid** $2\times4\times1$	1.5	$\begin{bmatrix} 27.95 & 0.00 & 0.00 \\ 0.00 & 29.90 & 0.00 \\ 0.00 & 0.00 & 25.05 \end{bmatrix}$	$a=2.5212$ $b=4.3181$ $c=1.4064$
 **Pyramid** $2\times2\times2$	500	$\begin{bmatrix} 12.85 & 0.00 & 0.00 \\ 0.00 & 12.85 & 0.00 \\ 0.00 & 0.00 & 8.50 \end{bmatrix}$	$a=1.0775$ $b=1.0775$ $c=0.7569$
 **Cube** $2\times2\times2$	10000	$\begin{bmatrix} 28.90 & 0.00 & 0.00 \\ 0.00 & 28.90 & 0.00 \\ 0.00 & 0.00 & 28.90 \end{bmatrix}$	$a=1.3201$ $b=1.3201$ $c=1.3201$

Note: d, diameter; h, height.

Table 11.2 The PT for a few objects presented in [2]. For each object, the normalized eigenvalues and the mean of the normalized eigenvalues are given. See [50] for more details

Object, B	**Dimension (cm)**	**The first order PT, m_B**	**Normalized eigenvalues**	**Mean**
Cone	$d = h = 3$	$10^{-5} \times \begin{bmatrix} 0.28 & 0 & 0 \\ 0 & 0.28 & 0 \\ 0 & 0 & 0.31 \end{bmatrix}$	0.9032 0.9032 1.0000	0.9355
Cube	$l = w = h = 3$	$10^{-5} \times \begin{bmatrix} 0.98 & 0 & 0 \\ 0 & 0.98 & 0 \\ 0 & 0 & 0.98 \end{bmatrix}$	1.0000 1.0000 1.0000	1.0000
Pyramid	$l = w = h = 3$	$10^{-5} \times \begin{bmatrix} 0.42 & 0 & 0 \\ 0 & 0.42 & 0 \\ 0 & 0 & 0.32 \end{bmatrix}$	0.7619 1.0000 1.0000	0.9206

Note: cm, centimetre; *d*, diameter; *h*, height; *l*, length; *w*, width.

In the series (11.4), the higher order terms decay more quickly than the lower, so it is easier to fit the dipole term accurately and this corresponds to the intuitive procedure we described in the EIT lab. Specifically, we are suggesting that when the fish senses the restriction of a dipole field to the part of its surface that is covered with receptors, it can estimate the dipole that would fit, and as it moves to different nearby locations it is effectively measuring the components of m for different $\nabla u_0(\mathbf{z})$. As the fish homes in on its prey and $|\mathbf{x} - \mathbf{z}|$ is smaller, the other terms in the expansion come into play, perhaps allowing the fish to characterize the object beyond a 'best fitting ellipsoid'. For example, for shapes such as cones and pyramids that do not have mirror symmetry normal to an eigenvector of m, the PT is not able to detect which way the shape is pointing, whereas the next order term in the GPT could provide this information.

In our previous studies [31,33,9], we have computed m for some objects used in experiments in [2] and shown that the electrosensing fish used in [2] need more time to discriminate between two objects that have a similar m. See Table 11.2 for the calculated tensors m and Table 11.3 for the training times required by fish to discriminate between these objects.

Recently, in [50], we have proposed several pairs of objects such that both objects have the same m. The pair consists of one ellipsoid and one other object, where the ellipsoid has been calculated to have the same m as the other object using our technique presented in [48]. Our suggested technique relies on two steps. First we calculate the PT of the given object, typically a regular geometric object.

Table 11.3 Results for the training in [2]. Here v is the difference between the mean normalized eigenvalue for each pair. The period is the time taken to train each fish. See [50] for more details

<table>
<tr><th>Fish</th><th>Accept</th><th>Reject</th><th>Period (days)</th><th>v</th></tr>
<tr><td>A</td><td rowspan="2">Pyramid</td><td rowspan="2">Cube</td><td>4</td><td rowspan="2">0.0794</td></tr>
<tr><td>B</td><td>7</td></tr>
<tr><td>C</td><td rowspan="2">Cone</td><td rowspan="2">Cube</td><td>8</td><td rowspan="2">0.0645</td></tr>
<tr><td>D</td><td>10</td></tr>
<tr><td>E</td><td>Cone</td><td>Pyramid</td><td>19</td><td>0.0149</td></tr>
</table>

This involves solving the integral equations (11.6) numerically by first meshing the surface of the object B with a triangular mesh and then discretizing the integral equations on that mesh. In [50], this is done using library functions from the Boundary Element Method code BEM++ [30] that give matrix representations for layer potential operators. Given the matrix $m(B, k)$, we need to solve (11.9) for a, b, and c for a suitable choice of conductivity k.

For further investigation, our suggestion here is that the experiments be repeated with such a pair of objects to test the hypothesis that the PT plays a role in discrimination. If so we might expect that either the fish would take longer to learn to distinguish between objects with the same PT, or could do so less reliably even after a long training.

So far we have treated the target as having a real conductivity. However, interesting targets for the fish are biological in origin and will display some dispersion effects so, their conductivity will be complex, $k = \sigma + i\omega\varepsilon$, and indeed σ and ε can vary with frequency [51]. For fish with a continuous wave electrical discharge, the complex conductivity at their fundamental frequency is the most important. For fish generating an electrical pulse, the complex conductivity at a range of frequencies is relevant and that might give these fish better discrimination.

As examples, black ghost knife fish use almost pure sine waves whereas Peters' elephantnose fish use a pulse with a wider range of frequencies (see [6]). The knife fish also use the frequency of electrical discharge to indicate their status in a hierarchy and negotiate their frequency band by fighting [7, Chapter 11]. Thus m, with complex conductivity that depends on frequency, might discriminate between objects of interest for different fish. In this case, for example, it might be worthwhile conducting experiments to test whether the Peters' elephantnose fish can distinguish objects that cannot be discriminated by the black ghost knife fish.

By considering an arbitrarily chosen complex conductivity $k = 1.5 + 0.1\,\omega i$ for ellipsoid with semi-principal axes 3 cm, 2 cm, and 1 cm, we now evaluate using

Table 11.4 m_f *for an ellipsoid at a few frequencies,* f *(Hz)*

Frequency, f (Hz)	$\lambda^r_{21}, \lambda^r_{31}$	$\lambda^i_{21}, \lambda^i_{31}$
20	0.3370	0.0927
	0.6778	0.3998
50	0.2821	0.0766
	0.6015	0.3524
120	0.2730	0.0740
	0.5880	0.3441
250	0.2715	0.0736
	0.5857	0.3427
500	0.2712	0.0735
	0.5852	0.3424

(11.8) its m at a few frequencies, f Hz denoted by m_f, where $\omega = 2\pi f$. Each m_f is now a complex matrix and thus can be expressed as $m_r + m_i\mathrm{i}$ such that the 3×3 symmetrical real matrices m_r and m_i are the real and imaginary parts of m_f.

For every f, the eigenvalues of m_r, $\lambda^r_1 \geq \lambda^r_2 \geq \lambda^r_3$ are normalized as $\lambda^r_{21} = \lambda^r_2/\lambda^r_1, \lambda^r_{21} = \lambda^r_3/\lambda^r_1$. Similarly m_i. In Table 11.4, the normalized eigenvalues of both m_r and m_i are shown for each f.

Finally, every experiment should be performed in such a way that the fish are at least a certain distance from the object. This is because the expansion is asymptotic in distance.

11.7 PT in metal detection by human technology

A low frequency electromagnetic field is usually used in metal detectors to locate a high conductivity target in a low conductivity background from electromagnetic induction data. By neglecting the displacement currents in the Maxwell's equations, metal detection can be modelled by the eddy current equations (see [37,40]). In this case, a mathematical justification for approximating the Maxwell's equations by the eddy current model is given in [52,53].

Recently, Marsh *et al.* [42,43] have used the polarization (or polarizability) tensor for metal detectors to describe the location, size, orientation, and material property of metallic targets for security screening. In this case the tensor was fitted empirically rather than calculated directly.

We assume the presence of an object in the form $B_\alpha = \mathbf{z} + \alpha B$ where B is a unit object centred at the origin, α denotes a scaling for B, and $\mathbf{z}$ denotes a point in the object as before. We introduce

$$\mu_\alpha = \begin{cases} \mu_* & \text{in } B_\alpha \\ \mu_0 & \text{in } \mathbb{R}^3 \backslash B_\alpha \end{cases}, \quad \sigma_\alpha = \begin{cases} \sigma_* & \text{in } B_\alpha \\ 0 & \text{in } \mathbb{R}^3 \backslash B_\alpha \end{cases} \tag{11.12}$$

where $\mu_0 = 4\pi \times 10^{-7}\mathrm{NA}^{-2}$ is the permeability of the free space while μ_* and σ_* denote the permeability and conductivity, respectively, of the inclusion B_α. The conductivity contrast between the object and the background has been assumed to be sufficiently high that the background can be approximated by a zero conductivity.

Let $\boldsymbol{E}_\alpha$ and $\boldsymbol{H}_\alpha$ be the time harmonic eddy current fields (electric and magnetic) in the presence of a conducting object B_α that result from a current source $\boldsymbol{J}_0$ located outside B_α. Suppose that $\nabla \cdot \boldsymbol{J}_0 = 0$ in $\mathbb{R}^3$. Both fields $\boldsymbol{E}_\alpha$ and $\boldsymbol{H}_\alpha$ satisfy the eddy current equations

$$\begin{aligned} \nabla \times \boldsymbol{E}_\alpha &= \mathrm{i}\omega\mu_\alpha \boldsymbol{H}_\alpha \quad \text{in } \mathbb{R}^3, \\ \nabla \times \boldsymbol{H}_\alpha &= \sigma_\alpha \boldsymbol{E}_\alpha + \boldsymbol{J}_0 \quad \text{in } \mathbb{R}^3, \\ \boldsymbol{E}_\alpha(\mathbf{x}) &= O(|\mathbf{x}|^{-1}), \boldsymbol{H}_\alpha(\mathbf{x}) = O(|\mathbf{x}|^{-1}) \quad \text{as } |\mathbf{x}| \to \infty, \end{aligned} \tag{11.13}$$

where ω is a fixed angular frequency of the current source. The depth of penetration of the magnetic field in the conducting object is described by its skin depth, $s = \sqrt{2/(\omega\mu_*\sigma_*)}$. On the other hand, without the object B_α, the fields $\boldsymbol{E}_0$ and $\boldsymbol{H}_0$ that result from the time varying current source satisfy

$$\begin{aligned} \nabla \times \boldsymbol{E}_0 &= \mathrm{i}\omega\mu_0 \boldsymbol{H}_0 \quad \text{in } \mathbb{R}^3, \\ \nabla \times \boldsymbol{H}_0 &= \boldsymbol{J}_0 \quad \text{in } \mathbb{R}^3, \\ \boldsymbol{E}_0(\mathbf{x}) &= O(|\mathbf{x}|^{-1}), \boldsymbol{H}_0(\mathbf{x}) = O(|\mathbf{x}|^{-1}) \quad \text{as } |\mathbf{x}| \to \infty. \end{aligned} \tag{11.14}$$

By introducing $\nu = 2\alpha^2/s^2$, the related asymptotic formula for the above model that describes the perturbation in the magnetic field at a position $\mathbf{x}$, away from $\mathbf{z}$, due to the presence of B_α when $\nu = O(1)$ and $(\mu_*/\mu_0) = O(1)$ as $\alpha \to 0$ is given by Ammari *et al.* [37] in the form

$$\begin{aligned} (\boldsymbol{H}_\alpha - \boldsymbol{H}_0)(\mathbf{x}) = &-\frac{\mathrm{i}\nu\alpha^3}{2}\sum_{i=1}^{3} \boldsymbol{H}_0(\mathbf{z})_i \Big(\int_B \boldsymbol{D}_{\mathbf{x}}^2 G(\mathbf{x},\mathbf{z})\boldsymbol{\xi} \times (\boldsymbol{\theta}_i + \boldsymbol{e}_i \times \boldsymbol{\xi})\mathrm{d}\boldsymbol{\xi}\Big) \\ &+ \alpha^3\left(1 - \frac{\mu_0}{\mu_*}\right)\left(\sum_{i=1}^{3} \boldsymbol{H}_0(\mathbf{z})_i \boldsymbol{D}_{\mathbf{x}}^2 G(\mathbf{x},\mathbf{z}) \int_B \left(\boldsymbol{e}_i + \frac{1}{2}\nabla \times \boldsymbol{\theta}_i\right)\mathrm{d}\boldsymbol{\xi}\right) \\ &+ \boldsymbol{R}(\mathbf{x}) \end{aligned} \tag{11.15}$$

where ξ is measured from the centre of B. Here, $G(\mathbf{x},\mathbf{z}) = (4\pi|\mathbf{x}-\mathbf{z}|)^{-1}$ is the free space Laplace Green's function and $\boldsymbol{R}(\mathbf{x}) = O(\alpha^4)$ is a small remainder term.

Furthermore, for $i = 1, 2, 3$, $\boldsymbol{e}_i$ is a unit vector for the ith Cartesian coordinate direction, $\boldsymbol{H}_0(\mathbf{z})_i$ denotes the ith element of $\boldsymbol{H}_0(\mathbf{z})$, and $\boldsymbol{\theta}_i$ is the solution to the transmission problem

$$\begin{cases} \nabla_\xi \times \mu^{-1}\nabla_\xi \times \boldsymbol{\theta}_i - \mathrm{i}\omega\sigma\alpha^2\boldsymbol{\theta}_i = \mathrm{i}\omega\sigma\alpha^2\boldsymbol{e}_i \times \boldsymbol{\xi} & \text{in } B \cup B^c, \\ \nabla_\xi \cdot \boldsymbol{\theta}_i = 0 & \text{in } B^c, \\ [\boldsymbol{\theta}_i \times \boldsymbol{n}]_{\partial B} = \mathbf{0} & \text{on } \partial B, \\ \left[\mu^{-1}\nabla_\xi \times \boldsymbol{\theta}_i \times \boldsymbol{n}\right]_{\partial B} = -2\left[\mu^{-1}\right]_{\partial B}\boldsymbol{e}_i \times \boldsymbol{n} & \text{on } \partial B, \\ \boldsymbol{\theta}_i(\boldsymbol{\xi}) = O(|\boldsymbol{\xi}|^{-1}) & \text{as } |\boldsymbol{\xi}| \to \infty \end{cases} \tag{11.16}$$

Based on (11.15) and (11.16), the *conductivity polarization tensor* (CPT) and the *magnetic* (or *permeability*) *polarization tensor* (MPT) have been introduced in [37].

Using this framework, Ledger and Lionheart [40] have applied tensor operations to combine both CPT and MPT as well as to reformulate (11.15) in the alternative form

$$(\boldsymbol{H}_\alpha - \boldsymbol{H}_0)(\mathbf{x}) = \boldsymbol{D}^2_{\mathbf{x}}G(\mathbf{x}, \mathbf{z})M\boldsymbol{H}_0(\mathbf{z}) + \boldsymbol{R}(\mathbf{x}) \tag{11.17}$$

where M is the new polarization tensor for a conducting and magnetic inclusion B_α. Importantly, they show that M is a complex symmetric rank 2 tensor defined by six complex coefficients after reducing (11.15)–(11.17). The *hp*-finite element method presented in [54] is also used in [40] to numerically compute M and some analysis for the errors in the method are also given. As again shown by [40], M is equivalent with the polarizability tensor for a metal detector in the engineering literature (for example, see [39,42–44]). The rank 2 tensor M is given by [40] as

$$M = N_{ji} - C_{ji} \tag{11.18}$$

where

$$N_{ji} = \alpha^3\left(1 - \frac{\mu_0}{\mu_*}\right)\int_B\left(\boldsymbol{e}_j \cdot \boldsymbol{e}_i + \frac{1}{2}\boldsymbol{e}_j \cdot \nabla \times \boldsymbol{\theta}_i\right)\mathrm{d}\boldsymbol{\xi} \tag{11.19}$$

and

$$C_{ji} = \left(-\frac{\mathrm{i}\nu\alpha^3}{4}\right)\boldsymbol{e}_j \cdot \int_B \boldsymbol{\xi} \times (\boldsymbol{\theta}_i + \boldsymbol{e}_i \times \boldsymbol{\xi})\,\mathrm{d}\boldsymbol{\xi} \tag{11.20}$$

for $j = 1, 2, 3$ and θ_i is the solution to the system (11.16) for $i = 1, 2, 3$. Note that N is an equivalent form of the original rank 2 magnetic polarization tensor in [37]. Furthermore, the rank 2 tensor C is a reduction given by [40] of the original rank 4 tensor in [37] using symmetries.

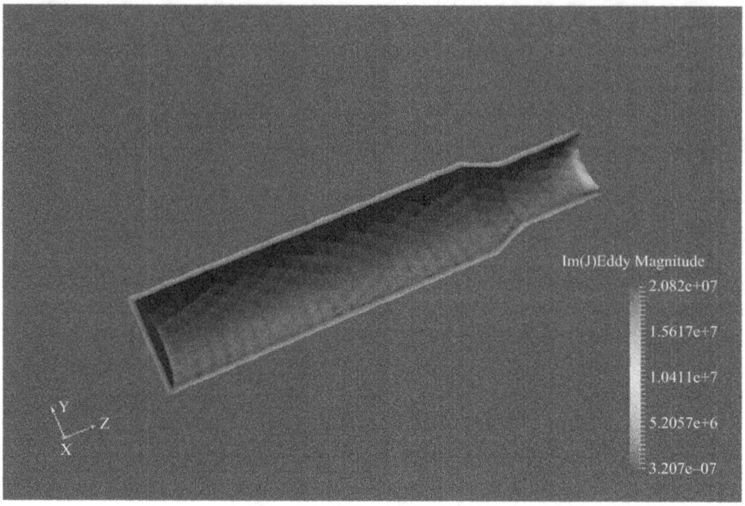

Figure 11.3 A mesh of a Remington .222 rifle cartridge. Metal cartridges cases are typical of metallic debris found in mine fields. Image courtesy of Paul Ledger. See [41] for more details

The coefficients of the rank 2 tensor M can be expressed as elements of a 3×3 complex matrix and if M is expressed as $M = M_r + M_i\mathrm{i}$, for real M_r and M_i each with three real eigenvalues. For a conducting and magnetic sphere, the analytical formula for M is also given in [40]. Many other properties of M have been explored by [33,35,40,41]. In their studies, Ledger and Lionheart [40] show that (11.18) reduces to the real symmetric $M = N$ when B_α has $\sigma_* = 0$, that is a magnetic but non-conducting object. They also explain that $M = N$ does not depend on frequency, ω, and reduces to the standard Pólya–Szegő PT. Therefore, in this case, $M = N$ can now be determined by solving boundary integral equations (11.5)–(11.7) where the parameter k is the contrast μ_*/μ_0 (or the relative permeability of B_α). The analytical formula of m for ellipsoids as given in (11.8) could also be used.

In locating minimal metal anti-personnel landmines, every small metal object located must be tested to see if it could be a mine. Mine fields unfortunately are often contaminated with metal objects including fragments of ammunition. A rifle cartridge casing is a typical example of battlefield scrap, see Figure 11.3. This particular Remington .222 calibre casing has been investigated experimentally by [39] and numerically by [41]. For frequency dependence of components of M, see Figure 11.4.

One experimental verification that the second rank tensor M represents the perturbed field accurately is that as the object is rotated through θ about an axis, the tensor transformation rule results in $\sin^2\theta$ and $\cos^2\theta$ and $\sin\theta\cos\theta$ terms. All of these are periodic with period 2. See Figure 11.5.

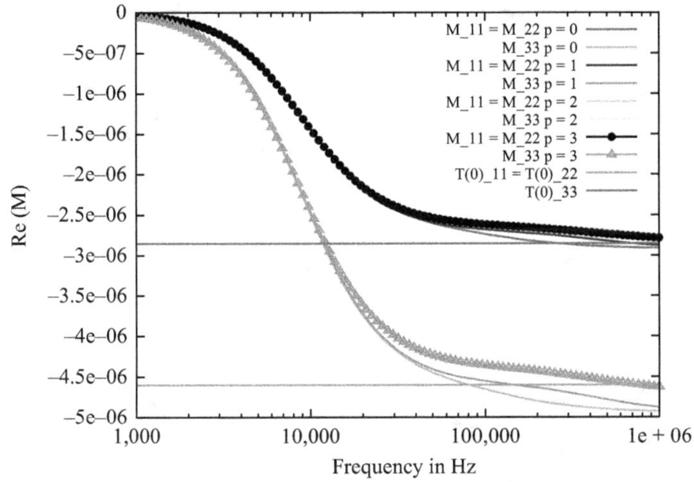

Re (M)

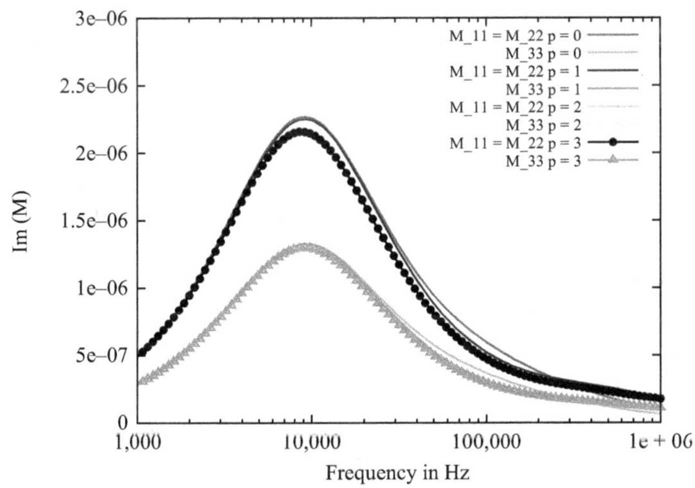

Im (M)

Figure 11.4 Frequency dependence of real and imaginary coefficients of M for the Remington .222 cartridge case. Results are in good agreement with measured data. See [41]

11.8 Conclusions

Polarization tensors provide a plausible low-dimensional representation of the response of an isolated object in an electric field that captures shape information about the object independently of position and transform naturally under rotation. It would certainly be a candidate for the basis of human-made algorithms to replicate

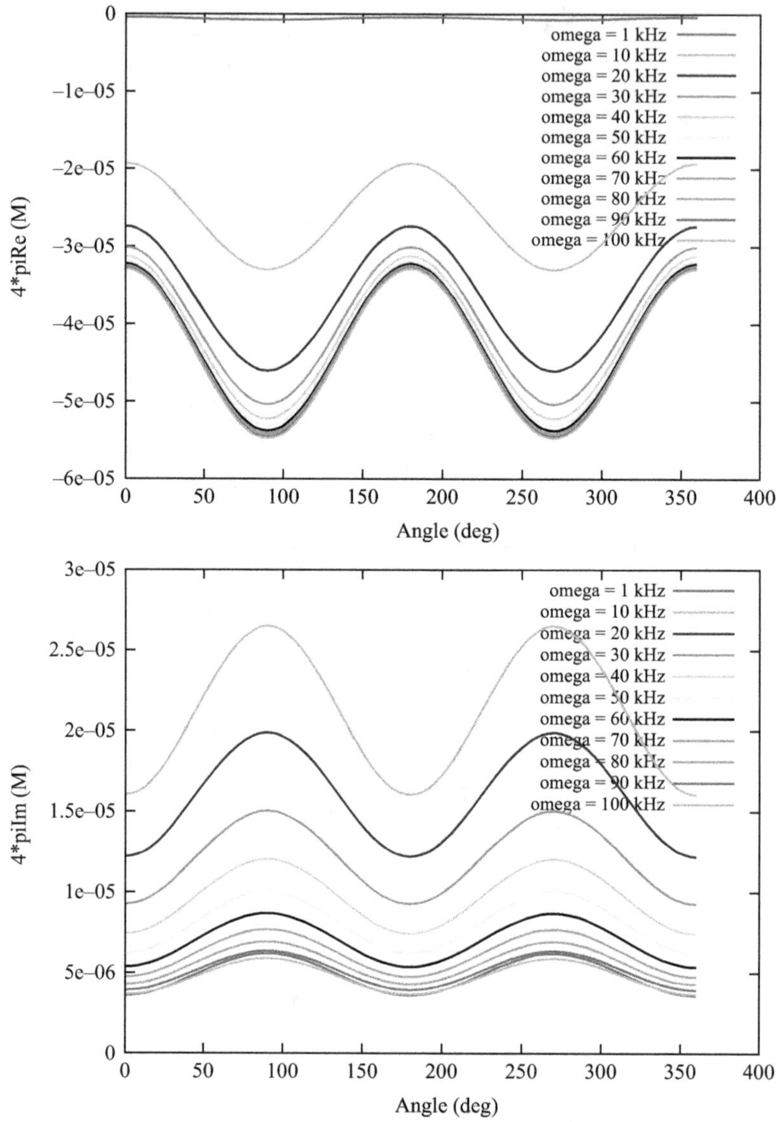

Figure 11.5 *A measurement with two coils will correspond to a component of M and as the object rotates about an axis the measurement will be sinusoidal with period 2 if represented accurately be a rank 2 tensor. This is a simulated measurement of the Remington rifle cartridge in Figure 11.3 at a range of frequencies. See [41] for more details. Reprinted with permissions from [41]*

some of the capabilities of electrosensing fish. Polarization tensors can also be used for locating and characterizing metal objects using eddy current inductive measurements, and similar theory has been developed for far-field radio and sound wave remote sensing methods.

References

[1] G. von der Emde. "Electroreception: object recognition in African weakly electric fish," in T. Hara and B. Zielinski (eds.). *Sensory Systems Neuroscience*. Academic Press Elsevier, Cambridge, MA, 2007.

[2] G. von der Emde and S. Fetz. "Distance, shape and more: recognition of object features during active electrolocation in a weakly electric fish," *The Journal of Experimental Biology*, 210, 3082–3095, 2007.

[3] G. von der Emde, K. Behr, B. Bouton, J. Engelmann, S. Fetz and C. Folde. "Three-dimensional scene perception during active electrolocation in a weakly electric pulse fish," *Frontiers in Behavioral Neuroscience*, 4(26), 1–13, 2010.

[4] D.J. Clarke, H.M. Whitney, G.P. Sutton and D. Robert. "Detection and learning of floral electric fields by bumblebees," *Science*, 340, 66–69, 2013.

[5] N.J. Cowan and E.S. Fortune. "The critical role of locomotion mechanics in decoding sensory systems," *The Journal of Neuroscience*, 27(5), 1123–1128, 2007.

[6] M.E. Nelson. "Biological smart sensing strategies in weakly electric fish," *Smart Structures and Systems*, 7(6), 539–549, 2011.

[7] P. Moller, *Electric Fishes: History and Behavior*, Chapman & Hall, New York, NY, 1995.

[8] M. Lewicki, B. Olshausen, A. Surlykke and C. Moss. "Scene analysis in the natural environment," *Frontiers in Psychology*, 5, 199, 2014.

[9] A. Adler, J.H. Arnold, R. Bayford *et al.* "GREIT: a unified approach to 2D linear EIT reconstruction of lung images," *Physiological Measurement*, 30, S35–S55, 2009.

[10] M. Soleimani and W.R.B. Lionheart. "Nonlinear image reconstruction for electrical capacitance tomography using experimental data," *Measurement Science and Technology*, 16, 1987–1996, 2005.

[11] H. Ammari, T. Boulier and J. Garnier. "Modeling active electrolocation in weakly electric fish," *SIAM Journal on Imaging Sciences*, 5, 285–321, 2013.

[12] A. Adler, R. Gaburro and W. Lionheart. "Electrical impedance tomography," in O. Scherzer (ed.). *Handbook of Mathematical Methods in Imaging*, Springer-Verlag, New York, NY, 2011.

[13] W. Lionheart, N. Polydordes and A. Borsic. "The reconstruction problem," in D.S. Holder (ed.). Part 1 of *Electrical Impedance Tomography: Methods, History and Applications*, Institute of Physics, Bristol, UK, pp. 3–64, 2005.

[14] M. Soleimani, W.R.B. Lionheart and O. Dorn. "Level set reconstruction of conductivity and permittivity from boundary electrical measurements using

experimental data," *Inverse Problems in Science and Engineering*, 14(2), 193–210, 2006.

[15] Z. Pizlo. "Perception viewed as an inverse problem," *Vision Research*, 41(24), 3145–3161, 2001.

[16] H. Ammari, T. Boulier, J. Garnier and H. Wang. "Shape recognition and classification in electrosensing," *Proceedings of the National Academy of Sciences of the USA*, 11, 11652–11657, 2014.

[17] G. Pólya. "A minimum problem about the motion of a solid through a fluid," *Proceedings of the National Academy of Sciences USA*, 33, 218–221, 1947.

[18] G. Pólya and G. Szegő. *Isoperimetric Inequalities in Mathematical Physics. Annals of Mathematical Studies Number 27*. Princeton University Press, Princeton, NJ, 1951.

[19] M. Schiffer and G. Szegő. "Virtual mass and polarization," *Transactions of the American Mathematical Society*, 67, 130–205, 1949.

[20] R.E. Kleinman and T.B. Senior. "Low frequency scattering by space objects," *IEEE Transactions Aerospace and Electronic Systems*, AES-11, 672–675, 1975.

[21] R.E. Kleinman and T.B. Senior. "Rayleigh scattering," in V.K. Varadan and V.V. Varadan (eds.). *Low and High Frequency Asymptotics*. Elsevier, Amsterdam, pp. 1–70, 1986.

[22] H. Ammari and H. Kang. *Polarization and Moment Tensors: With Applications to Inverse Problems and Effective Medium Theory*. Springer-Verlag, New York, NY, 2007.

[23] G. Milton. *Theory of Composites*, Cambridge University Press, New York, NY, 2002.

[24] R.E. Raab and O.L. De Lange. *Multipole Theory in Electromagnetism: Classical, Quantum and Symmetry Aspects, with Applications*. Cambridge University Press, New York, NY, 2002.

[25] H. Ammari, J. Garnier, V. Jugnon and H. Kang. "Direct reconstruction methods in ultrasound imaging of small anomalies," *Mathematical Modeling in Biomedical Imaging II, Lecture Notes in Mathematics*, 2035, 31–55, 2011.

[26] Y. Capdeboscq, A.B. Karrman and J.-C. Nedelec. "Numerical computation of approximate generalized polarization tensors," *Applicable Analysis*, 91(6), 1189–1203, 2012.

[27] T.K. Ahmad Khairuddin and W.R.B. Lionheart. "Some properties of the first order polarization tensor for 3D domains," *Matematika UTM*, 29(1), 1–18, 2013.

[28] T.K.A. Khairuddin and W.R.B. Lionheart. "Numerical comparisons for the approximated first order polarization tensor for ellipsoids," *Applied Mathematics and Computational Intelligence*, 4(1), 341–354, 2015.

[29] T.K. Ahmad Khairuddin and W.R.B Lionheart. "Computing the first order polarization tensor: welcome BEM++!," *Menemui Matematik*, 35(2), 15–20, 2013.

[30] W. Śmigaj, S. Arridge, T. Betcke, J. Phillips and M. Schweiger. "Solving boundary integral problems with BEM++," *ACM Transactions on Mathematical Software*, 41(2), 6, 2015.

[31] T.A. Khairuddin and W.R.B. Lionheart. *Do Electro-Sensing Fish Use the First Order Polarization Tensor for Object Characterization?* in 100 years of Electrical Imaging, 149, Presses des Mines, Paris, 2012.

[32] T.K.A. Khairuddin and W.R.B. Lionheart. "Do electro-sensing fish use the first order polarization tensor for object characterization? Object discrimination test," *Sains Malaysiana*, 43(11), 1775–1779, 2014.

[33] T.K. Ahmad Khairuddin and W.R.B. Lionheart. "Biological and engineering applications of the polarization tensor," in *Proceedings of the 10th IMT-GT ICMSA 2014*, October 14–16, 2014, Kuala Terengganu, Malaysia, 228–234, 2014.

[34] T.K. Ahmad Khairuddin and W.R.B. Lionheart. "Polarization tensor: between biology and engineering," *Malaysian Journal of Mathematical Sciences*, 10(S), 179–191, 2016.

[35] T.K.A. Khairuddin, P.D. Ledger and W.R.B. Lionheart. "Investigating the polarization tensor to describe and identify metallic objects," in *Proceedings of the World Congress on Engineering 2015*, Vol. I, WCE 2015, July 1–3, 2015, London, UK, 122–127, 2015.

[36] H. Ammari, M.S. Vogelius and D. Volkov. "Asymptotic formula for perturbations in the electromagnetic fields due to the presence of inhomogeneities of small diameter II. The full Maxwell equations," *Journal de Mathématiques Pures et Appliquées*, 8, 769–814, 2001.

[37] H. Ammari, J. Chen, Z. Chen, J. Garnier and D. Volkov. "Target detection and characterization from electromagnetic induction data," *Journal de Mathématiques Pures et Appliquées*, 101, 54–75, 2014.

[38] H. Ammari, J. Chen, Z. Chen, D. Volkov and H. Wang. "Detection and classification from electromagnetic induction data," *Journal of Computational Physics*, 301, 201–217, 2015.

[39] B. Dekdouk, L.A. Marsh, D.W. Armitage and A.J. Peyton. *Estimating Magnetic Polarizability Tensor of Buried Metallic Targets for Landmine Clearance*, Springer Science and Business Media, LLC, Berlin, pp. 425–432, 2014 (alternatively in *10th Conference on Ultra-Wideband, Short-Pulse Electromagnetics*, July 5–9, 2010, UWBSP, Ottawa, ON, Canada).

[40] P.D. Ledger and W.R.B. Lionheart. "Characterising the shape and material properties of hidden targets from magnetic induction data," *IMA Journal of Applied Mathematics* 80(6), 1776–1798, 2015.

[41] P.D. Ledger and W.R.B. Lionheart. "Understanding the magnetic polarizibility tensor," *IEEE Transactions on Magnetics*, 52(5), 1–16, 2016.

[42] L.A. Marsh, C. Ktistis, A. Järvi, D.W. Armitage and A.J. Peyton. "Three-dimensional object location and inversion of the magnetic polarizability tensor at a single frequency using a walk-through metal detector," *Measurement Science and Technology*, 24(4), 045102, 2013.

[43] L.A. Marsh, C. Ktistis, A. Järvi, D.W. Armitage and A.J. Peyton. "Determination of the magnetic polarizability tensor and three dimensional object location for multiple objects using a walk-through metal detector," *Measurement Science and Technology*, 25(5), 055107, 2014.

[44] O. Rehim, J.L. Davidson, L.A. Marsh, M.D. O'Toole, D.W. Armitage and A.J. Peyton. "Measurement system for determining the magnetic polarizibility tensor of small metal targets," in *Proceedings of IEEE Sensor Applications Symposium (SAS)*, April 13–15, 2015, Zadar, Croatia

[45] F. Watson and W. Lionheart. "Polarization tensors for ground penetrating radar: maximising distinguishability for landmine detection," in *Proceedings of the BAMC*, Cambridge, 2015.

[46] H. Ammari and H. Kang. *Reconstruction of Small Inhomogeneities from Boundary Measurements*, Springer-Verlag, Heidelberg, 2004.

[47] J. Osborn. "Demagnetizing factors of the general ellipsoid," *Physical Review*, 67, 351–357, 1945.

[48] T.K.A. Khairuddin and W.R.B. Lionheart. "Fitting ellipsoids to objects by the first order polarization tensor," *Malaya Journal of Matematik*, 4(1), 44–53, 2013.

[49] T.K. Ahmad Khairuddin. "Characterization of objects by fitting the polarization tensor," PhD Thesis, The University of Manchester, 2016 MIMS e-print 2016.55, http://eprints.ma.man.ac.uk/2514/, accessed on 24 February 2017.

[50] T.K. Ahmad Khairuddin and W.R.B. Lionheart. "Characterization of objects by electrosensing fish based on the first order polarization tensor," *Bioinspiration & Biomimetics*, 11, 055004, 2016.

[51] B.J. Roth. "The electrical conductivity of tissues," in J.D. Bronzino (ed.). *The Biomedical Engineering Handbook*. CRC Press LLC, Boca Raton, FL, 2000.

[52] H. Ammari, A. Buffa and J.-C. Nedelec. "A justification of eddy currents model for the Maxwell's equations," *SIAM Journal on Applied Mathematics*, 60(5), 1805–1823, 2000.

[53] A.A. Rodríguez and A. Valli. *Eddy Current Approximation of Maxwell Equations: Theory, Algorithm and Applications*. Springer-Verlag Italia, Italy, 2010.

[54] P.D. Ledger and S. Zaglmayr. "*hp*-Finite element simulation of three-dimensional eddy current problems on multiply connected domains," *Computer Methods in Applied Mechanics and Engineering*, 199, 49–52, 2010.

Postscript

This book has presented a number of related research activities that seek to either better understand echoic sensing of natural systems and/or exploit this understanding to improve synthetic sensors. It is inevitable that such a text can only scratch the surface of what might be possible. One thing we can say is that today's technology already allows very high levels of versatility such that synthetic sensors can closely mimic the key characteristics of their natural counterparts. This is only going to improve with time. This coming together provokes more and more questions around how the brilliance of nature can be captured, or at least part captured so that synthetic sensors can be ever more effective. An obvious and highly exciting area of future research is improving the understanding of the mammalian brain. In many ways, there is an increasingly sophisticated understanding of biological sensors such as the eye, ear, etc. However, it is equally clear that the brain is able to operate on sensed 'signals' and process them to provide a sufficiently valuable perception of the world that allows natural systems to operate and to thrive. Indeed, one only has to look out of any window to see how human perception has resulted in the construction of an extraordinary plethora of highly complex and interrelated objects that are a normal part of our everyday lives. Returning to the topic of our book and how this might inspire new sensor research, we can use the simple example of synthetic cognition. The idea of a perception–action cycle, augmented through the generation and use of memory, was introduced in Chapter 7. Here the authors showed simple sensitivity improvements possible by varying the PRF and pulse length of the transmitted waveform in a radar system. The use of memory has not yet been addressed but is an area that is so well exploited in nature and therefore seems to be of rich promise. Memories can be generated by a radar or sonar system such that previously visited areas can re-use knowledge of clutter variation to chart a more favourable trajectory for target detection. Synthetic systems need not confine themselves to generation of own memories. Using data from, say, a GIS could remove or suggest the removal of false targets due to location coincidence with roads. These are just two fairly trivial examples of a myriad that could be considered, potentially making memory a very fertile area for future research. More broadly, it is interaction across the biological and engineering disciplines that will spawn new ideas and concepts. Synthetic sensors seem very much in their infancy and so the possibilities for improvement are almost limitless. We look forward to exploring new ideas and hope that many of our readers will be inspired sufficiently to become a part of our growing community.

Index

absorption 113
accuracy 138
acoustic aposematism 122
acoustic signatures 91
acuity 209
air-coupled sonar 63, 65, 84
ambiguity function 11, 138
amplitude–latency trading 21
angle (horizontal and vertical) 210
arms-race 109
artificial cognition 155
attention 156
audio 189, 202
auditory periphery 39
auditory spectrogram 38
auditory system 37
azimuth angle 215

baseband 42
baseband receiver 38
Baseband Spectrogram Correlation and Transformation receiver (BSCT) 43
bats 109, 111, 137, 209
  adapting its pulse emission rate 15
  biosonar 61
  echolocation 61, 65, 68, 76, 78, 83, 85
  flying in close proximity 13
  -pollinated plants 90
Bayesian Cramér–Rao lower bound (BCRLB) 160, 182
Bayesian filtering 156
Bayesian information matrix (BIM) 160, 182
Bayes–Markov recursion 158
Bayes risk 158
beam pattern 64–71, 73–4, 76–81, 84
bell-shaped flowers 101
biomimetic sonar 61, 63–5, 79, 85
biosonar 109, 189
biosonar model 193
blind 209
Bottlenose dolphin 189
broadband 38, 189
buds 100
Butterworth low-pass filter 39

calyxes 101
click train 193
closely located scatterers 37
clutter 14
clutter interference 6
cochlear block 39–40
coevolution 128
cognition 155
cognitive radar 156
cognitive sensor/processor system framework 157
collision hazards 15
computational model of biosonar 16
conditional Bayes risk 158
conductivity 229
constant-Q filters 190
counter adaptations 125
Cramér–Rao Lower Bound (CRLB) 139

dechirps 21
decision-making 156
distance 209
distributed sensor management 168

dolphins 189
Doppler shift 89
Doppler signature 89

echolocating bats 7, 37
echolocation 61, 65, 68, 76, 78, 83, 85, 110, 189, 209
Electrical Resistivity Tomography (ERT) 230
electric fish 229
electrosensing 229
elephantnose fishes 229
elevation 218
environmental factors 212
estimate of delay 19
estimation 139
expected Fisher information matrix (EFIM) 161
exponential envelope 195

facial sense 210
facial vision 210
feed-forward radar 156
Fisher information matrix (FIM) 139, 161, 165, 183
focused vs defocused images 8
fore-active control 137
frequency 214
frequency-modulated (FM) signals 6, 38
fully adaptive radar 156

Gaussian envelope 195
Gaussian random processes 138

half-wave instead of full-wave rectification 11
harmonics 10
high range resolution profiles (HRRPs) 91
histogram 199
humans 209

inflorescence 98
insects 89, 109
intensity 214
interception 137
interferences 38

joint conditional PDF 159

Kalman filter 137, 179
knifefishes 229

lag time 193
latency difference 21
likelihood function 158, 165
Linear Frequency Modulated waveform 92
linear frequency modulation (LFM) 140, 174
Linear Quadratic Gaussian (LQG) controller 138

matched filter 38
material 210
maximum a posteriori penalty function (MAP-PF) 163, 179
memory 155
metal detection 235
micro-Doppler modulations 89
moths 112

narrowband 189
narrowband Complex Ambiguity Function (CAF) 139
nectar feeding bats 90
neural representation 19
neurons 38
nocturnal predators 109
nonlinear least square (NLLS) 196
novel deconvolution process 22

operating range of biosonar 12
optimal guidance 137
oral click 210

parallel bandpass filters 16
parallel frequency channels 19
passive electroreception 229

penalty function 164
perception 209
perception–action cycle 155
periodogram 197
permittivity 229
pitch 212
polarizable minerals 230
polarization tensors 229
posterior Cramér–Rao lower bound (PCRLB) 160, 162
posterior PDF 158
predator–prey interactions 109
predicted conditional Bayesian information matrix (PC-BIM) 160
predicted conditional Bayes risk 159–60
predicted conditional Cramér–Rao lower bound (PC-CRLB) 160, 162
predicted conditional MSE (PC-MSE) 160, 163
predicted PDF 158
prediction 156
prey 137
Principal Components Algorithm (PCA) 103
probability density function (PDF) 158
processor cost function 158
pulse-Doppler radar 174
pulse repetition frequency (PRF) 174

range resolution 38
ranging 212
redicted information matrix (PIM) 161
relative bandwidth 18
RELAXation method 197
rendezvous 137
repetition pitch 213

sensor cost function 158
sensor/processor system 155
shape 210
signal properties 209
Signal-to-Noise Ratio 139
size 210
software defined radar 174
sonar clicks 189
sound emissions 210
spatial resolution 209
spectral block 42
Spectrogram Correlation And Transformation (SCAT) 16, 38
spectrogram delays 25
spectrogram transformation block 39
spike-time representation 18
state equations 138
static electric fields 229

target kinematics 138
target shape images 25
temporal block 41
temporal processing block 38
terminal buzz 122
time–frequency replica 15
timing, or latency, of neural responses 19
tracking 155
training 209
transition PDF 158
tymbals 123
tympanal ears 116

ultrasonic 189
ultrasonic sonar 68, 78
unpollinated corollas 91

voltage sensing receptors 229

waveform 137
white noise 214
wideband FM biosonar 9
wideband pulsed sounds 189
wingbeat flutter 9

www.ingramcontent.com/pod-product-compliance
Lightning Source LLC
LaVergne TN
LVHW051051220926
848485LV00001BA/74

* 9 7 8 1 6 1 3 5 3 2 3 5 5 *